*The origins of angiosperms and their
biological consequences*

The origins of angiosperms and their biological consequences

EDITED BY

Else Marie Friis
Section of Palaeobotany, Swedish Museum of Natural History, Stockholm

William G. Chaloner FRS
Department of Biology, Royal Holloway and Bedford New College, London University

Peter R. Crane
Department of Geology, Field Museum of Natural History, Chicago

Published by the Press Syndicate of the University of Cambridge
The Pitt Building, Trumpington Street, Cambridge CB2 1RP
40 West 20th Street, New York, NY 10011-4211, USA
10 Stamford Road, Oakleigh, Victoria 3166, Australia

© Cambridge University Press 1987

First published 1987
First paperback edition 1989

Reprinted 1992
Printed at Interprint Ltd, Malta

British Library cataloguing-in-publication data
The Origins of angiosperms and their
biological consequences.
1. Angiosperms   2. Plants – Evolution
I. Friis, E. M.   II. Chaloner, W. G.
III. Crane, P. R.
582.13′0438   QK495 A1
75937
Library of Congress cataloguing-in-publication data
The Origins of angiosperms and their biological consequences.
'Most of the contributions were presented at a
symposium held during the Third International Congress
of Systematic and Evolutionary Biology at the
University of Sussex, Brighton, England, in August 1985' – Pref.
Includes index.
1. Angiosperms, Fossil – Congresses.   2. Angiosperms – Congresses.
3. Paleontology – Cretaceous – Congresses.   4. Paleontology – Tertiary – Congresses.
I. Friis, E. M.   II. Chaloner, W. G. (William Gilbert).   III. Crane, P. R.
IV. International Congress of Systematic and Evolutionary Biology (3rd: 1985:
University of Sussex)
QE980.O73 1987   561′.2   86-21561
75937
ISBN 0 521 32357 6 hardback
ISBN 0 521 31173 X paperback

UP

Contents

_____•_____

Contributors

———————————— • ————————————

W. G. Chaloner Department of Biology, Royal Holloway and Bedford New College, Huntersdale, Callow Hill, Virginia Water, Surrey GU25 4LN, England

M. J. Coe Animal Ecology Research Group, Oxford University, Oxford OX1 3AS, England

M. E. Collinson Department of Biology, King's College London, Kensington Campus, Campden Hill Road, London W8 7AH, England.

P. R. Crane Department of Geology, Field Museum of Natural History, Roosevelt Road at Lake Shore Drive, Chicago, Illinois 60605, USA

W. L. Crepet Department of Ecology and Evolutionary Biology, The University of Connecticut, 75 North Eagleville Road, Storrs, Connecticut 06268, USA

D. L. Dilcher Department of Biology, Indiana University, Bloomington, Indiana 47405, USA

M. J. Donoghue Department of Ecology and Evolutionary Biology, University of Arizona, Tucson, Arizona 85721, USA

J. A. Doyle Department of Botany, University of California, Davis, California 95616, USA

J. O. Farlow Department of Earth and Space Sciences, Indiana University–Purdue University, Fort Wayne, Indiana 46805, USA

E. M. Friis Section of Palaeobotany, Swedish Museum of Natural History, Box 50007, S-104 05 Stockholm, Sweden

J. J. Hooker Department of Palaeontology, British Museum of Natural History, Cromwell Road, London SW7 5BD, England

D. M. Jarzen Paleobiology Division, National Museum of Natural Sciences, Ottawa, Ontario K1A 0M8, Canada

J. T. Parrish Branch of Oil and Gas Resources, Denver Federal Center, US Geological Survey, Box 25046, MS 971, Denver, Colorado 80225, USA

vii

D. A. Russell Paleobiology Division, National Museum of Natural Sciences, Ottawa, Ontario K1A 0M8, Canada

B. H. Tiffney Department of Geological Sciences, University of California, Santa Barbara, California 93106, USA

G. R. Upchurch, Jr National Center for Atmospheric Research, PO Box 3000, Boulder, Colorado 80307–3000, USA

S. L. Wing Division of Paleobiology, National Museum of Natural History, Washington, DC 20560, USA

J. A. Wolfe Paleontology and Stratigraphy Branch, Denver Federal Center, US Geological Survey, Box 25046, MS 919, Denver, Colorado 80225, USA

Preface

———————————————————— • ————————————————————

The dramatic radiation of the angiosperms toward the end of the Early Cretaceous initiated major changes in terrestrial ecosystems throughout the world. During the Late Cretaceous and Early Tertiary, the ancient Mesozoic plant communities of gymnosperms, ferns, horsetails and lycopods were replaced by angiosperm-dominated vegetation. Jurassic dinosaur faunas dominated by sauropods were replaced by ornithopods, and these, in turn, were ultimately replaced by rapidly diversifying mammalian faunas during the Early Tertiary. During the same Late Cretaceous and Early Tertiary interval, birds and insects underwent remarkable radiation and many of the sophisticated pollination and dispersal systems that characterize extant angiosperms developed for the first time. Although the geological history of these plant and animal groups has been intensively studied for many years, studies of their co-evolutionary interactions have concentrated on living organisms, with little integration of the historical perspective provided by data from the fossil record.

This book provides a new interdisciplinary perspective on one of the most significant events in the evolution of terrestrial organisms, based principally on data that have accumulated over the last two decades. In nine integrated chapters, the authors review alternative hypotheses on angiosperm origins and relationships, the vegetational and faunal changes consequent on the rise of the angiosperms, and the time sequence of development of ecological interactions between angiosperms, pollination vectors, dispersal agents and herbivores. In addition to contributing broad reviews of the major subject areas, each chapter also presents new data and new interpretations. Most of the contributions were presented at a symposium held during the Third International Congress

ix

of Systematic and Evolutionary Biology at the University of Sussex, Brighton, England in August 1985. The theme of the symposium, 'Angiosperm origins and the biological consequences', was intended to update and develop further the published results from an earlier symposium on the 'Origin and early evolution of angiosperms' presented at the first International Congress of Systematic and Evolutionary Biology in 1973. The book is designed primarily for the advanced undergraduate and graduate student interested in ecology, evolutionary biology, and palaeobiology, and is intended to provide a factual background for understanding the historical development of past and present angiosperm-dominated ecosystems.

In view of the multidisciplinary nature of the book, and the diverse readership that we hope to attract, the introduction provides a general characterization of the angiosperms and a brief description of their position in the history of plants. There are also a glossary, a stratigraphic table that includes the Cretaceous and Tertiary stratigraphic units used throughout the book, and a conspectus of classification of the major animal and plant groups that includes every genus mentioned in the text.

We wish to thank all those who participated in any way in the preparation of the book, including the numerous typists and artists, particularly Mette Dybdahl for invaluable assistance, and the authors for their co-operation with our efforts to achieve an integrated coverage of the field. We are especially grateful to Dr R. A. Pellew, Cambridge University Press, for encouraging us to publish in this form, for his help and advice, but above all for his firm but kindly interpretation of the rights and obligations of editors.

March 1986

Else Marie Friis
University of Aarhus
William G. Chaloner
University of London
Peter R. Crane
Field Museum of Natural History, Chicago

1
·

Introduction to angiosperms

E. M. FRIIS, W. G. CHALONER AND
P. R. CRANE

Although the angiosperms, or flowering plants, are the most recently
evolved major group of plants (Figure 1.3), they now occupy a dominant
position in the world's vegetation. With the total number of extant
species thought to be in the range 240000 to 300000, they exceed the
combined diversity of algae, bryophytes, pteridophytes and gymnosperms (Sporne, 1974; Prance, 1977). With the exception of conifer
forest and moss–lichen tundra, angiosperms dominate all of the major
terrestrial vegetation zones, account for the majority of terrestrial
primary production, and exhibit a bewildering morphological diversity.
They range in habit from minute free-floating aquatics to herbs,
epiphytes, lianes, shrubs and large forest trees. Although most are auto-
trophs, some are parasites or saprophytes, and angiosperms are the only
group of vascular plants successfully to have colonized marine habitats
that are otherwise only occupied by algae.

What makes an angiosperm?

Angiosperms are distinguished from other seed plants, and united as a
group, by several features related to their unique reproductive system.
These are carpels enclosing the ovules, pollen-tube growth through the
sporophytic carpel tissue, double fertilization resulting in the formation
of a triploid endosperm, and the highly reduced male and female
gametophytes (Sporne, 1974; Crane, 1985; Doyle & Donoghue, this
volume, Chapter 2). However, several other characters that occur in
most angiosperms are not universally present within the group, and some
characteristic features of angiosperms also occur sporadically in
gymnosperms. Together with the enormous diversity of flowering

1

plants, the relative paucity of strict defining characters makes the group much more difficult to circumscribe than might appear at first sight. Even the characteristic angiosperm flower exhibits great variation in structure that defies unambiguous definition.

Vegetative features

The angiosperms exhibit an extreme range in habit including herbaceous forms not commonly encountered in other seed plants, and lianes otherwise known only in the Gnetales. In vegetative features, angiosperms are unique in that the conducting tissues of the phloem possess companion cells that are derived from the same mother cell as the sieve elements. Most angiosperms also possess vessel elements in the xylem, but the phylogenetic interpretation of this character is complicated by the existence of several angiosperms that lack vessels (e.g. Winteraceae) and the occurrence of vessels in some gymnosperms (Gnetales). The vessels of angiosperms typically have scalariform, reticulate or simple perforation plates (Figures 1.1 (*a*) to (*c*)), probably derived from tracheids with scalariform bordered pits, and ontogenetic studies indicate an evolutionary trend from elongated narrow vessels with tapering ends and scalariform perforation plates (Figure 1.1 (*a*)) to short and wide vessels with simple perforation plates (Figure 1.1 (*c*)). In the Gnetales the perforation plates are apparently mostly derived from circular bordered pits of typical gymnospermous nature (Bailey, 1944; Martens, 1971), but scalariform perforation plates have also been described from the Gnetales (Muhammad & Sattler, 1982). Whether the vessels of Gnetales and angiosperms have a common origin in a shared ancestor has therefore been the subject of much controversy. The vesselless condition in the angiosperms is most common within magnoliid angiosperms and associated with other vegetative and reproductive features that are generally considered to be primitive, but whether the lack of vessels in these taxa should be regarded as the primitive state or as a result of secondary loss remains unclear (Takhtajan, 1969; Young, 1981).

A further typical vegetative characteristic of the angiosperms is the presence of broad leaves, with the vascular bundles forming a reticulate branching pattern with a hierarchical system of successively thinner veins that often have free endings (Figure 1.1(*d*) to (*e*)). In the monocotyledons, the main veins are commonly parallel and the leaf differentiated into blade and sheath, while in the dicotyledons the main veins are usually pinnate or palmate (Figure 1.1 (*d*)) and the leaf differentiated into a petiole and blade. Broad leaves, reticulate venation and the

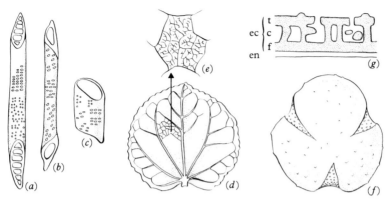

Figure 1.1. Characteristic features of angiosperms. (*a*) Vessel element with scalariform perforation plates. (*b*)–(*c*) Vessel elements with simple perforation plates. (*d*) Dicotyledonous leaf with palmate primary venation. (*e*) Detail of venation pattern. (*f*) Tricolpate pollen grain. (*g*) Section of pollen wall showing thin endexine (en), and footlayer (f), columellar layer (c) and tectum (t) of ectexine (ec).

presence of a petiole are not exclusive angiosperm features, although in extant non-angiospermous plants this combination is only present in the gymnospermous genus *Gnetum* (Gnetales).

Chemical characteristics of the angiosperms include the presence in many taxa of alkaloids derived from aromatic amino acids and hydro-lyzable tannins (Swain, 1976). In contrast, in non-angiosperms the occurrence of alkaloids as secondary metabolites is less common. Alkaloids of angiosperms are also different from those of gymnosperms, and gymnosperm tannins are condensed and non-hydrolyzable (Swain, 1976).

Pollen features

An apparently unique pollen characteristic of the angiosperms is the differentiation of the pollen wall into a non-laminate endexine, with the ectexine differentiated into a foot-layer, columellar layer and tectum (Figure 1.1(*g*)) (Muller, 1970; Walker & Doyle, 1975). Within the dicotyledons, members of the subclasses Hamamelididae, Caryophylli-dae, Dilleniidae, Rosidae and Asteridae are characterized by pollen with three or more apertures that are typically positioned equatorially (Figure 1.1 (*f*)). However, in the monocotyledons and dicotyledons of the subclass Magnoliidae, the pollen typically has a single distal furrow (monosulcate) and in this feature resembles the pollen grains of some gymnosperms included in the Cycadales, Bennettitales, Gnetales,

Ginkgoales and other groups. The male gametophyte of angiosperms consists of only three cells, compared with the more extensive four-, five- and six-celled male gametophytes of gymnosperms, and the female gametophyte is similarly reduced. In angiosperms the female gameto- phyte consists typically of only 8 to 16 cells, and no archegonia are differentiated. Although archegonia are also lacking in *Gnetum* and *Welwitschia* (Gnetales), the female gametophyte in these and all other gymnosperms is much more extensive than in angiosperms.

Floral features

To a large extent the diversity of angiosperm floral structure is a reflection of the sophisticated breeding systems that have evolved within the group, and apparently promote gene exchange by outcrossing. The angiosperm flower is typically hermaphroditic (bisexual), with carpels (female parts) and stamens (male parts) aggregated in the same flower, with the former borne above the latter. Outcrossing in hermaphroditic flowers is usually achieved by insects that transfer pollen from the stamens of one flower to the stigma of another. In such typical entomo- philous flowers, the reproductive organs are surrounded by a perianth of sterile appendages that are differentiated into an outer group of sepals and an inner group of petals that form the calyx and corolla, respectively (Figure 1.2(a)). The sepals are usually green and typically serve to protect the flower during its development, while the petals are generally colored, frequently function as a visual attractant and are often involved in the production of nectar or odors. These basic floral parts may be free or united both between themselves and to each other. The symmetry of flowers varies from radial (actinomorphic) to bilateral (zygomorphic) and may reflect specialization for pollination by various groups of insects (Proctor & Yeo, 1973; Faegri & van der Pijl, 1979). Other features typical of insect-pollinated flowers include the presence of nectar-producing tissue, ovaries with many ovules, and sticky pollen grains. In contrast, wind-pollinated (anemophilous) flowering plants typically have flowers protruding from the foliage (or formed earlier than the leaves) with a perianth that is inconspicuous or absent (Figure 1.2(b)–(d)). Large quantities of dry, more or less smooth pollen grains are produced typically, and the ovaries usually have one or a few ovules. Other characteristic features of many anemophilous plants are well-exposed reproductive organs aggregated into pendulous catkins, expanded feathery or papillose stigmatic surfaces, and the absence of scent and

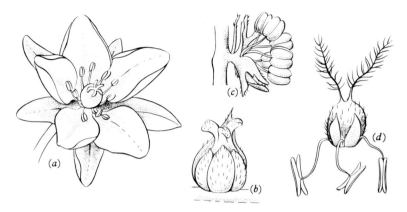

Figure 1.2. Floral features in angiosperms. (*a*) Entomophilous, hermaphroditic flower of a dicotyledon with showy perianth. (*b*)–(*c*) Anemophilous, unisexual flowers of an oak with reduced perianth: (*b*) female (pistillate) flower; (*c*) male (staminate) flower. (*d*) Anemophilous, hermaphroditic flower of a grass with feathery stigmas.

nectar production. These characters prevail in both dicotyledonous (Figure 1.2(*b*)–(*c*)) and monocotyledonous plants (Figure 1.2(*d*)).

Self-fertilization in hermaphroditic flowers may be kept at a low level by various incompatibility systems or by differences in the timing of maturation of stamens and ovaries. In protogynous flowers, the stigmas are receptive to pollination before the stamens mature and release their pollen, while in protandrous flowers, the pollen grains are released before the stigmas become receptive and are often only viable for a short period.

Although most angiosperm flowers are hermaphroditic, separation of the sexes into staminate (male) and pistillate (female) flowers occurs in some groups and this is a further means by which self-fertilization is apparently reduced. The pistillate and staminate flowers may be borne on the same individual (monoecious) or on separate plants (dioecious). Dioecious plants are most common in tropical areas and are generally pollinated by insects. Frequently they exhibit modifications for insect attraction comparable to those of hermaphroditic entomophilous flowers, but in dioecious plants the flowers are generally smaller than those of related hermaphroditic taxa (Bawa & Opler, 1975; Bawa, 1980). Monoecious angiosperms are most common in temperate regions and are less frequent at lower latitudes (Whitehead, 1983). They are typically wind pollinated, with simple flowers. Typical anemophilous features occur in many trees of temperate regions but are also shared by the grasses, which provide a good example of hermaphroditic anemophilous flowers (Figure 1.2(d)). The flowers of most grasses appear to be particularly well adapted to wind pollination in possessing versatile, exserted anthers and an expanded feathery stigma (Figure 1.2(d)).

Origin and form of the primitive angiosperms

Hypotheses about the nature of primitive angiosperms may be divided broadly into two opposing theories, developed over the contrasting floral characteristics associated with entomophily and anemophily (Doyle & Donoghue, this volume, Chapter 2). The most widely accepted view is the Euanthial, or Anthostrobilus, Theory (Arber & Parkin, 1907), in which the angiosperm flower is interpreted as being derived from an unbranched bisexual strobilus bearing spirally arranged ovulate and pollen organs, similar to the hermaphroditic reproductive structures of some extinct bennettitalean gymnosperms. The Bennettitales and the angiosperms were linked by those authors by hypothetical transitional hemiangiosperms that bore ovules on the margin of leaf-like structures resembling the megasporophylls of extant *Cycas*. Among living angiosperms, the strobiloid flower of the Magnoliaceae and related families, with numerous, free flower parts, were interpreted as being most similar to the basic archetype (Bessey, 1897, 1915). The flowers of the Magnoliaceae are typically solitary and large, with many spirally arranged carpels, stamens and perianth parts on an elongated axis. They are hermaphroditic, usually pollinated by beetles, and have a showy perianth. The stamens are often more or less laminar, and the gynoecium is apocarpous, with each carpel enclosing several ovules. According to the Euanthial Theory, more simple flowers, including the typical anemophilous forms, are derived from the basic *Magnolia* type, by reduction and fusion of parts.

Over the last two decades the fossil record has reinforced the view that extant Magnoliidae are among the most primitive living angiosperms, but the Euanthial Theory has also received some support, over a much longer period, from comparative studies with extant angiosperms. This interpretation of the angiosperm flower, developed by Bessey and by Arber and Parkin, also forms the basis of most modern phylogenetic classifications of flowering plants (Cronquist, 1968, 1981; Thorne, 1968; Hutchinson, 1969; Takhtajan, 1969, 1980; Walker & Walker, 1984).

The alternative, Pseudanthial Theory (Wettstein, 1907) interprets the angiosperm flower as being derived from unisexual gymnosperm reproductive structures, perhaps similar to those of the gnetalean gymnosperms. According to this concept, the small, simple, unisexual, anemophilous flowers of some Hamamelididae (Fagales, Casuarinales, Juglandales, Myricales, sometimes collectively termed the Amentiferae) and of the Piperales, retain the largest number of primitive floral characters among living angiosperms. The flowers are typically aggregated on elongated axes, and the gynoecium is unilocular, enclosing a single anatropous or orthotropous, unitegmic ovule. However, the occurrence in the Amentiferae of occasional, often imperfectly developed, bisexual flowers may suggest that the unisexual condition is secondary. This is also supported by the discovery of small bisexual flowers from the Upper Cretaceous that may have been produced by early primitive members of the Juglandales or Myricales (Friis, 1983). In addition, the fossil pollen record clearly demonstrates that triaperturate, particularly triporate grains, typical of the Amentiferae are not present in the initial radiation of the angiosperms, and are substantially predated in the fossil record by pollen resembling that of extant Magnoliidae (Muller, 1970; Doyle & Hickey, 1976). However, while the primitive status of the Amentiferae has been convincingly refuted, the phylogenetic position of the Piperales, themselves included within the Magnoliidae, remains of considerable interest. In particular, the Chloranthaceae have emerged as an especially critical group. While this family is often considered to be highly derived within the Magnoliidae (Walker & Walker, 1984), this may be in conflict with the abundance of chloranthoid pollen very early in the angiosperm fossil record (Muller, 1981; Walker & Walker, 1984). Futhermore, recent analyses of the relationships between major groups of seed plants have identified both the Gnetales and Bennettitales as gymnosperms that may be equally, and very closely, related to angiosperms (Crane, 1985; Doyle & Donoghue, this volume, Chapter 2).

A major problem in any theory of angiosperm origin is the difficulty of explaining the derivation of the closed angiosperm carpel from the ovuliferous structures in gymnosperms. In both the Bennettitales and Gnetales, the naked ovules are borne directly on an axis, with no evidence of sporophylls or other organs that might contribute to carpel formation. However, in several other groups of Mesozoic gymnosperms the ovules are borne on leaf-like structures, permitting several plausible possibilities for the derivation of the carpel. Attention has centered mainly on the Caytoniales, Corystospermales and Glossopteridales (Andrews, 1963; Stebbins, 1974; Doyle, 1978; Retallack & Dilcher, 1981; Crane, 1985),

8

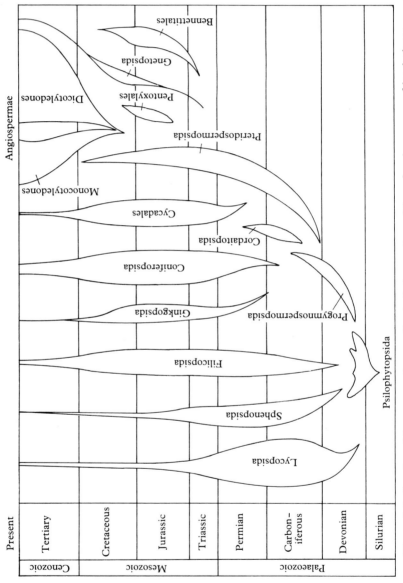

Figure 1.3. Stratigraphic range and generalized phylogenetic position of major groups of land plants.

all extinct taxa included in a highly heterogeneous group of extinct seed plants, the pteridosperms (seed ferns). The traditional interpretation of the position of angiosperms relative to other vascular plants which incorporates these interpretations of the angiosperm carpel, is summarized in Figure 1.3. The complex issues involved in interpreting the origins and relationships of the angiosperms are discussed in more detail by Doyle and Donoghue in Chapter 2. These authors adopt a numerical cladistic approach in attempting to resolve the relationships between major groups of seed plants, and also consider some of the possible evolutionary processes and ecological factors that may have played an important role in angiosperm origin.

Timing and pattern of the early angiosperm radiation

A long pre-Cretaceous history of the angiosperms has been postulated by several authors (Axelrod, 1952, 1960, 1970; Takhtajan, 1969; Stebbins, 1974), and although it is possible that the angiosperm lineage may have diverged from closely related gymnosperms very early in the Mesozoic (Crane, 1985; Doyle & Donoghue, this volume, Chapter 2), it is unlikely that the angiosperms were diverse during the Triassic and Jurassic. The available fossil evidence indicates instead that the initial major diversification of angiosperms took place during the late Early Cretaceous, following the first appearance of undoubted angiosperm fossils in the Barremian or slightly earlier in the Hauterivian (Figure 1.3). Palynological data suggest that the angiosperm radiation began at low palaeolatitudes (Brenner, 1976; Hughes, 1976), but, within a relatively short span of time during the mid-Cretaceous (Aptian–Turonian), angiosperms became established world wide as the ecologically dominant terrestrial plant group (see Upchurch & Wolfe, this volume, Chapter 4; Crane, this volume, Chapter 5).

The differentiation of the two major angiospermous groups, the dicotyledons and the monocotyledons, was established very early in angiosperm history. The earliest well-documented angiosperm fossils include dispersed monosulcate pollen grains of the form-genus *Clavatipollenites*, first described by Couper (1958) from Barremian and Aptian strata of southern England. These grains are believed to be angiospermous on the grounds that their sculptured exine shows a tectate structure (cf. Figure 1.1 (*g*)), coupled with a single distal furrow. *Clavatipollenites* grains are closely similar to pollen of the extant chloranthaceous genus *Ascarina* (Couper, 1958; Walker & Walker, 1984) and unequivocally indicate the early appearance of the Piperales. Other pollen types, leaves and floral structures with chloranthoid characters also occur in Aptian and Albian strata of eastern North America (Upchurch, 1984; Walker & Walker, 1984; Friis, Crane & Pedersen, 1986), and clearly establish the importance of this family in the initial angiosperm radiation. The oldest unequivocal fossils of the Magnoliales so far recorded are pollen grains related to the Winteraceae recorded from Aptian–Albian strata of Israel (Walker, Brenner & Walker, 1983) and leaves with features characteristic of Magnoliales from Aptian–Early Albian strata of eastern North America (Upchurch, 1984). Large magnolialean flowers with numerous free parts of the kind predicted by the classic formulation of the Euanthial Theory are not recorded until the Late Albian and Cenomanian (Dilcher & Crane, 1984; Crane & Dilcher, 1984).

Extant Chloranthaceae include both anemophilous and entomophilous taxa, and comparisons of fossil and Recent material indicate that a differentiation into insect and wind pollination systems was also established in the Early Cretaceous chloranthoid angiosperms (Friis *et al.*, 1986). Low pollen production in many entomophilous plants, combined with the unspecialized morphology of much magnolialean pollen, may explain the apparent paucity of the Magnoliales relative to the Chloranthaceae in the early palynological record of angiosperm diversification.

Triaperturate pollen grains diagnostic of non-magnoliid dicotyledons appear in the latest Barremian or earliest Albian (slightly later than monosulcate chloranthoid pollen) and increased dramatically both in diversity and in abundance during the mid-Cretaceous. The earliest triaperturate grains are small, with microreticulate ornamentation and three equatorial colpi. Albian floral structures with pollen *in situ* indicate that at least some of these are related to probable members of the Hamamelididae and Rosidae, which may have borne platanoid and *Sapindopsis* foliage, respectively (Crane, Friis & Pedersen, 1986). The very small size and microreticulate sculpture of the pollen grains in some of these triaperturate forms suggest entomophily.

Monosulcate pollen of possible monocotyledonous affinity is recorded during the Aptian and Albian (Kemp, 1968; Doyle, 1973; Walker & Walker, 1984). They are elongated in shape and have reticulate sculpture of the exine differentiated into areas with either a coarse or fine network. Similar features occur in extant entomophilous groups, whereas smooth monoporate pollen grains typical of anemophilous monocotyledons are not recorded until the Late Cretaceous (Muller, 1970; Doyle, 1973).

Monocotyledonous foliage with apically fusing veins and sheathing leaf bases (*Acaciaephyllum*) is also known from the Aptian. Although the fossil record has not yet documented conclusively the nature of the pollination systems in the earliest angiosperms, it provides strong support for insect pollination as a major factor in the early differentiation of the group, and indicates that anemophily developed independently within several monocotyledonous and dicotyledonous taxa. The chronological appearance of major floral features in the angiosperms is reviewed in Chapter 6 (Friis & Crepet), while Chapter 7 (Crepet & Friis) discusses in detail the palaeontological data bearing on the historical development of insect pollination.

Impact of the angiosperm radiation on Cretaceous and Early Tertiary ecosystems

As the timing and pattern of the angiosperm radiation has been clarified gradually by neontological and palaeobotanical work, new possibilities have emerged for assessing the biological consequences of the origins and diversification of this dominant group of terrestrial plants. It is now possible to begin to evaluate the vegetational, faunal and other ecological changes related to the rise of the angiosperms, and to begin to trace the historical development of co-evolutionary interactions between angiosperms and their pollinators, dispersal agents and herbivores.

The transition to angiosperm dominance during the mid-Cretaceous took place during a period of relative tectonic stability in which both sea-level and global temperature were relatively high. The latitudinal temperature gradient was low, and the Cretaceous to Eocene was probably the warmest interval during Phanerozoic time (Parrish, this volume, Chapter 3; Upchurch & Wolfe, this volume, Chapter 4). These climatic conditions undoubtedly played an important role in the initial development and geographical expansion of flowering plants between the mid-Cretaceous and Early Tertiary, and the climatic and geological background to the angiosperm radiation is reviewed in Chapter 3 (Parrish). Chapter 4 (Upchurch & Wolfe) discusses the extent to which climatic conditions influenced the characteristics and development of angiosperm-dominated vegetation throughout this period, and also considers the implications of the available palaeobotanical data for interpreting palaeoclimates accurately.

The appearance of the angiosperms was soon to be followed by break-up of the major continental masses (Laurasia and Gondwana). At the very time that the major new groups were appearing within the angiosperms and spreading over wide areas, they were also becoming separated on the moving continental plates and by the development of associated topographic barriers. Throughout the subsequent development of the angiosperms, climatic gradation from pole to equator became more marked, and their response, by migration and extinction, to the changing physical environment is a major feature of angiosperm history.

The rapid increase in both the diversity and abundance of angiosperms through the mid-Cretaceous was accompanied by widespread extinction in many groups that dominated the vegetation earlier in the Mesozoic. The Bennettitales, cycads and the so-called Mesozoic pteridosperms were the groups affected most but major systematic changes also occurred in pteridophytes and conifers through the same period. Ecologically, the angiosperms expanded from their initial role as early successional herbs and shrubs eventually to dominate the wide variety of habitats in which they occur today. The systematic changes involved in the transition from gymnosperm- to angiosperm-dominated vegetation, and their ecological implications are reviewed in Chapter 5 (Crane).

Major vegetational changes inevitably influence the herbivores, and carnivores that are ultimately dependent on terrestrial plant communities. More than any other group of land plants, the angiosperms show evidence of co-evolution with a variety of animal groups. Many angiosperms combine attractive fleshy fruits with tough seeds, as symptoms of adaptation to mammal and bird dispersal. Others have a range of adaptations to adhere to mammalian fur. In addition to these specific adaptations, it may be said fairly that from the mid-Cretaceous onwards all terrestrial animal life is largely dependent on angiosperms as primary producers. General aspects of the co-evolutionary interactions between angiosperms and both Late Cretaceous dinosaurs and Early Tertiary mammals are reviewed in Chapter 8 (Wing & Tiffney). Dinosaur faunas changed from sauropod-dominated communities in the Jurassic to ornithopod-dominated communities in the Cretaceous, and the whole group finally became extinct at the Cretaceous–Tertiary boundary. Possible effects of the angiosperm radiation on the history and biology of dinosaurs are considered in Chapter 9 (Coe et al.). Major changes in angiosperm vegetation also occurred during the Early Tertiary radiation of mammals. Chapter 10 (Collinson & Hooker) provides an integrated review of floral and mammalian faunal change in the Early Tertiary of southern England. These sediments provide one of the most complete and intensively sampled sequences in the world in which this kind of co-ordinated change can be directly assessed. Many of the data on which the reviews and interpretations in this book are based have emerged over the last two decades, and provide an indication of the rapidity with which new information is accumulating. Inevitably, however, the data are much less complete than would be desired, particularly with respect to geographical sampling in present-day tropics, the ancient tropics and the Southern Hemisphere. Nevertheless it is clear that certain general and well-corroborated patterns in the data are beginning to emerge, which now seem securely established. The following chapters review these patterns in the fossil record, explore their phylogenetic and ecological implications and, most importantly, highlight the diversity of questions that remain to be addressed with future research.

We are grateful to Mary Jane Spring for preparing the illustrations. Much of this work has been supported by the Royal Danish Academy of Sciences and Letters.

References

Andrews, H. N. (1963). Early seed plants. *Science*, **142**, 925–31.

Arber, E. A. N. & Parkin, J. (1907). On the origin of angiosperms. *Journal of the Linnean Society*, **38**, 29–80.

Axelrod, D. I. (1952). A theory of angiosperm evolution. *Evolution*, **6**, 29–60.

Axelrod, D. I. (1960). The evolution of flowering plants. In *The Evolution of Life*, ed. S. Tax, pp. 227–305. Chicago: University of Chicago Press.

Axelrod, D. I. (1970). Mesozoic paleogeography and early angiosperm history. *The Botanical Review*, **36**, 277–319.

Bailey, I. W. (1944). The development of vessels in angiosperms and its significance in morphological research. *American Journal of Botany*, **31**, 421–8.

Bawa, K. S. (1980). Evolution of dioecy in flowering plants. *Annual Review of Ecology and Systematics*, **11**, 15–39.

Bawa, K. S. & Opler, P. A. (1975). Dioecism in tropical trees. *Evolution*, **29**, 167–79.

Bessey, C. E. (1897). Phylogeny and taxonomy of the angiosperms. *Botanical Gazette*, **24**, 145–78.

Bessey, C. E. (1915). The phylogenetic taxonomy of flowering plants. *Annals of the Missouri Botanical Garden*, **2**, 109–64.

Brenner, G. J. (1976). Middle Cretaceous floral provinces and early migrations of angiosperms. In *Origin and Early Evolution of Angiosperms*, ed. C. B. Beck, pp. 23–47. New York: Columbia University Press.

E. M. Friis et al.

Crane, P. R. (1985). Phylogenetic analysis of seed plants and the origin of angiosperms. *Annals of the Missouri Botanical Garden*, **72**, 716–93.

Crane, P. R. & Dilcher, D. L. (1984). *Lesqueria*: an early angiosperm fruiting axis from the mid-Cretaceous. *Annals of the Missouri Botanical Garden*, **71**, 384–402.

Crane, P. R., Friis, E. M. & Pedersen, K. R. (1986). Angiosperm flowers from the Lower Cretaceous: fossil evidence on the early radiation of the dicotyledons. *Science*, **232**, 852–4.

Cronquist, A. (1968). *The Evolution and Classification of Flowering Plants.* London: Nelson.

Cronquist, A. (1981). *An Integrated System of Classification of Flowering Plants.* New York: Columbia University Press.

Couper, R. A. (1958). British Mesozoic microspores and pollen grains, a systematic and stratigraphic study. *Palaeontographica B*, **103**, 75–179.

Dilcher, D. L. & Crane, P. R. (1984). *Archaeanthus*: an early angiosperm from the Cenomanian of the Western Interior of North America. *Annals of the Missouri Botanical Garden*, **71**, 351–83.

Doyle, J. A. (1973). The Monocotyledons: their evolution and comparative biology. V. Fossil evidence on early evolution of Monocotyledons. *Quarterly Review of Biology*, **48**, 399–413.

Doyle, J. A. (1978). Origin of angiosperms. *Annual Review of Ecology and Systematics*, **9**, 365–92.

Doyle, J. A. & Hickey, L. J. (1976). Pollen and leaves from the Mid-Cretaceous Potomac Group and their bearing on early angiosperm evolution. In *Origin and Early Evolution of Angiosperms*, ed. C. B. Beck, pp. 139–206. New York: Columbia University Press.

Faegri, K. & van der Pijl, L. (1979). *Pollination Ecology*, 3rd edn. Oxford: Pergamon Press.

Friis, E. M. (1983). Upper Cretaceous (Senonian) floral structures of juglandalean affinity containing Normapolles pollen. *Review of Palaeobotany and Palynology*, **39**, 161–88.

Friis, E. M., Crane, P. R. & Pedersen, K. R. (1986). Floral evidence for Cretaceous chloranthoid angiosperms. *Nature*, **320**, 163–4.

Hughes, N. F. (1976). *Palaeobiology of Angiosperm Origins.* Cambridge: Cambridge University Press.

Hutchinson, J. (1969). *Evolution and Phylogeny of Flowering Plants.* London: Academic Press.

Kemp, E. M. (1968). Probable angiosperm pollen from British Barremian to Albian strata. *Palaeontology*, **11**, 421–34.

Martens, P. (1971). *Les Gnétophytes.* Handbuch der Pflanzenanatomie, vol. 12(2). Berlin: Gebrüder Borntraeger.

Muhammad, A. F. & Sattler, R. (1982). Vessel structure of *Gnetum* and the origin of angiosperms. *American Journal of Botany*, **69**, 1004–21.

Muller, J. (1970). Palynological evidence on early differentiation of angiosperms. *Biological Reviews*, 45, 417–50.

Muller, J. (1981). Fossil pollen records of extant angiosperms. *The Botanical Review*, 47, 1–142.

Prance, G. T. (1977). Floristic inventory of the tropics: Where do we stand? *Annals of the Missouri Botanical Garden*, 64, 659–84.

Proctor, M. & Yeo, P. (1973). *The Pollination of Flowers*. London: Collins.

Retallack, G. & Dilcher, D. L. (1981). Arguments for a glossopterid ancestry of angiosperms. *Paleobiology*, 7, 54–67.

Sporne, K. R. (1974). *The Morphology of Angiosperms*. London: Hutchinson.

Stebbins, G. L. (1974). *Flowering Plants. Evolution above the Species Level*. Cambridge, Massachusetts: Belknap Press.

Swain, T. (1976). Angiosperm–reptile co-evolution. *Linnean Society Symposia*, 3, 107–22.

Takhtajan, A. (1969). *Flowering Plants. Origin and Dispersal*. Edinburgh: Oliver & Boyd.

Takhtajan, A. J. (1980). Outline of the classification of flowering plants (Magnoliophyta). *The Botanical Review*, 46, 225–359.

Thorne, R. F. (1968). Synopsis of a putative phylogenetic classification of the flowering plants. *Aliso*, 6, 57–66.

Upchurch, G. R. (1984). Cuticle evolution in Early Cretaceous angiosperms from the Potomac Group of Virginia and Maryland. *Annals of the Missouri Botanical Garden*, 71, 522–50.

Walker, J. W., Brenner, G. J. & Walker, A. G. (1983). Winteraceous pollen in the Lower Cretaceous of Israel: early evidence of a Magnolialean angiosperm family. *Science*, 220, 1273–5.

Walker, J. W. & Doyle, J. A. (1975). The bases of angiosperm phylogeny: palynology. *Annals of the Missouri Botanical Garden*, 62, 664–723.

Walker, J. W. & Walker, A. G. (1984). Ultrastructure of Lower Cretaceous angiosperm pollen and the origin and early evolution of flowering plants. *Annals of the Missouri Botanical Garden*, 71, 464–521.

Wettstein, R. von (1907). *Handbuch der systematischen Botanik*, 2nd edn. Leipzig and Wien: Franz Deuticke.

Whitehead, D. R. (1983). Wind pollination: some ecological and evolutionary perspectives. In *Pollination Biology*, ed. L. Real, pp. 97–108. London: Academic Press.

Young, D. A. (1981). Are the angiosperms primitively vesselless? *Systematic Botany*, 6, 313–30.

2

The origin of angiosperms : a cladistic approach

J.A.DOYLE AND M.J.DONOGHUE

As the present volume illustrates, there has been renewed interest recently in the old problem of the origin and early evolution of angiosperms, stimulated in large part by studies of Cretaceous fossils. Although there is more agreement now than there was two decades ago concerning the timing and pattern of the early diversification of angiosperms, the question of their origin remains controversial, largely as a result of different ideas on how angiosperms are related to other groups of seed plants (gymnosperms). The purpose of this paper is to summarize a numerical cladistic analysis of seed plants, documented in detail elsewhere (Doyle & Donoghue, 1986b), and to explore its implications for evolutionary processes and ecological factors involved in the origin of angiosperms.

Previous ideas on angiosperm relationships

One of the first comprehensive theories concerning the relationships of angiosperms, commonly associated with the Englerian school of angiosperm systematics, was proposed by Wettstein (1907), who postulated that the angiosperms were derived from the gymnosperm order Gnetales, represented today by *Ephedra*, *Welwitschia*, and *Gnetum*. The Gnetales show more angiosperm-like features than does any other group of gymnosperms: vessels in the wood; compound strobili made up of minute flower-like units, with either a perianth and a whorl of more or less fused microsporophylls or a terminal ovule surrounded by an additional envelope; strong gametophyte reduction and cellular embryogeny in *Welwitschia* and *Gnetum*; dicotyledon-like leaves in *Gnetum* (cf. Martens, 1971). Wettstein homologized the compound strobili of Gnetales with the inflorescences of the wind-pollinated Amentiferae, and he interpreted the showy, insect-pollinated, bisexual flowers of *Magnolia* and other groups as pseudanthia derived by aggregation of unisexual units. This would imply that the first angiosperms were wind pollinated, and that insect pollination arose later within the group.

A competing view was proposed by Arber & Parkin (1907, 1908), stimulated by recognition that some members of the Mesozoic gymnosperm order Bennettitales had flower-like bisexual reproductive structures, with a perianth, pinnate microsporophylls, and a central ovuliferous receptacle. They homologized these structures with the flowers of *Magnolia* and argued that the flowers of the Amentiferae were secondarily reduced and aggregated during reversion to wind pollination. Seeds of Bennettitales were borne singly on simple stalks, interspersed with interseminal scales, rather than on structures resembling an angiosperm carpel, and the microsporophylls were whorled and usually fused basally. Therefore, rather than deriving angiosperms directly from Bennettitales, Arber & Parkin proposed that the two groups evolved from a hypothetical common ancestor with a 'pro-anthostrobilus' bearing pinnate microsporophylls and megasporophylls. In Bennettitales, the megasporophylls were presumably reduced, but in angiosperms the microsporophylls were. Like Wettstein, Arber & Parkin assumed that Gnetales were related to angiosperms, but they interpreted the flowers of Gnetales as reduced, like the flowers of Amentiferae, citing as support the presence of an abortive ovule in the staminate flowers of *Welwitschia*.

Both of these views have subsequently fallen into wide disfavor. Relationships between angiosperms and Gnetales, assumed by both theories, have come under special criticism. First, closer examination suggests that many of their common features arose independently. For example, some presumably primitive angiosperms, such as the magnoliid dicotyledon family Winteraceae, lack vessels in the wood, suggesting derivation from ancestors without vessels rather than ancestors with them. Furthermore, vessel members in the two groups appear to be derived from different kinds of tracheid: Gnetales have vessel members with perforations that intergrade with circular-bordered pits, whereas primitive angiosperms have tracheids with scalariform pitting or vessel members with scalariform perforations (Thompson, 1918; Bailey, 1944). In general, recognition of apparently primitive features in magnoliids (vesselless wood, gymnosperm-like monosulcate pollen, leaf-like carpels and stamens) and derived features in Amentiferae (advanced vessels, triporate pollen) has cast doubt on Wettstein's arguments that are based on similarities between Gnetales and Amentiferae. Most recently, these ideas on evolution within the angiosperms have been strengthened by palaeobotanical studies. The first recognizable Cretaceous angiosperm pollen is monosulcate, whereas pollen of the amentiferous type appears much later, after a long series of intermediates, and the leaf record reveals consistent trends (Doyle, 1969, 1978; Muller, 1970; Doyle & Hickey, 1976; Crane, this volume, Chapter 5).

In addition, several lines of evidence have led to the idea that the Gnetales are related to the coniferopsid gymnosperms (conifers, Palaeozoic cordaites, and ginkgos), characterized by linear-dichotomous leaves, pycnoxylic wood, and bilaterally symmetrical (platyspermic) seeds, whereas angiosperms are related to the cycadopsid gymnosperms (cycads, Bennettitales, and so-called seed ferns), characterized by pinnate leaves, manoxylic wood, and radially symmetrical (radiospermic) seeds. For example, Gnetales have circular-bordered pits in the metaxylem and even the protoxylem, like conifers and *Ginkgo*, whereas angiosperms and cycadopsids have scalariform metaxylem pitting (Bailey, 1944). Eames (1952) homologized the flowers of *Ephedra* with the axillary fertile short shoots of *Cordaites*, equating the perianth of the staminate flower and the envelope around the seed in *Ephedra* with the sterile scales on the short shoot. Although Eames rejected relationships between *Ephedra* and *Welwitschia* and *Gnetum*, others have noted that all three genera have similar wood anatomy and their reproductive structures can be interpreted as modifications of the same plan (Bailey, 1944; Bierhorst, 1971; Doyle, 1978). Conversely, the leaf-like carpels of magnoliids suggest relationships with cycadopsids, in which seeds are usually borne on obvious leaf homologs, and within seed plants scalariform secondary xylem pitting is largely restricted to angiosperms, Bennettitales, and some cycads.

It should be noted that the views just summarized are not universally accepted. For example, Meeuse (1963, 1972 *a*, *b*, *c*) has proposed a highly modified version of Wettstein's theory, in which angiosperm reproductive structures are derived from a polyaxial 'anthocorm' system, and Gnetales and Piperales are seen as links between 'higher cycadopsids' and typical angiosperms. Young (1981) challenged the view that the first angiosperms were vesselless, arguing on the basis of a cladistic analysis of primitive angiosperms that it is more parsimonious to assume that vessels were lost in several early lines. Muhammad & Sattler (1982) found scalariform perforations in vessel elements of *Gnetum* and suggested that angiosperms might be derived from Gnetales after all.

Reasons for the rejection of relationships between angiosperms and
Bennettitales are less clear, since the two groups have similar wood
anatomy, and the presence of primitive features in magnoliids supports
many of Arber & Parkin's ideas on angiosperm evolution. Recent authors
have tended to emphasize the morphological differences between the
parts making up the flowers of the two groups, particularly the carpels
of angiosperms and the isolated ovules of Bennettitales, and to regard
the similarities as convergent adaptations to insect pollination. Skepti-
cism has centered on the fact that Arber and Parkin's attempt to reconcile
these differences relied so heavily on a purely hypothetical prototype.
Takhtajan (1969) and Ehrendorfer (1976) have continued to argue that
angiosperms share a common ancestry with Bennettitales and that
Gnetales are highly modified bennettitalean derivatives, but most discussions have focused instead on seed ferns (Cronquist, 1968; Stebbins,
1974; Doyle, 1978), in which megasporophylls are less reduced.

   Among seed ferns, two Mesozoic families have attracted particular
attention. Caytoniaceae, a widespread Triassic–Cretaceous group, had
palmately compound leaves, simple reticulate venation, and reflexed
cupules borne in two rows along the rachis of a once-pinnate megasporophyll. Corystospermaceae, a Gondwana Triassic group, had fern-
like leaves and bipinnate megasporophylls. Early attempts to homologize
the cupules of these groups with carpels (e.g. Thomas, 1925) were un-
successful (the cupules appear to be modified leaflets rather than whole
sporophylls borne on a stem), but several authors have noted that
reduction to one ovule per cupule (seen in corystosperms) would result
in a structure like the anatropous, bitegmic ovule of angiosperms, with
the outer integument corresponding to the cupule wall (Gaussen, 1946;
Stebbins, 1974; Doyle, 1978). The carpel itself might then be derived
from the sporophyll rachis by expansion and folding to enclose the
ovules. Doyle (1978) pointed out that relationships with Mesozoic seed
ferns would be consistent with the concept of Takhtajan (1969) that
many of the conspicuous innovations of angiosperms can be interpreted
in terms of paedomorphosis (phylogenetic shifting of juvenile features
to later stages of ontogeny): simple leaves and stamens, scalariform
secondary xylem pitting, condensation of sporophylls into flowers,
closed carpels (suggesting unopened conduplicate leaves), the primordial
state of ovules at fertilization, the reduced, partly free-nuclear megagametophyte, and lack of archegonia. Doyle argued that the reduced size,
rapid functioning, and changes in relative proportions of angiosperm
structures relative to those of seed ferns suggest an origin through

progenesis (paedomorphosis resulting from precocious reproductive maturity, as opposed to neoteny, paedomorphosis resulting from retarded somatic development: Gould, 1977). Since Gould associates progenesis with selection for high reproductive rates (*r*-selection), this hypothesis would be consistent with the idea that early angiosperms were weedy colonizing species, as proposed by Stebbins (1974) and supported by Doyle & Hickey (1976) on Cretaceous fossil evidence (cf. Crane, this volume, Chapter 5).

Also much-discussed are the predominantly Permian glossopterids of Gondwana, considered to be coniferopsids by Schopf (1976) but seed ferns by Gould & Delevoryas (1977) and most other authors. Glossopterids had simple leaves with simple reticulate venation and fructifications consisting of a leaf bearing one or more cupule-like structures on its adaxial side. Stebbins (1974) and Retallack & Dilcher (1981) argued that the ovulate structures could be transformed into an angiosperm carpel by reduction to one ovule per cupule (seen in one glossopterid, *Denkania*) and folding of the leaf.

This brief review by no means exhausts current hypotheses on angiosperm relationships. For example, several authors support a polyphyletic origin of angiosperms (e.g. Meeuse, 1963, 1972 *a*, *b*, *c*; Hughes, 1976; Krassilov, 1977), and Krassilov and Hughes have suggested that some angiosperms were derived from the Mesozoic ginkgophyte order Czekanowskiales, which had seeds enclosed in bivalved capsules.

Previous cladistic studies

Cladistic analysis offers the most rigorous method for formulating and testing hypotheses on relationships of groups, and hence for elucidating evolutionary events and processes involved in their origin. Cladistic studies attempt to reconstruct the branching pattern of phylogeny from the distribution of shared derived character states (synapomorphies), and hence to identify clades or monophyletic groups in the strict sense of Hennig (1966). The most widely accepted method for determining ancestral versus derived character states (polarity) is outgroup comparison: if one character state is restricted to the group in question but another occurs in related groups, the latter is considered to be ancestral. Most commonly, hypotheses of relationship are judged and character conflicts resulting from convergence and reversal (homoplasy) are resolved using the criterion of parsimony; that is, the scheme is preferred that requires the fewest character state changes. It should be noted that parsimony analysis does not assume that evolution follows a most parsimonious path; it is simply a method of finding the hypothesis that is best supported by the totality of known characters (Farris, 1983). The first detailed cladistic analysis of seed plants, which considered extant groups only, was performed by Hill & Crane (1982). They presented several almost equally parsimonious arrangements but favored one in which angiosperms are the sister group of conifers plus Gnetales, and cycads plus *Ginkgo* are the sister group of the angiosperm–conifer–gnetalean clade. This breaks up both cycadopsids and coniferopsids and calls into question the cycadopsid affinities of angiosperms. Unfortunately, Hill and Crane's study suffered from problems in character analysis (Doyle & Donoghue, 1986b): questionable polarity decisions, often because fossil groups were not considered; redundant characters (e.g. siphonogamy and non-motile sperm); vaguely defined characters (e.g. strobili); questionable scoring of groups; and omission of many potentially informative characters (most aspects of leaf architecture, organization of the ovule-bearing structures, and pollen morphology). Furthermore, over half of the 50 characters used are either invariant in seed plants or advances of terminal groups (autapomorphies) or Gnetales, and are therefore not informative in determining relationships among major groups. In addition, Hill and Crane did not take advantage of numerical cladistic (computer-assisted) methods, which are all but essential when character conflicts are common and equally or more parsimonious alternative relationships are easy to overlook.

This analysis was subsequently completely redone by Crane (1985), taking into account fossil groups, correcting most of the problems in character analysis seen in Hill & Crane (1982), and using numerical techniques. He concluded that coniferopsids were derived from platyspermic seed ferns, as proposed by Rothwell (1982), rather than from *Archaeopteris*-like progymnosperms (Meeuse, 1963; Beck, 1971, 1981), but he separated Gnetales from the coniferopsids and grouped them with angiosperms, Bennettitales, and *Pentoxylon* in a clade related to Mesozoic seed ferns, which are also platyspermic. He emphasized possible homologies of the cupules of Mesozoic seed ferns, the outer integument of the angiosperm ovule, and the layer that surrounds the ovule in *Pentoxylon* and some Bennettitales. As Crane noted, these results reconcile Arber & Parkin's (1907, 1908) views on relationships of angiosperms, Bennettitales, and Gnetales and more recent comparisons between angiosperms and Mesozoic seed ferns. However, this analysis is still unsatisfying in several respects. Some characters were coded in ways that bias the results toward particular hypotheses (e.g. one functional megaspore, integument, and micropyle were coded separately, and platyspermic seeds were coded as derived from radiospermic, thus favoring a single origin of the seed). Several characters were omitted that have been cited as evidence for alternative relationships (e.g. similarities in branching pattern in *Archaeopteris* and coniferopsids, anatomical similarities between Gnetales and coniferopsids). In some cases, the ancestral state was assumed to be present in fossil groups for which information is lacking (e.g. angiosperms are linked with Gnetales on one character, siphonogamy, but, as Crane recognizes, there is no reason to assume that siphonogamy did not exist in Bennettitales and *Pentoxylon*). For these reasons, it is difficult to judge how much stronger is the support for Crane's scheme than for alternatives.

In an attempt to overcome these problems, we undertook our own numerical cladistic study, which differs from previous analyses in several respects. First, we made a great effort to amass as many potentially informative (non-autapomorphic) characters as possible from all parts of the plant body, and to code them in ways consistent with major competing morphological theories. In addition, in order to assess the robustness of our results and the relative merits of alternative hypotheses, we adopted an experimental approach, asking the computer not only to produce the most parsimonious tree(s) but also to determine the lengths of alternative trees. One preliminary analysis was described in a discussion of relationships between angiosperms and Gnetales (Doyle & Donoghue, 1986*a*); results of that study differ from those summarized here (Doyle & Donoghue, 1986*b*) largely in grouping angiosperms, Bennettitales, and Gnetales with cycads and *Medullosa* rather than with *Caytonia* and glossopterids. This change is due largely to the recoding of sporophyll characters in cycads and ovule and cupule characters in Bennettitales, *Pentoxylon*, and angiosperms, on the basis of data and arguments of Crane (1985) and recognition of subtle biases in our previous character codings.

Methods

Choice of taxa for our analysis (Table 2.1) was necessarily a compromise based on: (1) a desire to represent all major seed plant groups and to recognize only monophyletic groups, (2) the variable quantity and quality of information available on different groups, and (3) potential relevance to major problems of seed plant phylogeny. Thus we treated *Ephedra*, *Welwitschia*, and *Gnetum* as separate taxa because we hoped to

Table 2.1. *Terminal taxa used by Doyle & Donoghue (1986b), with abbreviations used in Figure 2.1*

Aneurophyton s. lat., including *Triloboxylon* and *Eospermatopteris*	An
Archaeopteris s. lat., including *Svalbardia*	Ar
Early Carboniferous protostelic lyginopterids with multiovulate cupules	ML
'Higher' lyginopterids, including *Heterangium* and *Lyginopteris*	HL
Medullosa, not including *Quaestora* and *Sutcliffia*	Md
Callistophyton	Ca
Glossopteridales	Gl
Peltaspermum (*Lepidopteris, Antevsia*)	Pl
Corystospermaceae (*Dicroidium, Rhexoxylon, Umkomasia, Pteruchus*)	Cs
Caytonia (*Sagenopteris, Caytonanthus*)	Ct
Cycadales, including Nilssoniales	Cy
Bennettitales (= Cycadeoidales)	Bn
Pentoxylon	Pn
Euramerican cordaites, including *Cordaites, Cordaianthus*, and *Mesoxylon*	Cd
Ginkgoales, including *Baiera, Karkenia*, and *Ginkgo*	Go
Coniferales, including Lebachiaceae, Podocarpaceae, and Taxaceae	Cn
Ephedra	Ep
Welwitschia	We
Gnetum	Gn
Angiosperms	Ag

test the recurrent suggestion that Gnetales are polyphyletic. In order to test the hypothesis that cycadopsids and coniferopsids are derived independently from progymnosperms (Meeuse, 1963; Beck, 1971, 1981), we included two groups of progymnosperms (*Aneurophyton, Archaeopteris*) as well as seed plants. We split up many conventional taxa, such as seed ferns, because they are thought to be paraphyletic; that is, grade taxa, some members of which are more closely related to various 'higher' groups than to each other. Other groups may also be paraphyletic but contain many members that are too incompletely known to be sorted into assuredly monophyletic taxa; in such cases, we selected one particularly well-reconstructed or phylogenetically critical member for analysis (e.g. *Aneurophyton, Peltaspermum, Medullosa*). Where the amount of information on various organs is marginal, our decision on whether or not to include groups was often based on potential relevance to angiosperm relationships: thus we included *Caytonia* and glossopterids but not two interesting Permian coniferopsid groups, *Buriadia* (Pant & Nautiyal, 1967) and Angaran 'cordaites' (Meyen, 1984).

Czekanowskiales were excluded because so many key aspects of their morphology are unknown or obscure, and because most characters that are available can be interpreted in ginkgoalean terms (Meyen, 1984).

While this procedure resulted in many small taxa, several large, diverse groups such as conifers and angiosperms were left undivided because they have so many apomorphies that they can be assumed safely to be monophyletic. When characters vary within such groups, we attempted to identify the basic conditions, which were usually clear from comparison with any of several plausible outgroups (e.g. monosulcate pollen in angiosperms), probable relationships within the group, and the stratigraphic record. Thus we used magnoliid dicotyledons and pre-Albian fossils as guides in coding angiosperms and Permo-Carboniferous Lebachiaceae in coding conifers. Since many authors have questioned whether angiosperms are monophyletic, it is worth noting that they are united by at least nine apomorphies: sieve tubes and companion cells derived from the same initials, stamens with two lateral pairs of pollen sacs, a closed carpel with stigmatic pollen germination, a hypodermal endothecium in the anther, lack of a laminated endexine, a megaspore wall without sporopollenin, a three-nuclear male gametophyte with neither prothallials nor a sterile cell, a megagametophyte with only eight nuclei (or various related conditions), and double fertilization associated with endosperm formation. There are some exceptions to these characters, but they occur in taxa that seem to be well-enough nested within angiosperm groups to assume that they represent secondary reversals or elaborations. In order to support a polyphyletic origin of angiosperms, it must be shown that these characters are outweighed by synapomorphies between particular angiosperm subgroups and different gymnosperm groups.

Choice of characters involves preliminary hypotheses on homology, by which we mean any trait inherited from a common ancestor and its subsequent transformations. Potential homology may be recognized by analysis of positional and developmental relationships of structures (Kaplan, 1984), but the ultimate test of homology is congruence with the totality of evidence on phylogenetic relationships. In seed plants, several alternative derivations have been proposed for many structures, with varying degrees of plausibility, and it would be easy to reject possibilities prematurely because of unconscious preference for one or another morphological theory or some overly strict criterion of similarity.

In numerical methods of the sort used here, characters are coded in binary form: 0 (which we used uniformly for the ancestral state), 1 (the derived state), and X (missing data, which takes on the value that gives the most parsimonious result for a given position of a taxon). A multistate character with a linear series of three states is binary-coded as 00, 10, 11; independent origin of two derived states is coded as 00, 10, 01. In our analysis we used 62 binary characters, defined briefly in the Appendix (p. 42; for details, see Doyle & Donoghue, 1986*b*). Toward our goal of testing alternative hypotheses of seed plant evolution in as unbiased a way as possible, we allowed relatively wide leeway in treating similarities as potential homologies. In certain cases where the morphology of structures is particularly obscure or controversial (e.g. ovule symmetry in angiosperms, 'sporophylls' of glossopterids), or where there is controversy over which of two states is primitive within a group (e.g. presence or absence of vessels in angiosperms), we attempted to avoid bias by scoring groups X. In compiling characters, we consistently excluded autapomorphies; although these are important in determining that the groups used are monophyletic, they contribute nothing to the understanding of relationships between groups and give a false sense of the amount of information present in the matrix. We also attempted to eliminate redundant (developmentally correlated) characters, usually identified as such because they changed simultaneously on cladograms obtained in preliminary analyses and could be plausibly attributed to the same morphogenetic factors (e.g. extension of circular-bordered pitting into both the metaxylem and the protoxylem in conifers, ginkgos, and Gnetales).

Polarities were determined by outgroup analysis (Maddison, Donoghue & Maddison, 1984), which necessitated a preliminary analysis of relationships among vascular plants as a whole (Doyle & Donoghue, 1986*b*). We concluded that progymnosperms and seed plants form a monophyletic group, united by possession of both secondary xylem and phloem, periderm, and cortical fiber strands (lost in more advanced seed plants). Several other advances of living seed plants in organs that are not preserved in progymnosperms (several apical cells, loss of neck canal cells in the archegonium, a free-nuclear stage in early embryogenesis, and an embryo with shoot apex, root apex, and suspensor in a row) could be synapomorphies of progymnosperms and seed plants or of seed plants alone. Cladoxylales, sphenopsids, and ferns are the closest outgroup(s) of progymnosperms and seed plants, trimerophytes, the next outgroup. Comparison with these groups indicates that *Aneurophyton* retains the ancestral state for the group in all characters considered;

hence, it can be used as a functional outgroup for assessment of polarities in the remaining taxa (the functional ingroup: Watrous & Wheeler, 1981).

On this basis, deciding on polarity and the resulting coding was often relatively easy. The greatest difficulties arose in coding multistate characters, where the ancestral state is clear but there are alternative interpretations of relationships of the derived states. A good example concerns major categories of leaf morphology. Most workers agree that the pinnately compound leaves of seed ferns were derived from branch systems bearing dichotomous leaves, as in Devonian progymnosperms (Meeuse, 1963; Beck, 1971, 1981; Rothwell, 1982; Meyen, 1984), and the once-pinnately organized leaves of cycads, Bennettitales, and angiosperms are clearly derived relative to fronds. Progymnosperms can thus be coded 000, seed ferns 100, and once-pinnate groups 110. However, there is disagreement over the homologies of the linear-dichotomous leaves of coniferopsids. On the basis of similarities in branching patterns and anatomy, Beck (1971, 1981) suggested that coniferopsids were independently derived from *Archaeopteris*-like progymnosperms. The leaves of coniferopsids, which are *Archaeopteris*-like in groups such as *Ginkgo*, would thus be directly homologous with those of progymnosperms, although at least one change would have to be postulated: coniferopsids with dichotomous leaves also have cataphylls (scale leaves), like other seed plants, but progymnosperms do not. More recently, however, on the basis of the discovery of coniferopsid characters such as platyspermic seeds and saccate pollen in the Late Carboniferous seed fern *Callisto-phyton*, Rothwell (1982) proposed instead that coniferopsids were derived from seed ferns, perhaps by heterochrony: suppression of the fronds and production of cataphylls throughout the life of the plant (cf. also Meyen, 1984). Coding coniferopsids 001, one step from progymnosperms, would be appropriate under the Beck hypothesis but would bias against the Rothwell hypothesis by making derivation from seed ferns take two steps, whereas coding coniferopsids 101, one step from seed ferns, would bias against the Beck hypothesis. We attempted to avoid these biases by coding coniferopsids X01, one step from either progymnosperms (where X = 0) or seed ferns (where X = 1). We coded the linear leaves of *Ephedra* and *Welwitschia* XX1 in order to allow for a still broader range of prototypes, since it has been postulated that Gnetales are related to groups coded both X01 (coniferopsids) and 110 (Bennettitales, angiosperms). Subtle biases can be introduced by this technique, since it adds an extra step (or steps) when an X-coded state is positioned between two other states, but these are unlikely to be serious in the present case (Doyle & Donoghue, 1986*b*).

We used similar systems to code sporophyll morphology and seed symmetry, which also show different pathways of derivation under the Beck and Rothwell hypotheses. We also used X to code autapomorphies that may be derived either directly from the ancestral state or from another derived state (e.g. the three-nucleate microgametophyte of angiosperms versus the more complex basic condition and the partially reduced condition in *Welwitschia* and *Gnetum*).

For our analyses, we used two basically similar programs: the Wagner parsimony algorithm in PHYSYS (Mickevich & Farris, 1982) and the Mixed Method Parsimony algorithm with the Wagner option in PHYLIP (Felsenstein, 1985). Both programs attempt to find the branching diagram that minimizes the total number of character state transitions, treating forward changes and reversals equally. We attempted to evaluate alternative hypotheses by adding 'dummy' synapomorphies to the matrix to force particular taxa together and then subtracting the corresponding numbers of steps after analysis, or by employing the user tree option in PHYLIP, which allows one to specify whole trees and determine their length.

Because of the large number of possible trees, present methods cannot guarantee finding the most parsimonious tree(s) with large data sets. In Wagner algorithms, taxa are added sequentially to the analysis in the most parsimonious position, and what trees are found depends in part on what taxa have already been entered. PHYSYS determines the order of entry based on an advancement index, but in PHYLIP the order of entry is specified by the user. Finding most parsimonious trees with both programs required considerable experimentation and familiarity with possible alternative relationships, much of it gained from preliminary analyses. PHYSYS improves the chances of finding most parsimonious trees by global branch-swapping, but the shortest trees that we obtained with PHYSYS (124 steps) were found by forcing together taxa using dummy characters. Some of our shortest trees (123 steps) were obtained with PHYLIP, by judicious shuffling of the order of entry (entering taxa roughly in order of advancement but placing possible 'linking' taxa in various arrangements before specialized and problematical ones) and with the user tree option; others were found by W. E. Stein (personal communication), using the PAUP program of D. L. Swofford.

Results

One of our most parsimonious cladograms is shown in Figure 2.1; others differ in reversing Bennettitales and *Pentoxylon* and/or in rearranging *Callistophyton*, coniferopsids, corystosperms, and cycads in various ways. Arranged in pectinate fashion from the base are the progymnosperms *Aneurophyton* and *Archaeopteris*, two groups of Carboniferous lygino-pterid seed ferns, the Late Carboniferous seed fern *Medullosa*, and a major clade that includes all extant groups of seed plants. Seed plants are thus a monophyletic group, initially united by cataphylls and pinnately compound leaves, multiseriate rays, fused pollen sacs, radiospermic seeds with a lagenostome, and probably axillary branching (all modified or reversed in some members). *Medullosa* and higher seed plants are linked by loss (or transformation into a new integument) of the lygino-pterid cupule, reduction of the lagenostome to a 'normal' pollen chamber, and bilateral pollen.

Excluding later reversals, the clade above *Medullosa* is united by normal eusteles, platyspermic seeds, saccate pollen, and linear megaspore tetrads. The representative that has retained the most primitive charac-ters is the Late Carboniferous seed fern *Callistophyton*. Coniferopsids are either basal or nested within the clade, supporting the hypothesis that they were derived from *Callistophyton*-like platyspermic, saccate seed ferns (Rothwell, 1982) rather than *Archaeopteris*-like progymno-sperms (Beck, 1971, 1981). However, trees with coniferopsids linked with *Archaeopteris* are only one step longer (124 steps), implying that the Beck hypothesis is still a viable alternative. A sulcus and pollen tube and abaxial microsporangia arise independently within coniferopsids and below *Callistophyton*. The remaining groups, which include Permian and Mesozoic seed ferns, are initially united by multilacunar nodes (a convergence with medullosans) and secondarily free microsporangia (also seen in coniferopsids). Unexpectedly, cycads are linked with the Permo-Triassic seed fern *Peltaspermum* (on secondary loss of saccate pollen), not with medullosans as is often suggested; this means that cycads are secondarily radiospermic, which is consistent with the fact that seeds of Permian taeniopterids (considered as primitive cycads by Mamay (1976)) appear to be flattened and *Cycas* seeds have bilateral symmetry (Meyen, 1984).

The angiosperms also belong in the platyspermic group, forming a clade with Bennettitales, *Pentoxylon*, and Gnetales, initially united by leaves with simply pinnate organization, scalariform pitting in the

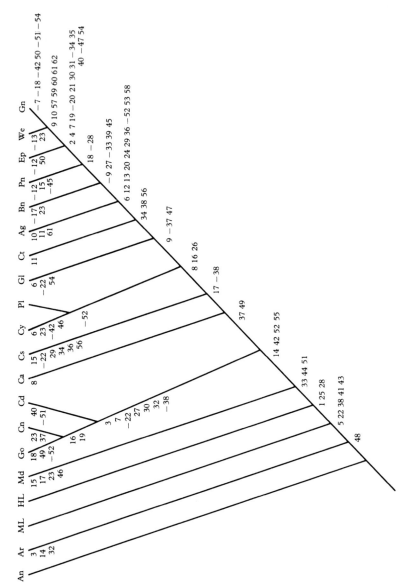

Figure 2.1. Representative most parsimonious 123-step cladogram of progymnosperms and seed plants (Doyle & Donoghue, 1986b). Taxa are defined in Table 2.1, characters in the Appendix. Minus signs before characters indicate reversals.

secondary xylem, once-pinnate microsporophylls, one ovule per cupule, secondarily non-saccate pollen, granular exine structure, and syndetocheilic stomata. Other possible synapomorphies are a tunica layer in the apical meristem, siphonogamy, and lignin chemistry (Mäule reaction), characters seen in angiosperms and Gnetales but not preserved in fossils. Since all four groups show strong aggregation of sporophylls into flower-like reproductive structures, we will refer to this clade as the anthophytes. It may be objected that the term anthophyte is already in use as a synonym for angiosperms, but we have deliberately extended it in order to emphasize the conclusion that flowers are not a unique advance of angiosperms but rather an older feature of the larger group to which they belong. This contrasts sharply with recent suggestions that flowers originated within the angiosperms (Meeuse, 1963, 1972*a*, *b*, *c*; Krassilov, 1977; Dilcher, 1979; Meyen, 1984). Recognition of the anthophyte clade is a striking agreement with Crane (1985), although our scheme differs slightly in relating anthophytes most closely to *Caytonia* and glossopterids, whereas he interpolated corystosperms between *Caytonia* and anthophytes. Glossopterids are linked with *Caytonia* and anthophytes on reticulate venation (later lost below Bennettitales), a shift away from abaxial microsporangia, and a thick nucellar cuticle; *Caytonia* is linked with anthophytes on anatropous cupules (presumably homologous with the glossopterid cupule), secondary formation of microsynangia, and reduction of the megaspore wall. The cupules of glossopterids and *Caytonia* presumably represent enrolled leaflets bearing laminar ovules, since the original lyginopterid cupule was lost several nodes below.

Within anthophytes, our results imply that angiosperms are the sister group of the remaining taxa. Bennettitales, *Pentoxylon*, and Gnetales share several advances over angiosperms: erect, solitary ovules (cupules), whorled microsporophylls, and possibly a micropylar tube (if lost in *Pentoxylon*). This scheme again differs slightly from that of Crane (1985), who groups angiosperms with Gnetales (as proposed by Arber & Parkin, 1908) and Bennettitales with *Pentoxylon*; as noted above, Crane links angiosperms and Gnetales on siphonogamy, which may be basic in anthophytes. The three genera of Gnetales are in turn united by multiple axillary buds, opposite-decussate leaves, vessels, loss of scalariform secondary xylem pitting, extension of circular-bordered pitting into the protoxylem, one-veined microsporophylls, a single terminal ovule, loss of the cupule, origin of a new outer integument from the perianth of the ovulate flower, compound strobili, reduction of the nucellar cuticle, and

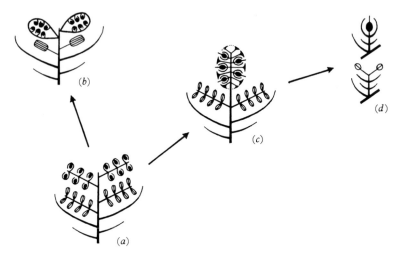

Figure 2.2. Major transformations in reproductive structures of anthophytes inferred from Figure 2.1. (a) Hypothetical common ancestor; (b) angiosperms; (c) Bennettitales; (d) Gnetales.

probably linear leaves and striate pollen (assuming reversals in *Gnetum*). Of these, linear leaves, the pitting characters, reduced microsporophylls, and compound strobili are convergences with coniferopsids. Within Gnetales, *Welwitschia* and *Gnetum* share additional advances, many of them convergences with angiosperms: vein anastomoses, interpolated higher-order veins, reduction of the male gametophyte, a tetrasporic megagametophyte with free-nuclear eggs, cellular embryogeny, and a feeder in the embryo. Although Gnetales are thus the closest living relatives of angiosperms, most of the commonly cited similarities between the two groups (vessels, dicotyledon-like leaves, simple stamens, embryology) arose independently, and most of the homologies are rather cryptic (siphonogamy, tunica–corpus, lignin chemistry, reduced megaspore wall, granular exine). As we discuss elsewhere (Doyle & Donoghue, 1986a), this goes far toward explaining why angiosperm–gnetalean relationships have been widely rejected (including by one of us: Doyle, 1978).

Inferences concerning floral evolution in the anthophytes are summarized in Figure 2.2. Although our results indicate that Gnetales are the closest living relatives of angiosperms, they in no way support derivation of angiosperms from Gnetales, as in Wettstein's (1907) Pseudanthial Theory. Rather, they are more consistent with the views of Arber & Parkin (1907, 1908), that angiosperms, Bennettitales, and Gnetales were derived from a common ancestor with bisexual flowers and pinnate megasporophylls and microsporophylls, and that the flowers of Gnetales were secondarily reduced and aggregated in response to wind pollination, like the flowers of Amentiferae within angiosperms. In retaining relatively leaf-like carpels, angiosperms are primitive, but they are advanced in having simplified stamens (with two pairs of pollen sacs representing synangia of the *Caytonia* or bennettitalean type). Bennettitales are primitive in retaining pinnate microsporophylls, but megasporophylls were reduced to single ovules in the common ancestor of Bennettitales and Gnetales, and Gnetales went on to simplify the microsporophylls as well. Whether anthophyte flowers were originally bisexual is equivocal but consistent with present data (Doyle & Donoghue, 1986*b*). We find the agreements with Arber & Parkin (1907, 1908) all the more significant because our results were obtained with a large set of characters, many of which were not known to them, and without their speculative assumption that the clade had bisexual strobili originally.

Experiments

These results go far toward bridging the gap between angiosperms and other seed plants, since they allow angiosperm flowers, carpels and bitegmic ovules to be homologized with specific structures in related groups. The same is true of the wood anatomy, stomatal structure, and granular monosulcate pollen of primitive angiosperms, all of which resemble Bennettitales. However, there are several reasons for concern. Even our most parsimonious trees include a large amount of homoplasy: the presence of 123 steps with 62 binary characters means, on average, almost one convergence or reversal per character. When homoplasy is common, many almost equally parsimonious arrangements can be expected, corresponding to different concepts of which shared advances are homologies and which are convergences, and the results may be unstable (sensitive to addition, deletion, or reinterpretation of characters). Furthermore, although our results are generally consistent with the stratigraphic record, derivation of angiosperms and Bennettitales plus Gnetales from a common ancestor indicates that the line leading to angiosperms existed at least as far back as Bennettitales (Late Triassic), but convincing angiosperm remains are not known until the Cretaceous. The loss and reappearance of reticulate venation within the anthophytes may also seem implausible. These problems led us to undertake a series of computer experiments, designed to test the robustness of the results and the relative merits of alternatives. In general, these experiments indicate that several major clades are quite stable (i.e. hard to break up without adding a large number of steps), but they can be placed almost equally parsimoniously in several different positions.

Experiments relating to non-angiospermous groups are discussed in detail by Doyle & Donoghue (1986*b*); here we will concentrate on experiments that bear on angiosperm relationships. Our primary result is that the position of the anthophyte clade is somewhat unstable, but the group itself is robust, implying that its unity can be assumed with some confidence and used as a basis for further inferences. This can be seen by comparing the lengths of trees in which anthophytes as a whole, and particular subgroups alone, are moved to alternative positions. For example, moving all anthophytes into a clade with cycads and *Medullosa* (as in Doyle & Donoghue, 1986*a*) adds four steps (127), whereas associating Bennettitales and Gnetales with cycads and *Medullosa* but leaving the angiosperms linked with *Caytonia* and glossopterids adds nine steps.

The least securely included group in the anthophytes appears to be the Gnetales; only four extra steps are needed to move them into the coniferopsids (linked with ginkgos on primary xylem pitting, two-trace nodes, and non-saccate pollen). However, trees only two steps longer than our best trees are obtained when the anthophytes as a whole are moved to the same position, with Gnetales as the sister group of *Pentoxylon*, Bennetttitales, and angiosperms. This rather disconcerting result suggests that the last three groups originated from coniferopsids via Gnetales-like intermediates, as envisioned for angiosperms by Wettstein (1907); however, the fact that Bennettitales and *Pentoxylon* are interpolated between angiosperms and Gnetales would suggest that the angiosperm flower originated by elaboration rather than by aggregation of simple units into pseudanthia. Since ginkgos are the sister group of anthophytes in this scheme, it recalls the suggestion of Krassilov (1977) that angiosperms are derived from Czekanowskiales, which we interpret as ginkgophytes (cf. Meyen, 1984). We find such 'neo-englerian' trees highly implausible in morphological terms, since they require first drastic reduction and then re-elaboration of leaves and sporophylls into a pinnate pattern convergent with that of Mesozoic seed ferns and cycads, a shift from cycadopsid to coniferopsid anatomical features and back again, and an origin *de novo* of the cupule or outer integument. It is easier to imagine that the similarities between the simple appendages of Gnetales and coniferopsids are due to independent reduction. These results show that earlier authors were right in seeing evidence for relationships between the Gnetales and coniferopsids, but the links between the Gnetales and other anthophytes are stronger. The most parsimonious alternative positions of the anthophytes are found with different arrangements of cycads and the various platyspermic taxa. Relationships among these groups are highly unstable, undoubtedly reflecting the lack of data on many characters in Permian and Mesozoic fossils. We obtained a large number of 124-step trees with cycads interpolated between *Medullosa* and the platyspermic clade, in which case their commonly cited similarities with medullosans (multilacunar nodes, secretory canals, nucellar vasculature) could be homologies, and the ancestor of all higher groups would be more like *Medullosa* than *Callistophyton*. Most of these trees associate coniferopsids with glossopterids and/or *Peltaspermum*, which implies that the lack of a sulcus in cordaites and primitive conifers is a reversal rather than a primitive retention. Some trees link anthophytes with corystosperms alone, with *Caytonia* and corystosperms (in either order) but not glossopterids, or with all three groups in various orders (cf. Crane, 1985). Others diverge more fundamentally in linking *Caytonia* directly with angiosperms, on the basis of reticulate venation and flat guard cells; however, this is not a major challenge to the unity of the anthophyte clade, since the closest relative of the anthophytes is simply shifted inside the group. These variations have generally consistent morphological implications, since all three seed fern groups have cupules and sporophylls that can be homologized with angiosperm structures in similar ways.

Many of these uncertainties might be resolved by new data on missing characters in fossil groups. For example, the presumed homology between bitegmic angiosperm ovules and the cupules of Mesozoic seed ferns implicitly requires that both structures are derived from circinately enrolled leaflets with ovules on their *adaxial* surface. This is based not on the fact that angiosperm ovules are borne on the adaxial side of the carpel but rather on the positions of the nucellus plus inner integument (equivalent to the original unitegmic ovule), the funicle (the basal part of the leaflet), and the micropyle (its reflexed tip) relative to the whole carpel (Doyle, 1978, p. 384). We did not include adaxial versus abaxial ovule position in the data matrix because relevant information is lacking in many critical groups. Harris (1940) presented indirect evidence that the cupules of *Caytonia* are oriented adaxially, but this is questioned by G. Retallack (personal communication). One problem is that ovule position appears to be abaxial in other platyspermic groups (e.g.

Callistophyton, peltasperms; Meyen, 1984). If the cupules of glossopterids were borne facing the subtending blade, as reconstructed by Gould & Delevoryas (1977), ovule position would also be an obstacle to a close relationship between angiosperms and glossopterids. Retallack & Dilcher (1981) circumvented this problem by comparing angiosperms with the glossopterid genus *Denkania*, which had orthotropous, uniovulate cupules, and by interpreting angiosperms as being primitively orthotropous. However, this may have been unnecessary, since Pant & Nautiyal (1984) have reported that ovules in the glossopterid fructification *Ottokaria* were oriented adaxially.

A final reason for confidence in relationships among angiosperms, Bennettitales and Gnetales is that they share several features excluded from the original data set because of problems in interpretation that may, with hindsight, be additional synapomorphies. One of these, flower-like strobili, has already been discussed. Anthophytes also show several striking parallel trends; these cannot be used as synapomorphies, but they may reflect shared genetic advances (cf. Cantino, 1985). Ehrendorfer (1976) noted that Gnetales differ from other gymnosperms and resemble angiosperms in having relatively small chromosomes, less repetitive DNA, and extensive polyploidy. Gnetales also show fusion of the second sperm nucleus with one or another nucleus of the megagametophyte (Martens, 1971), a possible precursor of double fertilization of the angiosperm type. Angiosperms and Gnetales also show parallel trends for acceleration of the life cycle and associated paedomorphic structural features (i.e. progenesis), and some tendency in this direction may be suspected in Bennettitales and *Caytonia*, on the basis of their small seed size. Anthophytes also show strong tendencies for adaptation to hot and/or dry conditions. This is clearest for angiosperms and Gnetales, which were most abundant and diverse in the Early Cretaceous tropics (Africa–South America, southern China), associated with evidence of aridity (Brenner, 1976; Doyle, Jardiné & Doerenkamp, 1982). However, it is also true for Bennettitales, which had a predominantly low-latitude distribution and were one of the dominant groups in southern Eurasia during the Late Jurassic, a time of widespread aridity (Vakhrameev, 1970). There is also evidence that early angiosperms and Gnetales both tended to occupy disturbed flood-plain habitats (Doyle & Hickey, 1976; Doyle *et al.*, 1982; Upchurch & Crane, 1985; Crane, this volume, Chapter 5), supporting the idea that they were colonizing species, as proposed for early angiosperms by Stebbins (1974). The remarkable vegetative similarities between angiosperms and *Gnetum* are also easier to understand if they represent parallel responses to similar selection pressures acting on plants with a relatively recent common ancestor rather than on members of very distantly related clades.

Within anthophytes, the strength of different links varies considerably. It is highly unparsimonious to force angiosperms into Gnetales, as the sister group of *Welwitschia* and *Gnetum* (133 steps) or *Gnetum* alone (136 steps), as implied by suggestions that angiosperms are derived from Gnetales (e.g. Muhammad & Sattler, 1982). However, only two extra steps (125) are needed to link angiosperms directly with the Bennettitales, in which case the sister group of anthophytes is corystosperms rather than *Caytonia*. Three extra steps are required to link angiosperms with Gnetales, as proposed by Crane (1985).

General implications

Some further implications of our results for the evolution of angiosperms and related groups may be discussed with reference to Figure 2.3, which shows three phylogenetic trees and associated character changes plotted against time, with *Pentoxylon* omitted for clarity.

As already noted, our most parsimonious cladograms (Figures 2.1, 2.3(*a*)) imply that simplified leaves, syndetocheilic stomata, non-saccate granular pollen, and aggregation of sporophylls into flowers are homologies of angiosperms and other anthophytes, and that reduced megasporophylls, whorled microsporophylls, and a micropylar tube are further advances of Bennettitales and Gnetales. A problem is that this arrangement entails a gap in the record of the angiosperm line from the Late Triassic to the Early Cretaceous. The idea that angiosperms underwent much of their radiation before the Cretaceous (Axelrod, 1952, 1970) conflicts with present evidence for rapid diversification during the Cretaceous, but the earlier existence of primitive angiosperms cannot be ruled out (Doyle, 1969, 1978; Muller, 1970). Late Triassic monosulcate pollen with tectal perforations and columellae (Cornet, 1977) and Jurassic dicotyledon-like leaves (e.g. *Phyllites* Seward: Crane, 1985) could represent such plants. Alternatively, many advances (autapomorphies) of the angiosperms may not have evolved until the Cretaceous. Our scheme predicts that early members of the angiosperm line (like our hypothetical pro-anthophyte, Figure 2.2(*a*)) might be indistinguishable from Bennettitales in most characters commonly preserved in fossils – leaf morphology, stomata, wood anatomy, and pollen morphology – that is, in most respects other than nodal anatomy,

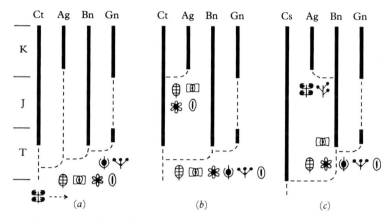

Figure 2.3. Alternative scenarios for the evolution of angiosperms and related groups. (*a*) Angiosperms as the sister group of other anthophytes, as in the most parsimonious cladograms (Figure 2.1, 123 steps); (*b*) angiosperms linked with *Caytonia* (124 steps); (*c*) angiosperms derived from some member of Bennettitales (consistent with some 125-step cladograms). T = Triassic, J = Jurassic, K = Cretaceous, Ct = *Caytonia*, Ag = angiosperms, Bn = Bennettitales, Gn = Gnetales, Cs = corystosperms. Characters indicated (reading left to right and down in (*a*)): solitary orthotropus cupulate ovule; whorled microsporophylls; once-pinnate leaf; syndetocheilic stomata; sporophylls aggregated into flowers; granular monosulcate pollen; pinnate megasporophyll with anatropous cupulate ovules.

multicupulate megasporophylls, and spirally arranged microsporophylls (another idea anticipated by Arber & Parkin, 1907). This highlights the need for a search for reproductive structures associated with the many 'bennettitalean' genera that are known only as leaves. If the record of other adaptive radiations is any guide, a great variety of experimental anthophyte lines probably existed in the Triassic and Jurassic, of which the known groups are only the most common.

In almost equally parsimonious (124-step) cladograms with angiosperms linked with *Caytonia* (Figure 2.3 (*b*)), the reticulate venation and flat guard cells of the two groups are homologous rather than convergent, but simplified leaves, syndetocheilic stomata, granular pollen, and flowers originate independently in angiosperms and the bennettitalean–gnetalean line (or were lost in *Caytonia*). Presumably insect pollination, the ability to colonize unstable and/or arid habitats, and acceleration of the life cycle also arose independently in angiosperms and other anthophytes. A weakness of this scheme is that most of the characters that potentially unite the anthophyte-*Caytonia* clade are unknown in *Caytonia*: not only presence of a tunica, lignin chemistry and siphonogamy, which are also undocumented in Bennettitales and *Pentoxylon*, but also scalariform pitting. However, this tree has the advantage of eliminating the stratigraphic gap in figure 2.3(*a*): since *Caytonia* is more primitive than angiosperms in all known characters, angiosperms could have evolved from some species of *Caytonia* at any time up to the Cretaceous.

As noted above, the next-best arrangement within the anthophytes links angiosperms directly with Bennettitales (125 steps). A closely related hypothesis that deserves special attention (Figure 2.3(*c*)) is that angiosperms were directly derived from some member of the Bennettitales, making Bennettitales a paraphyletic group. This would imply that angiosperms reverted to megasporophylls with several anatropous bitegmic ovules (cupules), spirally arranged microsporophylls, and no micropylar tube, and lost the few supposed autapomorphies of Bennettitales (unilacunar nodes, secretory canals, interseminal scales). However, this hypothesis would eliminate the stratigraphic gap in Figure 2.3(*a*). Since Bennettitales were a diverse group, recognition of special derived similarities between angiosperms and particular Bennettitales could make this alternative more parsimonious. Reinterpretations of basic conditions within angiosperms that would have the same effect are also conceivable. For example, we scored angiosperms as having multilacunar or trilacunar nodes, but nodal anatomy is diverse in magnoliids, and there is controversy over whether unilacunar, trilacunar or multilacunar nodes are ancestral (Bailey, 1956; Benzing, 1967; Takhtajan, 1969). Likewise, orthotropous ovules are typical of one magnoliid order, Piperales, including Chloranthaceae, which have pollen and leaves similar to some of the oldest Cretaceous angiosperms (Walker, 1976; Burger, 1977; Muller, 1981; Upchurch, 1984). The most implausible change required is derivation of carpels from stalked ovules, but S. V. Meyen (personal communication) suggested a speculative mechanism, gamoheterotopy, that might produce this result, parallel to the hypothesis that the ear of maize originated from the staminate inflorescence of teosinte, rather than the two-ranked pistillate inflorescence, by a regulatory mutation (Iltis, 1983). Some Bennettitales had leaf-like microsporophylls with microsynangia borne in two rows on the adaxial surface; extension of the developmental program for such

structures to the ovuliferous receptacle could result in flat megasporophylls with cupules substituted for microsynangia, like the hypothetical
primitive carpel.

Since Gnetales have many of the same features that constitute obstacles
to a bennettitalean origin of angiosperms, similar reinterpretations might
also increase the plausibility of a direct relationship between angiosperms
and Gnetales (Crane, 1985), now three steps less parsimonious than our
best arrangement, especially if any additional 'gnetalean' features were
reinterpreted as basic within angiosperms. Again, candidates for such
characters exist in the Chloranthaceae (opposite leaves; swollen, two-
trace nodes; inflorescences composed of small, apetalous flowers with
uniovulate carpels). Gnetales have so many additional advances that
derivation of angiosperms from Gnetales is highly unlikely. However,
the two groups might both be derived from a bennettitalean line that
began a trend for leaf simplification and floral reduction, which was
continued in Gnetales but reversed in angiosperms.

These schemes help to put in clearer perspective possible causal factors
in the origin of angiosperms. The scheme in which angiosperms are
directly linked with *Caytonia* (Figure 2.3(*b*)) implies that the origin of
flowers and acceleration of the life cycle occurred in the angiosperm line
after its separation from its closest sister group and were therefore
intimately tied with the origin of angiosperms, as assumed by Stebbins
(1974), Doyle & Hickey (1976), and Doyle (1978, 1984). This would be
consistent with suggestions that the origin of angiosperms from their
immediate ancestors can be explained by paedomorphosis (Takhtajan,
1969), more specifically progenesis (Doyle, 1978). However, if angiosperms are the sister group of Bennettitales plus Gnetales (Figure 2.3(*a*)),
which we consider most likely, or if angiosperms are derived from
Bennettitales (Figure 2.3(*c*)), these same features originated earlier in the
evolution of the anthophyte clade. Here the possible role of progenesis
is more complex: some paedomorphic traits (gametophyte reduction,
simplified stamens, carpel closure) arose in the angiosperm line, but
others are basic for anthophytes as a whole (scalariform pitting, aggregation of sporophylls, small seeds, reduced megaspore wall). This
suggests that factors favoring progenesis were operating on the anthophytes since their origin, although they were continued and intensified
in angiosperms and Gnetales. This is consistent with the fact that all
three major anthophyte groups show evidence of adaptation to arid
and/or disturbed environments, but that this tendency is strongest in
early angiosperms and Gnetales. Similarly, insect pollination might be

proposed as a causal factor in the origin of angiosperms under the scheme in Figure 2.3(*b*), but under the others (Figure 2(*a*) and (*c*)) it was presumably established much earlier.

Under all three schemes, the closed carpel, stigmatic pollen germination, and double fertilization with endosperm formation arose only in angiosperms, whereas broad leaves with reticulate venation, vessels, simple microsporophylls, and reduced gametophytes evolved independently in angiosperms and Gnetales. All of these changes must be explained if the origin of angiosperms is to be understood. Where similar traits arose independently in angiosperms and Gnetales, comparisons between the two groups may clarify how and why such traits evolved. Thus the concentration of Early Cretaceous angiosperms and Gnetales in presumed seasonally arid tropical areas supports the idea that vessels originated as an adaptation to aridity, as suggested for Gnetales by Carlquist (1975). Once evolved, vessels may have been a preadaptation for evolution of large, undissected leaves in the tropics, since such leaves would tend to overheat even in wet tropical conditions unless vessels are present to allow rapid transpiration (Doyle *et al.*, 1982).

The implications of these results for factors in the rise of angiosperms are explored in detail by Doyle & Donoghue (1986*b*), on the basis of the idea that the relative diversity of clades is a function of speciation and extinction rates, and that some traits may have incidental effects on speciation rate, independent of why they evolved (Gould & Eldredge, 1977; Stanley, 1979; Vrba, 1983). Insect pollination might lead to higher speciation rates, by making possible pollinator-mediated isolating mechanisms, and to lower extinction rates, by allowing species to maintain more dispersed distributions of individuals and thus to escape herbivores and pathogens (Janzen, 1970; Regal, 1977; Stanley, 1979; Burger, 1981; Doyle *et al.*, 1982; Doyle, 1984; but see Stebbins, 1981). However, insect pollination probably existed in Bennettitales, Gnetales, and the angiosperm line itself since the Triassic. Possession of vessels and intercalary meristems (Stebbins, 1974, 1981), allowing leaf expansion, may have helped angiosperms to exploit tropical habitats more effectively, but the same features also arose in Gnetales. Some uniquely angiospermous trait(s) must be invoked to explain the Cretaceous expansion of angiosperms and their replacement of other anthophytes. We suggest that closure of the carpel may have played a key role, by raising speciation rates still higher. Carpel modifications for dispersal might lead to more frequent establishment of isolated populations, while stigmatic pollen germination might increase the probability that mutations would result in incompatibility with pollen from partly differentiated populations. Since every new species represents an evolutionary experiment, high speciation rate alone might lead in the long term to occupation of more and more niches by angiosperms and their piecemeal replacement of other groups (Stanley, 1979; Doyle, 1984), independent of any superior features common to all angiosperms.

Although these scenarios are necessarily speculative and leave many questions unanswered, the analysis on which they rest greatly reduces the number of hypotheses on the origin of angiosperms that need to be seriously considered and makes a great number of detailed predictions that can be tested by future work. On the neontological front, a fruitful approach may be study of appropriate DNA sequences, which should show closer relationships between Gnetales and angiosperms than between any of the other major extant seed plant groups. On the palaeobotanical front, critical evidence could come from data on cryptic characters in Bennettitales (meristem type, lignin chemistry, embryology), stem anatomy and cupule orientation in *Caytonia* and other Mesozoic seed ferns, and morphological diversity within the anthophyte clade.

Appendix

Characters used by Doyle & Donoghue (1986*b*). 0 is used for the presumed ancestral state, 1 for the derived state, and X for missing information (see pp. 25–8). When only one state is listed, it is the derived state.

1. 0 = branching apical; 1 = axillary.
2. 0 = axillary buds single; 1 = multiple.
3. Leaves on (homologs of) progymnosperm penultimate order branches.
4. 0 = phyllotaxy spiral; 1 = opposite-decussate or whorled.
5–7. 000 = simple, dichotomous leaves only; 100 = pinnately compound leaves and cataphylls; 110 = simple (or dissected) pinnately veined leaves and cataphylls; X01, XX1 = simple, one-veined leaves only, or linear or dichotomous leaves and cataphylls.
8. 0 = rachis regularly bifurcate; 1 = usually or always simple.
9, 10. 00 = one order of laminar venation, open; 10 = one order of laminar venation, reticulate; 11 = two or more orders of laminar venation, at least one order reticulate.
11. 0 = poles of guard cells raised; 1 = level with aperture.
12. 0 = stomata entirely haplocheilic; 1 = some or all syndetocheilic.
13. Apical meristem with differentiation of tunica and corpus.
14, 15. 00 = protostele (including vitalized); 10 = eustele usually with external

secondary xylem only; X1 = eustele with regular internal secondary xylem.

16. 0 = some or all stem bundles mesarch or exarch; 1 = all endarch.

17, 18. 00 = leaf traces from one stem bundle or protoxylem strand (one-trace unilacunar); 10 = from more than two bundles (multilacunar); X1 = from two adjacent bundles (two-trace unilacunar).

19. 0 = some scalariform pits in metaxylem; 1 = no scalariform metaxylem, circular-bordered pits in protoxylem.

20. 0 = circular-bordered pitting or perforations only in the secondary xylem; 1 = at least some scalariform.

21. Vessels in the secondary xylem.

22. 0 = rays uniseriate, rarely biseriate; 1 = at least some multiseriate.

23. Secretory canals.

24. Mäule reaction.

25–27. 000 = dichotomous megasporangiate appendages (cupules) on radial axis; 100 = pinnately compound megasporophyll; 110 = once-pinnate megasporophyll, with two rows of simple leaflets or cupules bearing ovules; X01, XX1 = ovule on one-veined sporophyll or sessile.

28–30. 000 = dichotomous microsporangiate appendages on radial axis; 100 = pinnately compound microsporophyll; 110 = once-pinnate microsporophyll, with two rows of simple leaflets or stalks bearing pollen sacs; X01, XX1 = one-veined microsporophyll.

31. 0 = ovule on lateral appendage; 1 = terminal.

32. 0 = homologs of progymnosperm fertile branchlets on homologs of lower-order axes; 1 = on homologs of last-order axes.

33–35. 000, 010 = ovule(s) in radial cupule; 100 = ovules directly on more or less laminar sporophyll; 110 = ovules in anatropous cupule, or anatropous and bitegmic; X01 = ovule with second integument derived from two appendages lower on axis.

36. 0 = several ovules per anatropous cupule or potential homolog; 1 = one.

37. 0 = microsporangia terminal, marginal, or adaxial; 1 = abaxial.

38. 0 = microsporangia free; 1 = more or less fused into microsynangia.

39. 0 = microsporophylls spirally arranged; 1 = whorled.

40. 0 = strobili on undifferentiated axes, or only female aggregated into compound strobili; 1 = both male and female strobili aggregated.

41, 42. 00 = no seeds; 10 = radiospermic seeds; X1 = platyspermic seeds.

43, 44. 00 = megasporangium with unmodified apex; 10 = lagenostome with central column; 11, X1 = pollen chamber without central column.

45. Micropylar tube.

46. Nucellar vasculature.

47. 0 = nucellar cuticle thin; 1 = thick, maceration resistant.

48. Heterospory.

49, 50. 00 = tetrad scar, no sulcus/pollen tube; 10 = sulcus/pollen tube; 11 = pollen tube but no sulcus.

51. 0 = pollen radially symmetrical or mixed; 1 = strictly bilateral.

52. 0 = pollen non-saccate or subsaccate; 1 = saccate.

53. 0 = infratectal structure alveolar; 1 = granular or columellar.

54. Pollen striate.

55. 0 = megaspore tetrad tetrahedral; 1 = linear.

56. 0 = megaspore wall thick; 1 = thin or lacking sporopollenin.

57. 0 = microgametophyte with prothallial(s) and sterile cell; 1 = prothallial but no sterile cell.

58. 0 = motile sperm; 1 = siphonogamy, non-motile sperm.

59. 0 = megagametophyte monosporic; 1 = tetrasporic.

60. Apex of megagametophyte free-nuclear or multinucleate; wall formation irregular, resulting in polyploid cells; egg a free nucleus.

61. 0 = early embryogenesis free-nuclear; 1 = entirely cellular.

62. Embryo with feeder.

Data matrix (Doyle & Donoghue, 1986*b*).

	1	2	3	4	5	6
Aneurophyton	0X00000X00XXX0000000000X000000X0000X000000000XX000000000X00XXX
Archaeopteris	0X10000X00XXX100000000XX000000X1000X000000000XX100000000X00XXX
Multiov. lygin.[a]	XX0010000000X0000000010X00000000000X010X101000X10000X000X000XX
Higher lygin.[b]	100010000000X0000000010X1001000000X010X101000X10000X000X000XX
Medullosa	100010000000XX101000011X10010000100X010X101101X1001000X0X0X0XX
Callistophyton	1000100100X0X1000000010X10010000100011100X1100X1101100X00X00XX
Glossopterids	10X0110X10X0X10XXX0000XXXX0XX000X000000XX1X10011101101X0XXX0XX
Peltaspermum	XX0X100100X0XXXXXXXXXXXX11010000100100XX111XX011010X0X0XXXXXX
Corystosperms	XX0010000000XX101000000X10011000110110000XX111XX011011X0X1XXXXXX
Caytonia	1000100X1010XXXXXXXXXX110100001100X10XX110011101100X1XXXXXX
Cycads	XXX0110X00000101100011101X01X00X100X10X10X1010110100010000000
Bennettitales	10X0110X0001X101000101XXX11100X010X0110XXX110X110101011XX00XX
Pentoxylon	10X0110X0000XX1XX101010XXX10100X010X0X10X1X100111010X0X1XXXXXX
Cordaites	1010X0X0000X1000000000XX01X0101X00X0001X1X10XX1000100X000X0XX
Ginkgos	1010X01X00000101X1100000X01X0101X00XX000X1X1000110100010000000
Conifers	1010X01X0000010100100010X01X0101X00X1000X1X1000100110010000000
Ephedra	11X1XX1X00001101X1101101XX1XX11XX01X1X1X1001111011110010000
Welwitschia	11X1XX1X11010101X1101111XX1XX11XX01X011X1X11001110110011X1111111
Gnetum	11X1110X11011101101X1101XX1XX11XX01X0X1110X11001110010X1111111
Angiosperms	10X0110X11111011001X1011101100X1101X100XXXX001110101011X10010

[a] Multiovulate lyginopterids.
[b] Higher lyginopterids.

References

Arber, E. A. N. & Parkin, J. (1907). On the origin of angiosperms. *Journal of the Linnean Society, Botany*, **38**, 29–80.

Arber, E. A. N. & Parkin, J. (1908). Studies on the evolution of the angiosperms. The relationship of the angiosperms to the Gnetales. *Annals of Botany*, **22**, 489–515.

Axelrod, D. I. (1952). A theory of angiosperm evolution. *Evolution*, **6**, 29–60.

Axelrod, D. I. (1970). Mesozoic paleogeography and early angiosperm history. *The Botanical Review*, **36**, 277–319.

Bailey, I. W. (1944). The development of vessels in angiosperms and its significance in morphological research. *American Journal of Botany*, **31**, 421–8.

Bailey, I. W. (1956). Nodal anatomy in retrospect. *Journal of the Arnold Arboretum*, **37**, 269–87.

Beck, C. B. (1971). On the anatomy and morphology of lateral branch systems of *Archaeopteris*. *American Journal of Botany*, **58**, 758–84.

Beck, C. B. (1981). *Archaeopteris* and its role in vascular plant evolution. In *Paleobotany, Paleoecology, and Evolution*, vol. 1, ed. K. J. Niklas, pp. 193–230. New York: Praeger.

Benzing, D. H. (1967). Developmental patterns in stem primary xylem of woody Ranales. II. Species with trilacunar and multilacunar nodes. *American Journal of Botany*, **54**, 813–20.

Bierhorst, D. W. (1971). *Morphology of Vascular Plants*. New York: Macmillan.

Brenner, G. J. (1976). Middle Cretaceous floral provinces and early migrations of angiosperms. In *Origin and Early Evolution of Angiosperms*, ed. C. B. Beck, pp. 23–47. New York: Columbia University Press.

Burger, W. C. (1977). The Piperales and the monocots. Alternative hypotheses for the origin of monocotyledonous flowers. *The Botanical Review*, **43**, 345–93.

Burger, W. C. (1981). Why are there so many kinds of flowering plants? *BioScience*, **31**, 572, 577–81.

Cantino, P. D. (1985). Phylogenetic inference from nonuniversal derived character states. *Systematic Botany*, **10**, 119–22.

Carlquist, S. (1975). *Ecological Strategies of Xylem Evolution*. Berkeley: University of California Press.

Cornet, B. (1977). Angiosperm-like pollen with tectate–columellate wall structure from the Upper Triassic (and Jurassic) of the Newark Supergroup, USA. *American Association of Stratigraphic Palynologists 10th Annual Meeting* (Tulsa), *Abstracts*, 8–9.

Crane, P. R. (1985). Phylogenetic analysis of seed plants and the origin of angiosperms. *Annals of the Missouri Botanical Garden*, **72**, 716–93.

Cronquist, A. (1968). *The Evolution and Classification of Flowering Plants*. Boston: Houghton Mifflin.

Dilcher, D. L. (1979). Early angiosperm reproduction: an introductory report. *Review of Palaeobotany and Palynology*, **27**, 291–328.

Doyle, J. A. (1969). Cretaceous angiosperm pollen of the Atlantic Coastal Plain and its evolutionary significance. *Journal of the Arnold Arboretum*, **50**, 1–35.

Doyle, J. A. (1978). Origin of angiosperms. *Annual Review of Ecology and Systematics*, **9**, 365–92.

Doyle, J. A. (1984). Evolutionary, geographic, and ecological aspects of the rise of angiosperms. *Proceedings of the 27th International Geological Congress* (Moscow), vol. 2, pp. 23–33. Utrecht: VNU Science Press.

Doyle, J. A. & Donoghue, M. J. (1986 a). Relationships of angiosperms and Gnetales: a numerical cladistic analysis. In *Systematic and Taxonomic Approaches in Palaeobotany*, ed. B. A. Thomas & R. A. Spicer, pp. 177–98. Oxford: Oxford University Press.

Doyle, J. A. & Donoghue, M. J. (1986 b). Seed plant phylogeny and the origin of angiosperms: an experimental cladistic approach. *The Botanical Review*, **52**, 321–431.

Doyle, J. A. & Hickey, L. J. (1976). Pollen and leaves from the mid-Cretaceous Potomac Group and their bearing on early angiosperm evolution. In *Origin and Early Evolution of Angiosperms*, ed. C. B. Beck, pp. 139–206. New York: Columbia University Press.

Doyle, J. A., Jardiné, S. & Doerenkamp, A. (1982). *Afropollis*, a new genus of early angiosperm pollen, with notes on the Cretaceous palynostratigraphy and paleoenvironments of Northern Gondwana. *Bulletin des Centres de Recherches Exploration–Production Elf-Aquitaine*, **6**, 39–117.

Eames, A. J. (1952). Relationships of the Ephedrales. *Phytomorphology*, **2**, 79–100.

Ehrendorfer, F. (1976). Evolutionary significance of chromosomal differentiation patterns in gymnosperms and primitive angiosperms. In *Origin and Early Evolution of Angiosperms*, ed. C. B. Beck, pp. 220–40. New York: Columbia University Press.

Farris, J. S. (1983). The logical basis of phylogenetic analysis. In *Advances in Cladistics*, vol. 2, ed. N. I. Platnick & V. A. Funk, pp. 7–36. New York: Columbia University Press.

Felsenstein, J. (1985). Confidence limits on phylogenies: an approach using the bootstrap. *Evolution*, **39**, 783–91.

Gaussen, H. (1946). *Les Gymnospermes, Actuelles et Fossiles*, vol. 2, sect. 1. Toulouse: Laboratoire Forestier.

Gould, R. E. & Delevoryas, T. (1977). The biology of *Glossopteris*: evidence from petrified seed-bearing and pollen-bearing organs. *Alcheringa*, **1**, 387–99.

Gould, S. J. (1977). *Ontogeny and Phylogeny*. Cambridge, Massachusetts: Harvard University Press.

Gould, S. J. & Eldredge, N. (1977). Punctuated equilibria: the tempo and mode of evolution reconsidered. *Paleobiology*, **3**, 115–51.

Harris, T. M. (1940). *Caytonia*. *Annals of Botany*, New Series, **4**, 713–34.

Hennig, W. (1966). *Phylogenetic Systematics*. Urbana: University of Illinois Press.

Hill, C. R. & Crane, P. R. (1982). Evolutionary cladistics and the origin of angiosperms. In *Problems of Phylogenetic Reconstruction*, ed. K. A. Joysey & A. E. Friday, pp. 269–361. London: Academic Press.

Hughes, N. F. (1976). *Palaeobiology of Angiosperm Origins*. Cambridge: Cambridge University Press.

Iltis, H. H. (1983). From teosinte to maize: the catastrophic sexual transmutation. *Science*, 222, 886–94.

Janzen, D. H. (1970). Herbivores and the number of tree species in tropical forests. *American Naturalist*, 104, 501–28.

Kaplan, D. R. (1984). The concept of homology and its central role in the elucidation of plant systematic relationships. In *Cladistics: Perspectives on the Reconstruction of Evolutionary History*, ed. T. Duncan & T. F. Stuessy, pp. 51–70. New York: Columbia University Press.

Krassilov, V. A. (1977). The origin of angiosperms. *The Botanical Review*, 43, 143–76.

Maddison, W. P., Donoghue, M. J. & Maddison, D. R. (1984). Outgroup analysis and parsimony. *Systematic Zoology*, 33, 83–103.

Mamay, S. H. (1976). Paleozoic origin of the cycads. *US Geological Survey Professional Paper*, 934, 1–48.

Martens, P. (1971). *Les Gnétophytes. Handbuch der Pflanzenanatomie*, vol. 12(2). Berlin: Gebrüder Borntraeger.

Meeuse, A. D. J. (1963). From ovule to ovary: a contribution to the phylogeny of the megasporangium. *Acta Biotheoretica*, 16, 127–82.

Meeuse, A. D. J. (1972*a*). Facts and fiction in floral morphology with special reference to the Polycarpicae. 1. A general survey. *Acta Botanica Neerlandica*, 21, 113–27.

Meeuse, A. D. J (1972*b*). Facts and fiction in floral morphology with special reference to the Polycarpicae. 2. Interpretation of the floral morphology of various taxonomic groups. *Acta Botanica Neerlandica*, 21, 235–52.

Meeuse, A. D. J. (1972*c*). Facts and fiction in floral morphology with special reference to the Polycarpicae. 3. Consequences and various additional aspects of the anthocorm theory. *Acta Botanica Neerlandica*, 21, 351–65.

Meyen, S. V. (1984). Basic features of gymnosperm systematics and phylogeny as evidenced by the fossil record. *The Botanical Review*, 50, 1–112.

Mickevich, M. F. & Farris, J. S. (1982). Phylogenetic analysis system (PHYSYS) (FORTRAN V software system of cladistic and phenetic algorithms).

Muhammad, A. F. & Sattler, R. (1982). Vessel structure of *Gnetum* and the origin of angiosperms. *American Journal of Botany*, 69, 1004–21.

Muller, J. (1970). Palynological evidence on early differentiation of angiosperms. *Biological Reviews of the Cambridge Philosophical Society*, 45, 417–50.

Muller, J. (1981). Fossil pollen records of extant angiosperms. *The Botanical Review*, **47**, 1–142.

Pant, D. D. & Nautiyal, D. D. (1967). On the structure of *Buriadia heterophylla* (Feistmantel) Seward & Sahni and its fructification. *Philosophical Transactions of the Royal Society, ser. B*, **252**, 27–48.

Pant, D. D. & Nautiyal, D. D. (1984). On the morphology and structure of *Ottokaria zeilleri* sp. nov. – a female fructification of *Glossopteris*. *Palaeontographica B*, **193**, 127–52.

Regal, P. J. (1977). Ecology and evolution of flowering plant dominance. *Science*, **196**, 622–9.

Retallack, G. & Dilcher, D. L. (1981). Arguments for a glossopterid ancestry of angiosperms. *Paleobiology*, **7**, 54–67.

Rothwell, G. W. (1982). New interpretations of the earliest conifers. *Review of Palaeobotany and Palynology*, **37**, 7–28.

Schopf, J. M. (1976). Morphologic interpretation of fertile structures in glossopterid gymnosperms. *Review of Palaeobotany and Palynology*, **21**, 25–64.

Stanley, S. M. (1979). *Macroevolution: Pattern and Process*. San Francisco: W. H. Freeman.

Stebbins, G. L. (1974). *Flowering Plants: Evolution above the Species Level*. Cambridge, Massachusetts: Harvard University Press.

Stebbins, G. L. (1981). Why are there so many species of flowering plants? *BioScience*, **31**, 573–7.

Takhtajan, A. L. (1969). *Flowering Plants: Origin and Dispersal*. Washington: The Smithsonian Institution.

Thomas, H. H. (1925). The Caytoniales, a new group of angiospermous plants from the Jurassic rocks of Yorkshire. *Philosophical Transactions of the Royal Society, ser. B*, **213**, 299–363.

Thompson, W. P. (1918). Independent evolution of vessels in Gnetales and angiosperms. *Botanical Gazette*, **65**, 83–90.

Upchurch, G. R. (1984). Cuticle evolution in Early Cretaceous angiosperms from the Potomac Group of Virginia and Maryland. *Annals of the Missouri Botanical Garden*, **71**, 522–50.

Upchurch, G. R. & Crane, P. R. (1985). Probable gnetalean megafossils from the Lower Cretaceous Potomac Group of Virginia. *American Journal of Botany*, **72**, 903, Abstract.

Vakhrameev, V. A. (1970). Yurskie i rannemelovye flory. In *Paleozoyskie i Mezozoyskie Flory Yevrazii i Fitogeografiya Etogo Vremeni*, ed. V. A. Vakhrameev, I. A. Dobruskina, Ye. D. Zaklinskaya & S. V. Meyen, pp. 213–81. Moscow: Nauka.

Vrba, E. S. (1983). Macroevolutionary trends: new perspectives on the roles of adaptation and incidental effect. *Science*, **221**, 387–9.

Walker, J. W. (1976). Evolutionary significance of the exine in the pollen of primitive angiosperms. In *The Evolutionary Significance of the Exine*, ed. I. K. Ferguson & J. Muller, pp. 1112–37. London: Academic Press.

Watrous, L. E. & Wheeler, Q. D. (1981). The out-group comparison method of character analysis. *Systematic Zoology*, **30**, 1–11.

Wettstein, R. R. von (1907). *Handbuch der systematischen Botanik*, 2nd edn. Leipzig: Franz Deuticke.

Young, D. A. (1981). Are the angiosperms primitively vesselless? *Systematic Botany*, **6**, 313–30.

3

·

Global palaeogeography and palaeoclimate of the Late Cretaceous and Early Tertiary

JUDITH TOTMAN PARRISH

The critical early evolution and dispersion of the angiosperms took place at a time of relative stability in the tectonic history of the Earth. The major plate tectonic developments, i.e. the formation of the Atlantic and Indian Oceans and the movement of India toward Asia, had already been initiated. By the end of the Early Cretaceous, the Atlantic Ocean basin was open along most of its length and the Indian Ocean was also open. No major collisions took place and the most significant new plate tectonic event to occur during the Late Cretaceous–Eocene was the separation of Australia and Antarctica.

The Late Cretaceous–Eocene was also a time of climatic quiescence, despite intense volcanism associated with rapid subduction of the Pacific Plate and an apparent large-bolide impact at the Cretaceous–Tertiary boundary (Alvarez *et al.*, 1980). Sea level was high, with only minor fluctuations (Vail, Mitchum & Thompson, 1977), and Cretaceous and early Tertiary global temperature was warm (Savin, 1977); this was perhaps the warmest interval in the history of the Earth (Frakes, 1979). Climatic patterns followed the zonal component of atmospheric circulation more closely than at any time in the Phanerozoic since the Early Carboniferous (Raymond, Parker & Parrish, 1985; Rowley *et al.*, 1985; Parrish, Parrish & Ziegler, 1986; Parrish & Doyle, 1984; Barron & Washington, 1984).

Angiosperms underwent their greatest diversification within this environment of stability. The purpose of this paper is to provide a global context for the studies of angiosperm evolution and ecology that appear in this volume. I will outline the palaeoclimatic history of the Late Cretaceous to early Tertiary and describe the palaeogeography of the

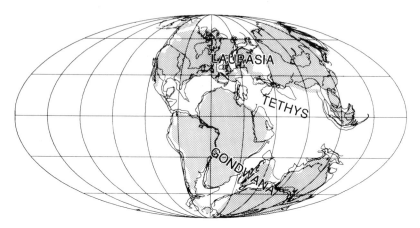

Figure 3.1. Palaeogeography in the Barremian (Early Cretaceous; from Parrish, 1985). By this time, the breakup of Gondwana had commenced. Land area is stippled.

Earth during this interval, especially the persistence of land bridges between continents.

Global palaeogeography and its palaeoclimatic consequences

In the early part of the Mesozoic, global climate was dominated by highly seasonal, monsoonal circulation (Robinson, 1973; Parrish *et al.*, 1986), which was caused by Pangaea, the vast supercontinent that straddled the equator. Unlike the present Asian monsoon, the Pangaean monsoon strongly affected the zonal circulation in both hemispheres. This circulation resulted in a dry region that included the equatorial land as well as the subtropics. In addition, the Tethyan coasts in low mid-latitudes, e.g. in southern Laurasia, experienced abundant, but strongly seasonal, rainfall. By the Cretaceous, the North Atlantic basin was well open (Phillips & Forsyth, 1972), causing the breakdown of the monsoon in the Northern Hemisphere, and the climatic consequences, including drying of southern Laurasia, had already been realized (Parrish & Doyle, 1984; Hallam, 1984). During the Early Cretaceous (Figure 3.1), the South Atlantic and Indian Ocean basins began opening (Sclater, Hellinger & Tapscott, 1977; Larson, 1977), destroying the southern monsoon as well (Parrish & Doyle, 1984). Thus, by the beginning of the Late Cretaceous, global palaeogeography had entered a mode of maximum continental dispersion, and the monsoonal circulation that had so dominated palaeoclimatic history during the previous 250 million years had disappeared. From this time until the collision of India and Asia re-established strong monsoonal circulation during the Miocene, continental palaeoclimates were characterized by the zonal circulation that is the fundamental component of atmospheric circulation on Earth (Parrish, 1982).

Temperature

Several quantitative methods exist for estimating palaeotemperatures; unfortunately, none can be confidently extended very far into the Cretaceous. One successful method is the use of transfer functions, a statistical method for analyzing the similarity of ancient faunas to recent ones and inferring palaeotemperature. This method requires relatively large sample sizes, however, and loses its effectiveness once the number of modern taxa in the ancient faunas drops below a certain percentage. Therefore, it has proved useful for only the Quaternary (CLIMAP, 1976; Flessa *et al.*, 1979). A similar, less rigorous method, is the so-called 'nearest-living relative' method, which extrapolates known physiological requirements of modern, usually terrestrial, taxa to related taxa in the past (e.g. Axelrod, 1981; Colbert, 1964); perhaps the most familiar example is the use of crocodilians to infer latitudinal temperature minima (Colbert, 1964; Estes & Hutchison, 1980). The effectiveness of this method has been questioned recently on a number of grounds. Specifically, some modern taxa are relicts of groups that were previously much more diverse in their adaptations (e.g. cycads), so that modern taxa are not necessarily good guides for the physiological requirements of their fossil relatives (e.g. modern reptiles versus dinosaurs; Thomas & Olson, 1978). Similarly, some modern taxa can be shown to have had different climatic requirements earlier in their history (e.g. *Glyptostrobus*; Wolfe, 1979).

Ratios of the stable isotopes of oxygen (Lowenstam & Epstein, 1954) and leaf-margin analysis (Bailey & Sinnott, 1916; Wolfe, 1979) are the most successful quantitative methods for estimating palaeotemperature that can be extended reasonably far back into the past. The advantage of stable isotope analysis is that it has been performed on hundreds of samples from different latitudes through the latest Cretaceous and Tertiary (Savin, 1977). The disadvantage is that it has proved most useful in open-ocean carbonates; analyses on epeiric sea organisms appear to have some unreliability, the source of which is not clear but is probably diagenetic (Spaeth, Hoefs & Vetter, 1971) or the result of an inconsistent vital effect (Pratt, 1985). Therefore, the use of stable

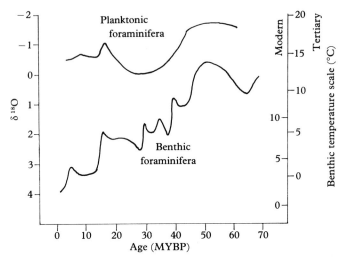

Figure 3.2. Palaeotemperature curves by Savin (1977), based on stable isotopes of oxygen in the tests of benthic and planktonic foraminifera from several Deep Sea Drilling Project sites in the North Pacific Ocean. The change in temperature scales corrects for isotopic changes in seawater brought about by the formation of the permanent ice; for isotopic temperature of planktonic foraminifera, add approximately 2.5 degC to the appropriate part of the temperature scale.

isotopes with regard to terrestrial organisms is limited to giving an overall view of global climate based on ocean temperatures. Figure 3.2 is generalized from the well-known palaeotemperature curve of Savin (1977), who also reviewed numerous palaeoclimatic studies for the Cretaceous and Tertiary. Savin's curve and nearly every other curve constructed, regardless of the basis (e.g. palaeobotanical data; Wolfe & Hopkins, 1967), show that the Cretaceous and early Tertiary were warm relative to the later Tertiary, particularly at mid- and high latitudes (see also Shackleton & Boersma, 1981), and that a marked drop in temperature occurred at about the Eocene–Oligocene boundary. In equatorial waters, benthic foraminifers record a much stronger decline in temperature than do planktonic foraminifers, whereas in high latitudes, the temperature curves for both groups are parallel (Savin, 1977). The temperatures recorded by the benthics are considered to be indicative of the temperature of high-latitude waters, based on the assumption that bottom waters were generated at high latitudes, as they are today (Savin, Douglas & Stehli, 1975). Therefore, the warmth of the Cretaceous and early Tertiary is more strongly expressed in high palaeo-latitudes than in low. Confirming this assumption has been difficult because few isotopic measurements have been made at high latitudes. Moreover, Cretaceous bottom waters may not have been generated near the poles but, rather, in marginal marine evaporite basins in low latitudes (Brass, Southam & Peterson, 1982). A global temperature decrease of about 15 to 20 degC may have occurred since the middle Cretaceous (Savin, 1977).

Leaf-margin analysis has the advantage of directly reflecting continental palaeotemperatures (Wolfe, 1979). The major disadvantage is that large sample sizes (> 20–30 leaf forms) and a number of sample localities are required to overcome taphonomic and local effects (Wolfe, 1971). In addition, leaf-margin analysis is potentially less reliable in the early phase of angiosperm history (mid-Cretaceous) because angiosperm evolution was so rapid then (Upchurch & Wolfe, this volume, Chapter 4). The physiological basis for the relationship between leaf-margin morphology and temperature is not well understood, so the influence of evolutionary instability is difficult to analyze. Nevertheless, leaf-margin analysis gives similar results for global temperature change as does the isotope record, showing that the Cretaceous and early Tertiary were warmer than the present world. On land, this difference was 5 to 10 degC at low mid-latitudes (30° N; Upchurch & Wolfe, this volume, Chapter 4) and about 30 degC at high latitudes (80° N; Spicer & Parrish, 1986; cf. modern surface air temperature curve of Barron (1983)).

Rainfall

Parrish, Ziegler & Scotese (1982) constructed maps of predicted global rainfall patterns based on the circulation maps of Parrish & Curtis (1982) for several geologic ages, including the Cenomanian (early Late Cretaceous; Figure 3.3), Maastrichtian (latest Cretaceous, Figure 3.4), and Lutetian (middle Eocene, Figure 3.5). These rainfall maps were successful for explaining the distributions of evaporites and coals. However, Parrish *et al.* (1982) noted that, for those three intervals, the circulation–rainfall maps were no more successful in predicting the distribution of those sediments than was a general zonal model of precipitation. They interpreted this as evidence that, as expected, climatic patterns would have been more nearly zonal when the geographic conditions for a monsoonal circulation were absent. The zonal model, used by palaeoclimatologists for many years (e.g. Briden &

Irving, 1964), predicts high rainfall between 15° N and 15° S and between 45° and 70° north and similarly south, and low rainfall in the intervening latitudes (Parrish *et al.*, 1982).

Frakes (1979, Fig. 9-1) constructed a mean global precipitation curve indicating that the Cretaceous was very dry and that precipitation increased dramatically toward the Eocene. This undocumented curve was based apparently on the relative abundances of coals and evaporites. In contrast, Hallam (1984) regarded the Late Cretaceous as humid, including in his data ironstones, bauxites, and semiquantitative estimates of evaporite and coal abundance. Although Parrish *et al.*'s (1982) data were restricted to the individual geologic ages, whereas Hallam's (1984) data covered the whole epoch, their data appear to be in general agreement – at least, the number of evaporite deposits decreases slightly between the Cenomanian and Maastrichtian. The agreement between the distribution of Hallam's (1984) climatic indicators and Parrish *et al.*'s (1982) rainfall predictions is good. Coals are more abundant in the middle Eocene than in the Cretaceous, seeming to support Frakes' curve, but so are evaporites (Parrish *et al.*, 1982). Parrish *et al.* (1982) pointed out that loci of coal and evaporite deposition may change through time, but that global drying or wetting trends can be difficult to document. Similarly, Parrish *et al.*'s (1982) method only gives relative rainfall, and the contrast between wet and dry regions may have differed through time, as this contrast is dependent on the magnitude of such variables as land–sea temperature contrast that cannot be treated qualitatively. The deposition of coals and evaporites is dependent on so many conditions other than climate that any quantitative estimate of rainfall based on those lithologies would be suspect.

Regional palaeoclimatology

The climate of the young South Atlantic is generally considered to have been arid (Arthur & Natland, 1979) once the rift valley formed (Hay *et al.*, 1982; Doyle, Jardiné & Doerenkamp, 1982). Flood-plain and lacustrine deposits in eastern Brazil gave way in the Barremian (late Early Cretaceous) to fluvial and aeolian deposits (Petri & Campanha, 1981; Bigarella, 1973) and subsequently to thick Aptian–Albian evaporites (Franks & Nairn, 1973; Asmus & Ponte, 1973). Low rainfall is indicated also by palynofloras in the Cape and Angola Basins (McLachlan & Pieterse, 1978; Morgan, 1978) and from Aptian–Albian rocks in the northern South Atlantic (Doyle *et al.*, 1977, 1982). A narrow wet belt lay along the palaeoequator and just north of the palaeoequator in

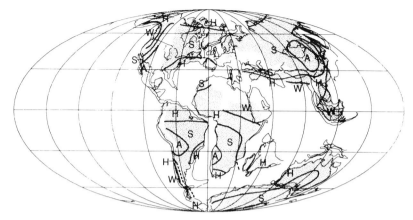

Figure 3.3. Palaeogeography (from Ziegler *et al.*, 1983) and rainfall in the Cenomanian (early Late Cretaceous; adapted from Parrish *et al.*, 1982). 'Arid' (A) is roughly equivalent to desert; 'semi-arid' (S) is roughly equivalent to steppe, taiga, or Mediterranean climate; 'humid' (H) is roughly equivalent to humid subtropical or marine climate; and 'wet' (W) is roughly equivalent to tropical rainforest, monsoonal, or tropical savanna. Land area is stippled.

northwestern Africa, outside the rift zone (Doyle *et al.*, 1982). Aptian–Albian palynofloras of southern Laurasia indicate humid conditions (Brenner, 1976), although seasonality of the rainfall may be indicated by the persistence of scattered evaporites in southwestern Europe. In addition, Chamley (1979) interpreted changes in clay mineralogy from the North Atlantic as indicating a change from dry to seasonally wet conditions during the Early Cretaceous.

By the Cenomanian (earliest Late Cretaceous, Figure 3.3), northwestern Africa was far enough north to have penetrated the dry subtropics, as indicated by evaporites in Algeria and Libya. Lignite and swamp vegetation of Cenomanian age are found in northeastern Brazil (Petri & Campanha, 1981). A striking feature of the early Late Cretaceous is the decline in the number of evaporite deposits not only in Gondwana but the rest of the world as well (Parrish *et al.*, 1982). The Tethyan subtropical high pressure cell is predicted to have existed year-round by the Cenomanian (Parrish & Curtis, 1982), carrying rainfall to eastern North America and southwestern Europe and reversing the Late Jurassic drying trend that had resulted from the breakdown of the northern monsoon (Parrish & Doyle, 1984; Doyle & Parrish, 1984). Evaporites disappeared from Laurasian rocks of Cenomanian age and were replaced

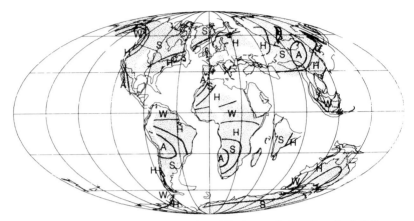

Figure 3.4. Palaeogeography (from Ziegler *et al.*, 1983) and rainfall
in the Maastrichtian (latest Cretaceous; adapted from Parrish *et al.*,
1982). Symbols as in Figure 3.3.

by bauxites in Apulia (Italy, Yugoslavia, and parts of Hungary, Romania,
and Greece) and southwestern Europe (Parrish *et al.*, 1982; Ronov &
Balukhovskii, 1982) and Cenomanian coals occur all along the southern
margin of Laurasia (Parrish *et al.*, 1982). Evaporites do occur just behind
the mountain front in the southern USSR, which comprises the present-
day Tien Shan and Kunlung Mountains, the Hindu Kush and Pamirs,
and the Elburz and Caucasus Mountains, but coals occur farther north,
reflecting the year-round presence of the subtropical high-pressure cell,
which would tend to make the region more humid (Parrish *et al.*, 1982).
    Maastrichtian (Figure 3.4) rainfall patterns in the circum-South
Atlantic region are predicted to have been very similar to those in the
Cenomanian, the major difference being the generally wetter equatorial
regions permitted by the width of the Central Atlantic, reflected in
laterite, coal, and kaolinite deposits of Central Africa (Ronov & Balukhovskii, 1982). The distribution of early Tertiary palaeoclimatic indicators is as expected, zonal with minor deviations caused by mountains.
South-central Laurasia remained humid, as indicated by the presence of
numerous coal deposits, laterites, and bauxites of Paleocene through
Oligocene age. Paleocene and Eocene evaporites are common in the
subtropical regions of eastern and western Laurasia, in what is now
China and southern North America, and a few evaporite deposits of
Eocene age occur in Spain, near the developing eastern Atlantic upwelling zone (Parrish *et al.*, 1982; Ronov & Balukhovskii, 1982).

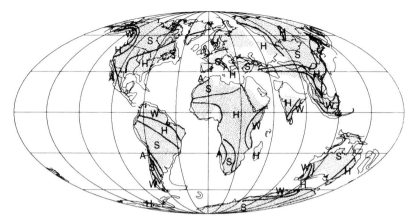

Figure 3.5. Palaeogeography (from Ziegler *et al.*, 1983) and rainfall in the Lutetian (middle Eocene; adapted from Parrish *et al.*, 1982). Symbols as in Figure 3.3.

Regional palaeogeography

For the distribution of angiosperms, the aspects of Late Cretaceous and early Tertiary palaeogeography of greatest concern are land connections among the drifting continents, particularly between North America and Europe and North America and Asia. Other connections include Africa–South America, South America–Antarctica–Australia, and South America-North America. India did not collide with Asia until the late Tertiary (Ziegler, Scotese & Barrett, 1983). The reader should be aware that in a treatment of this length, one cannot begin to do justice to the complexity of the history of each connection and that discussion must be confined to bare conclusions.

North America–Europe

North America and Europe remained in tectonic contact via Greenland until the Miocene, when the Mid-Atlantic Ridge had extended through to the Arctic Ocean (Pitman, Larson & Herron, 1974). Before the Miocene, the exchange of terrestrial organisms would have been hindered only by shallow seaways. According to the reconstructions of Ziegler, Scotese & Barrett (1983) and Thiede (1979, 1980) for the Late Cretaceous, a narrow but persistent seaway over continental crust existed through the end of the Cretaceous. Sea level remained high throughout the Late Cretaceous (Vail *et al.*, 1977), so it seems unlikely that a land bridge existed at any time during the epoch. Two land bridges were possible once rifting had been inititated (Eldholm & Thiede, 1980), one between Greenland and Svalbard and one along the Iceland–Faeroe Ridge. According to Ziegler *et al.*'s reconstruction, by the Lutetian (middle Eocene), rifting between Svalbard and Greenland was well advanced, and only the Iceland–Faeroe Ridge could have functioned as a land bridge. However, the initial movement between Scandinavia and Europe was apparently along a transform (Talwani & Udintsev, 1976; Talwani & Eldholm, 1977; Eldholm & Thiede, 1980), so that a land connection could have persisted later than the Eocene. According to Vail *et al.*'s (1977) curves, sea level stood lower than its Lutetian level at the Cretaceous–Tertiary boundary and at the end of the Danian (early Paleocene) and Ypresian (early Eocene) ages. These sea-level drops were significant and may have left the Iceland–Faeroe Ridge and the Svalbard–Greenland region emergent for a short time. Ziegler's (1982) palaeogeographic map for the Paleocene–Eocene of northwestern Europe indicates the presence of shallow-marine to deltaic sediments on the Iceland–Faeroe Ridge north of Scotland, and subaerially weathered, middle Eocene basalt was drilled on the ridge flank farther to the northwest (Talwani, Udintsev, *et al.*, 1976). Talwani & Udintsev (1976) estimated that at the time the basalt was formed, the ridge crest was 900 m above sea level and that the oldest part of the ridge would have subsided below sea level, creating the first breach, in late Oligocene–early Miocene time. However, as Talwani & Udintsev (1976) pointed out, this estimate was based on a number of simplifying assumptions, which Thiede (1980) and Eldholm & Thiede (1980) rejected on the grounds that marine sediments were being deposited on nearby topographic highs. These workers showed the Iceland–Faeroe Ridge as consisting of a series of islands, although they also suggested that more land area could have been exposed. Talwani & Udintsev (1976) and Talwani & Eldholm (1977) estimated, on tectonic and palaeomagnetic grounds, a breach of the Greenland–Svalbard connection in Oligocene time, with which Eldholm & Thiede (1980) seemed to concur. Eldholm & Thiede (1980) noted an apparent discrepancy between a loss of terrestrial faunal continuity between North America and Europe before or during the early Eocene (McKenna, 1972; West, Dawson & Hutchison, 1977) and the later evolution of the Greenland Sea. The best explanation is the high sea level of the Eocene, which could have kept the continental region submerged. By the time the sea level dropped in the Oligocene (Vail *et al.*, 1977), the rift was deep enough to maintain a water barrier; indeed, deep-water exchange was initiated in the Oligocene to early Miocene (Thiede, 1980).

However, it should be noted that McKenna (1972) and Simpson (1946) both described a drastic drop-off in faunal similarity after the early Eocene; this does not accord with the drop in sea level at the end of the early Eocene, which would have left more land exposed. Perhaps this discrepancy implies that this sea-level drop was relatively minor. The conclusion, then, is that a very narrow seaway existed between Europe and North America from the time of the Cretaceous sea-level rise onward, with a brief connection established during low sea-level stands in the Paleocene.

North America–Asia

McKenna (1972), in describing the faunal exchange between Europe and North America across the Svalbard–Greenland connection, also pointed out that dispersal could not have occurred via the Bering Sea because eastern Asia was isolated from Europe in the Paleocene and early Eocene but not from North America. Nevertheless, he did regard the Bering route as a more efficient filter than the Svalbard–Greenland connection, attributing the difference to a cooler climate. However, climate does not seem to be a likely explanation, because the two regions were at about the same latitude in the Eocene.

Ziegler *et al.* (1983) portrayed the land area between Alaska and northeasternmost USSR as continuous throughout the Late Cretaceous and early Tertiary (Figures 3.3 to 3.5), based on reconstructions by Vinogradov (1967, 1968) for the USSR and various data sources for Alaska. In the early Late Cretaceous, a prolific sediment source existed to the west of the North Slope of Alaska, in the region of the present Bering Strait, and thick deltaic sequences were deposited in the western and central North Slope (Huffman, 1985). Non-marine deposition continued on the western North Slope in the late Late Cretaceous. Tertiary rocks are absent on the western North Slope (Chapman & Sable, 1960) and uncommon elsewhere in the Bering Strait region. Paleogene rocks are non-marine, consisting largely of volcanics and coal-bearing sediments; marine rocks appear in the Neogene (Fisher, Patton & Holmes, 1981). St Lawrence Island has Oligocene plant-bearing rocks, and wells drilled northeast of the Bering Strait reached probable Cenozoic, coal-bearing rocks. On the basis of seismic studies, Fisher *et al.* (1981) believed that the offshore basin stratigraphy was the same as that onshore. The tectonic history lends further support to the reconstruction of the Bering Sea region as emergent. In the middle Cretaceous, the Bering Sea region was intruded by granitic plutons. In the Late Cretaceous, the region underwent compression, uplift, and substantial erosion (Fisher *et al.*, 1981), followed by extension, which formed basins that filled with volcanics and presumed alluvial fans and associated non-marine rocks. Regional subsidence did not occur until the Neogene. Thus, in contrast to the mammalian evidence, geologic evidence indicates that the Bering Strait region was emergent apparently until late in the Tertiary.

North America–Africa

The opening of the North Atlantic Ocean and Tethys created a southward motion of Gondwana relative to Laurasia, the northern continent consisting of North America and Eurasia. This sense of movement was reversed when the South Atlantic opened and Africa began moving northward toward Europe. The closing of Tethys between Africa and Europe has a very complicated history, consisting of several cycles of collision and rifting. For example, Apulia collided with Europe in the Late Cretaceous (Burchfiel & Royden, 1982), but was simultaneously rifted from northern Africa. A connection between Europe and Africa was possible then. Additional land connections between Africa and Europe were probably not established until Oligocene or Miocene time (Berggren & Hollister, 1974).

North America–South America

The history of the connections between North and South America also is complicated, with virtually every aspect the subject of controversy. The most recent, comprehensive, and coherent model of Caribbean plate evolution is that by Pindell & Dewey (1982). By the early Late Cretaceous, Mexico, which had been displaced westward relative to its present position (Pindell & Dewey, 1982; Scotese *et al.*, 1979), was in place. Subduction along the eastern end of the Antillean arc raised a chain of now-extinct volcanoes on the present Aves Ridge, which lies just west of the Lesser Antilles (Fox & Heezen, 1975). Throughout the Late Cretaceous the Caribbean plate moved northeastward. The western end of the Antillean arc collided with southern Yucatan, breaking the arc into blocks that eventually constituted Jamaica, the southern Haitian peninsula, and two central blocks of Hispaniola, one of which also contained Puerto Rico (Pindell & Dewey, 1982). Near the close of the Cretaceous, back-arc spreading split the Antillean arc lengthwise, opening the Yucatan Basin. The southern part of the split block constituted what is now the Cayman Ridge and the southern peninsula of Haiti (Pindell & Dewey, 1982). North of the rift lay what was to become western Cuba. A connection between North and South America through Yucatan, the Antillean arc, and the Aves Swell in Campanian time is supported by the geology of the various islands, which includes continental and very shallow marine deposits (Pindell & Dewey, 1982, and references therein; papers in Nairn & Stehli, 1975). Indeed, a connection at that time is required by the Late Cretaceous migration of hadrosaurs into South America from North America (Casamiquela, 1980; Brett-Surman, 1979).

During the early Tertiary, the Yucatan Basin opened and, in the Eocene, the Antillean arc collided with the Blake–Bahama Plateau. In addition, abundant subaerial volcanics of Eocene age were extruded on the Lesser Antilles (Tomblin, 1975) and an ?Eocene subaerial weathering surface (Edgar, Saunders, *et al.*, 1973) and shallow-water carbonates (Fox & Heezen, 1975) are found on Aves Ridge. These geologic features suggest that a connection between North America and South America most likely existed during the Eocene period. The length of time this connection persisted is not known, but the Lesser Antillean volcanism decreased after the Eocene and many parts of the Caribbean subsided several hundred meters, beginning in the Miocene. Therefore, no connection between North and South America after the early Miocene was likely to have existed until the formation of the Isthmus of Panama in the Pliocene.

South America–Africa

Sclater *et al.* (1977) presented a palaeobathymetric reconstruction for the South Atlantic in the Aptian, about 110 million years before present (MYBP). According to the reconstruction, a gap approximately 50 km wide existed between the 200 m depth contours off Brazil and Ghana, and they and Thiede (1977) postulated a 2000 m depth contour in that gap. A 110 MYBP opening was favored also by Rabinowitz & LaBrecque (1979) and Pindell & Dewey (1982). To the south, Sclater *et al.*'s (1977) reconstruction shows the Rio Grande Rise–Walvis Ridge complex as a wide corridor above 200 m palaeodepth between South America and Africa in the central South Atlantic. By the Cenomanian, about 95 MYBP, water depths between northern Brazil and Ghana had increased considerably, up to 4000 m deep in some basins. No evidence exists that topographically high fracture-zone ridges (not figured by Sclater *et al.* but discussed by them) stood above sea level. The results of Sclater *et al.*'s (1977) work and the lack of geologic evidence supporting a land bridge in the equatorial region suggest that the latest direct connection between South America and Africa was likely to have been in the central South Atlantic, along the Rio Grande Rise and Walvis Ridge.

The Walvis Ridge–Rio Grande Rise connection was still very shallow in the Cenomanian, mostly less than 200 m, but Sclater *et al.*'s (1977) reconstruction for the Campanian Stage (80 MYBP) shows water depths of 2000 m for parts of the Rio Grande Rise–Walvis Ridge complex. Sclater *et al.*'s (1977) palaeodepth predictions did not take sea level changes into account and they noted a possible error in palaeodepth reconstruction of as much as 300 m. With the exception of a drop in global sea level in the mid-Cenomanian, which did not go to pre-Aptian levels, sea level remained high throughout the Cretaceous and Paleogene (Vail *et al.*, 1977), with only one significant drop (still above present sea level) at the Cretaceous–Tertiary boundary. On the basis of drill samples of shallow-water indicators, Thiede (1977) estimated that the Rio Grande Rise finally subsided below the surface of the ocean in the late Oligocene. No similar analysis has been done for Walvis Ridge, but its time of submergence can be estimated and is consistent with the data on the Rio Grande Rise. The time when the Rio Grande Rise–Walvis Ridge complex was severed as a continuous land connection probably was mid-Early Cretaceous. Island-hopping across widening water barriers would have been the only possible dispersal mode for terrestrial organisms across the central South Atlantic after mid-Early Cretaceous time (Cox, 1980).

The faunal evidence for a northern connection between South America and Africa is disputed, with some workers favoring a rather late connection across the Gulf of Guinea (Turonian), in conflict with most of the geologic data. Some of the faunal evidence was summarized by Reyment, Pengtson & Tait (1976), who reviewed the faunas in Atlantic equatorial basins in Africa and South America: terrestrial microfloras (Jardiné, Kieser & Reyre, 1974) formed one large province in Africa and South America until the end of the Cenomanian; Niger Republic and South American dinosaurs were similar into the Coniacian (Buffetaut & Taquet, 1975), although the similarities do not require a connection in the Late Cretaceous (Sues & Taquet, 1979); pelecypods and gastropods were similar on both continents until the Cenomanian (Nicklés, 1950); and West African and Brazilian ostracodes showed strong affinities into the earliest Tertiary (Neufville, 1973). In addition, hadrosaurs dispersed to South America from North America during the Late Cretaceous (Casamiquela, 1980; Brett-Surman, 1979). The absence of hadrosaurs in Africa is negative evidence for the lack of a land bridge, but in the light of their large numbers and ubiquity elsewhere, not entirely without value. Freshwater fish taxa and a crocodilian are common to South America and Africa in Aptian time but not later (Taquet, 1978).

Data on marine faunas are conflicting (Premoli-Silva & Boersma, 1977; Förster, 1978; Kennedy & Cooper, 1975; Reyment, 1973). Equatorial currents would have assured that a flow of terrestrial organisms between South America and Africa by rafting, for example, was likely to have been predominantly east to west. The weight of the palaeogeographic and geophysical data is on the side of a late Albian severance of the last connection between South America and Africa. The faunal evidence does not strongly favor a later separation.

South America–Antarctica–Australia

The timing of the connection between South America and Australia via Antarctica, so vital to mammalogists (Tedford, 1974), is particularly difficult to outline because of the paucity of information from Antarctica. A rectilinear cordillera connecting the Andes and the Antarctic Peninsula might have existed at least until the Late Cretaceous (Dalziel & Elliot, 1973), although this model has recently been renounced in favor of sea-floor spreading between South America and Antarctica (Craddock, 1982; Dalziel, 1982). The modern configuration of the Scotia Sea was achieved during the Paleocene or early Eocene (Dalziel & Elliot, 1973) or as late as the Oligocene (Barker, Dalziel, *et al.*, 1976; Craddock, 1982). Rifting between Australia and Antarctica occurred during the Late Cretaceous (Cande & Mutter, 1982), although faunal evidence suggests some land connection between Australia and Antarctica as late as the early Eocene (Marshall, 1980). The final separation of South America from Antarctica permitted the development of the circum-Antarctic Current, to which has been attributed the global cooling event observed at the end of the Eocene (Kennett, 1977).

Summary

The potential barriers to dispersal for angiosperms are temperature and oceans. The apparent warmth of the entire globe suggests that temperature barriers are not likely to have been significant. Determining the history of emergent connections between continents is difficult for a variety of reasons, including lack of data, particularly on submerged volcanic edifices and fragments of continental crust; existence of two or more possible connections between any two continents; and controversy

Table 3.1. *Histories of emergent connections between continents during the Late Cretaceous and Tertiary*

Continents	Times of connection
North America/Europe	Paleocene; early Eocene?
North America/Asia	Late Cretaceous through early Tertiary
North America/Africa via Europe	Briefly during Late Cretaceous; Oligocene or Miocene
North America/South America	Campanian; Eocene–Oligocene?
South America/Africa	None (final severance at end of Albian)
South America/Antarctica	Late Cretaceous?
Antarctica/Australia	early Late Cretaceous; to Eocene?

among palaeontologists regarding the significance of the relationships between different faunas or floras. Nevertheless, a review of the literature results in relatively narrow time ranges for the possible connections, which are summarized in Table 3.1.

References

Alvarez, L. W., Alvarez, W., Asaro, R. & Michel, H. V. (1980). Extraterrestrial cause for the Cretaceous–Tertiary extinction. *Science*, **208**, 1095–108.

Arthur, M. A. & Natland, J. H. (1979). Carbonaceous sediments in the North and South Atlantic: the role of salinity in stable stratification of Early Cretaceous basins. In *Deep Drilling Results in the Atlantic Ocean: Continental Margins and Paleoenvironment*, ed. M. Talwani, W. Hay & W. B. F. Ryan, Maurice Ewing Series 3, pp. 375–401. Washington, D.C,: American Geophysical Union.

Asmus, H. E. & Ponte, F. C. (1973). The Brazilian marginal basins. In *The Ocean Basins and Margins*, vol. 1, *The South Atlantic*, ed. A. E. M. Nairn & F. G. Stehli, pp. 87–133. New York: Plenum Press.

Axelrod, D. I. (1981). Altitudes of Tertiary forests estimated from paleotemperature. In *Geological and Ecological Studies in Qinghai–Xizang Plateau*, vol. 1, *Geology, Geological History and Origin of Qinghai–Xizang Plateau*, Proceedings of Symposium on Qinghai–Xizang (Tibet) Plateau, pp. 131–7. Beijing and New York: Science Press and Gordon and Breach, Science Publishers.

Bailey, I. W. & Sinnott, E. W. (1916). The climatic distribution of certain types of angiosperm leaves. *American Journal of Botany*, **3**, 24–39.

Barker, P., Dalziel, I. A. W. *et al.*, with Harris, W. & Sliter, W. V. (1976). Evolution of the southwestern Atlantic Ocean basin: results of Leg 36,

Palaeogeography and palaeoclimate   67

Deep Sea Drilling Project. In *Initial Reports of the Deep Sea Drilling Project*, vol. 36, P. Barker, I. A. W. Dalziel *et al.*, pp. 993–1014. Washington, DC: US Government Printing Office.

Barron, E. J. (1983). A warm, equable Cretaceous: the nature of the problem. *Earth-Science Reviews*, **19**, 305–38.

Barron, E. J. & Washington, W. M. (1984). The role of geographic variables in explaining paleoclimates: results from Cretaceous climate model sensitivity studies. *Journal of Geophysical Research*, **89**, 1267–79.

Berggren, W. A. & Hollister, C. D. (1974). Paleogeography, paleobiogeography, and the history of circulation in the Atlantic Ocean. In *Studies in Paleoceanography*, ed. W. W. Hay, Society of Economic Paleontologists and Mineralogists, Special Publication 20, pp. 126–86.

Bigarella, J. J. (1973). Geology of the Amazon and Parnaiba Basins. In *The Ocean Basins and Margins*, vol. 1, *The South Atlantic*, ed. A. E. M. Nairn & F. G. Stehli, pp. 25–86. New York: Plenum Press.

Brass, G. W., Southam, J. R. & Peterson, W. H. (1982). Warm saline bottom water in the ancient ocean. *Nature*, **296**, 620–3.

Brenner, G. J. (1976). Middle Cretaceous floral provinces and early migrations of angiosperms. In *Origin and Early Evolution of Angiosperms*, ed. C. B. Beck, pp. 23–47. New York: Columbia University Press.

Brett-Surman, M. K. (1979). Phylogeny and palaeobiogeography of hadrosaurian dinosaurs. *Nature*, **277**, 560–2.

Briden, J. C. & Irving, E. (1964). Paleolatitude spectra of sedimentary paleoclimatic indicators. In *Problems in Palaeoclimatology*, ed. A. E. M. Nairn, pp. 199–224. London: John Wiley and Sons.

Buffetaut, E. & Taquet, P. (1975). Les vertébrés du Crétacé et la dérive des continents. *La Recherche*, **55**, 379–81.

Burchfiel, B. C. & Royden, L. (1982). Carpathian Foreland fold and thrust belt and its relation to Pannonian and other basins. *American Association of Petroleum Geologists*, **66**, 1179–95.

Cande, S. C. & Mutter, J. C. (1982). A revised identification of the oldest sea-floor spreading anomalies between Australia and Antarctica. *Earth and Planetary Science Letters*, **58**, 151–60.

Casamiquela, R. M. (1980). Considérations écologiques et zoogéographiques sur les vertébrés de la zone littorale de la mer du Maestrichtien dans le Nord de la Patagonie. *Memoires de la Societé Géologique de France, N.S.* **139**, 53–5.

Chamley, H. (1979). North Atlantic clay sedimentation and paleoenvironment since the Late Jurassic. In *Deep Drilling Results in the Atlantic Ocean: Continental Margins and Paleoenvironment*, ed. M. Talwani, W. Hay & W. B. F. Ryan, Maurice Ewing Series 3, pp. 342–61. Washington, DC: American Geophysical Union.

Chapman, R. M. & Sable, E. G. (1960). Geology of the Utukok–Corwin region, northwestern Alaska. *United States Geological Survey Professional Paper* **303-C**, 47–167.

CLIMAP (1976). The surface of the ice-age earth. *Science*, **191**, 1131–44.

Colbert, E. H. (1964). Climatic zonation and terrestrial faunas. In *Problems in Palaeoclimatology*, ed. A. E. M. Nairn, pp. 617–39. London: John Wiley and Sons.

Cox, C. B. (1980). An outline of the biogeography of the Mesozoic world. *Memoires de la Société Géologique de France, N.S.* **139**, 75–9.

Craddock, C. (1982). Antarctica and Gondwanaland. In *Antarctic Geoscience*, ed. C. Craddock, pp. 3–13. International Union of Geological Sciences Symposium. Madison: University of Wisconsin Press.

Dalziel, I. W. D. (1982). The early (pre-Middle Jurassic) history of the Scotia Arc region: a review and progress report. In *Antarctic Geoscience*, ed. C. Craddock, International Union of Geological Sciences Symposium, pp. 111–26. Madison: University of Wisconsin Press.

Dalziel, I. W. D & Elliot, D. H. (1973). The Scotia Arc and Antarctic margin. In *The Ocean Basins and Margins*, vol. 1, *The South Atlantic*, ed. A. E. M. Nairn & F. G. Stehli, pp. 171–246. New York: Plenum Press.

Doyle, J. A., Biens, P., Doerenkamp, A. & Jardiné, S. (1977). Angiosperm pollen from the pre-Albian Lower Cretaceous of equatorial Africa. *Bulletin des Centres de Recherche Exploration–Production Elf-Aquitaine*, **1**, 451–73.

Doyle, J. A., Jardiné, S. & Doerenkamp, A. (1982). *Afropollis*, a new genus of early angiosperm pollen, with notes on the Cretaceous palynostratigraphy and paleoenvironments of Northern Gondwana. *Bulletin des Centres de Recherche Exploration–Production Elf-Aquitaine*, **6**, 39–117.

Doyle, J. A. & Parrish, J. T. (1984). Jurassic–Early Cretaceous plant distributions and paleoclimatic models (abstract). *Abstracts, International Organization of Paleobotany Conference, Edmonton, August 18–26, 1984.*

Edgar, N. T., Saunders, J. B. *et al.* (1973). Site 148. In *Initial Reports of the Deep Sea Drilling Project*, vol. 15, N. T. Edgar, J. B. Saunders *et al.*, pp. 217–75. Washington, DC: US Government Printing Office.

Eldholm, O. & Thiede, J. (1980). Cenozoic continental separation between Europe and Greenland. *Palaeogeography, Palaeoclimatology, Palaeoecology*, **30**, 243–59.

Estes, R. & Hutchison, J. H. (1980). Eocene lower vertebrates from Ellesmere Island, Canadian Arctic Archipelago. *Palaeogeography, Palaeoclimatology, Palaeoecology*, **30**, 325–47.

Fisher, M. A., Patton, W. W. & Holmes, M. L. (1981). Geology and petroleum potential of the Norton Basin area, Alaska. *United States Geological Survey Open-File Report* 81–1316.

Flessa, K. W., Barnett, S. G., Cornue, D. B., Lomaga, M. A., Lombardi, N., Miyazaki, J. M. & Murer, A. S. (1979). Geologic implications of the relationship between mammalian faunal similarity and geographic distance. *Geology*, 7, 15–18.

Förster, R. (1978). Evidence for an open seaway between northern and southern proto-Atlantic in Albian times. *Nature*, 272, 158–9.

Fox, P. J. & Heezen, B. C. (1975). Geology of the Caribbean crust. In *The Ocean Basins and Margins*, vol. 3, *The Gulf of Mexico and the Caribbean*, ed. A. E. M. Nairn & F. G. Stehli, pp. 421–66. New York: Plenum Press.

Frakes, L. A. (1979). *Climates Throughout Geologic Time*. Amsterdam: Elsevier.

Franks, S. & Nairn, A. E. M. (1973). The equatorial marginal basins of west Africa. In *The Ocean Basins and Margins*, vol. 1, *The South Atlantic*, ed. A. E. M. Nairn & F. G. Stehli, pp. 301–50. New York: Plenum Press.

Hallam, A. (1984). Continental humid and arid zones during the Jurassic and Cretaceous. *Palaeogeography, Palaeoclimatology, Palaeoecology*, 47, 195–223.

Hay, W. W., Behensky, J. F., Jr, Barron, E. J. & Sloan III, J. L. (1982). Late Triassic–Liassic paleoclimatology of the proto-central North Atlantic rift system. *Palaeogeography, Palaeoclimatology, Palaeoecology*, 40, 13–30.

Huffman, A. C., Jr (ed.) (1985). Geology of the Nanushuk Group and related rocks, North Slope, Alaska. *United States Geological Survey Bulletin* 1614.

Jardiné, S., Kieser, G. & Reyre, Y. (1974). L'individualisation progressive du continent africain vue à travers les données palynologiques de l'ere secondaire. *Bulletin Sciences Géologiques Strasbourg*, 27, 69–85.

Kennedy, W. J. & Cooper, M. (1975). Cretaceous ammonite distributions and the opening of the South Atlantic. *Journal of the Geological Society of London*, 131, 283–8.

Kennett, J. P. (1977). Cenozoic evolution of Antarctic glaciation, the circum-polar Antarctic Ocean, and their impact on global paleoceanography. *Journal of Geophysical Research*, 82, 3843–60.

Larson, R. L. (1977). Early Cretaceous breakup of Gondwanaland off western Australia. *Geology*, 5, 57–60.

Lowenstam, H. A. & Epstein, S. (1954). Paleotemperatures of the post-Aptian Cretaceous as determined by the oxygen isotope method. *Journal of Geology*, 62, 207–48.

McKenna, M. C. (1972). Eocene final separation of the Eurasian and Greenland-North American landmasses. *24th International Geological Congress*, section 7, pp. 275–81.

McLachlan, I. R. & Pieterse, E. (1978). Preliminary palynological results: Site 361, Leg 40, Deep Sea Drilling Project. In *Initial Reports of the Deep Sea Drilling Project*, vol. 40, Bolli, H. M., Ryan, W. B. F. *et al.*, pp. 857–81. Washington, DC: US Government Printing Office.

Marshall, L. G. (1980). Marsupial paleobiogeography. In *Aspects of Vertebrate History: Essays in Honor of Edwin Harris Colbert*, ed. L. L. Jacobs, pp. 345–86. Flagstaff: Museum of Northern Arizona Press.

Morgan, R. (1978). Albian to Senonian palynology of Site 364, Angola Basin. In *Initial Reports of the Deep Sea Drilling Project*, vol. 40, Bolli, H. M., Ryan, W. B. F. *et al.*, pp. 915–51. Washington, DC: US Government Printing Office.

Nairn, A. E. M. & Stehli, F. G., eds. (1975). *The Ocean Basins and Margins*, vol. 3, *The Gulf of Mexico and the Caribbean*. New York: Plenum Press.

Neufville, E. M. H. (1973). Upper Cretaceous–Palaeogene Ostracoda from the South Atlantic. *Publications from the Palaeontological Institution of the University of Uppsala. Special Volume 1*.

Nicklés, M. (1950). Mollusques testaces marins de la côte occidentale d'Afrique. *Manuels Ouest-Africains*, vol 2. Paris: Lechevalier.

Parrish, J. M., Parrish, J. T. & Ziegler, A. M. (1986). Permian-Triassic paleogeography and paleoclimatology and implications for therapsid distribution. In *The Ecology and Biology of Mammal-like Reptiles*, ed. N. Hotton III, P. D. McLean, J. J. Roth & C. E. Roth, pp. 109–32. Washington, DC: Smithsonian Press.

Parrish, J. T. (1982). Upwelling and petroleum source beds, with reference to the Paleozoic. *American Association of Petroleum Geologists Bulletin*, **66**, 750–74.

Parrish, J. T. (1985). Global paleogeography, atmospheric circulation, and rainfall in the Barremian Age (late Early Cretaceous). *United States Geological Survey Open-File Report 85–728*.

Parrish, J. T. & Curtis, R. L. (1982). Atmospheric circulation, upwelling, and organic-rich rocks in the Mesozoic and Cenozoic. *Palaeogeography, Palaeoclimatology, Palaeoecology*, **40**, 31–66.

Parrish, J. T. & Doyle, J. A. (1984). Predicted evolution of global climate in Late Jurassic-Cretaceous time. *Abstracts, International Organization of Paleobotany Conference, Edmonton, August 18–26, 1984*.

Parrish, J. T., Ziegler, A. M. & Scotese, C. R. (1982). Rainfall patterns and the distribution of coals and evaporites in the Mesozoic and Cenozoic. *Palaeogeography, Palaeoclimatology, Palaeoecology*, **40**, 67–101.

Petri, S. & Campanha, V. A. (1981). Brazilian continental Cretaceous. *Earth Science Reviews*, **17**, 69–86.

Phillips, J. D. & Forsyth, D. (1972). Plate tectonics, paleomagnetism, and the opening of the Atlantic. *Geological Society of America Bulletin*, **83**, 1579–600.

Pindell, J. & Dewey, J. F. (1982). Permo-Triassic reconstruction of western Pangaea and the evolution of the Gulf of Mexico/Caribbean region. *Tectonics*, **1**, 1179–211.

Pitman, W. C., III, Larson, R. L. & Herron, E. M. (1974). The age of the ocean basins. *Geological Society of America Map and Chart Series*, **MC–6**.

Pratt, L. M. (1985). Isotopic studies of organic matter and carbonate in rocks of the Greenhorn marine cycle. In *Fine-Grained Deposits and Biofacies of*

the Cretaceous Western Interior Seaway: Evidence of Cyclic Sedimentary
Processes, ed. L. M. Pratt, E. G. Kauffman & F. B. Zelt, pp. 38–48.
Society of Economic Paleontologists and Mineralogists Field Trip
Guidebook no. 4, 1985 Midyear Meeting.

Premoli-Silva, I. & Boersma, A. (1977). Cretaceous planktonic
foraminifers – DSDP Leg 39 (South Atlantic). In *Initial Reports of the Deep
Sea Drilling Project*, vol. 39, P. R. Supko, K. Perch-Nielsen *et al.*, pp.
615–41. Washington, DC: US Government Printing Office.

Rabinowitz, P. D. & LaBrecque, J. (1979). The Mesozoic South Atlantic
Ocean and evolution of its continental margins. *Journal of Geophysical
Research*, **84**, 5973–6002.

Raymond, A., Parker, W. C. & Parrish, J. T. (1985). Phytogeography and
paleoclimate of the Early Carboniferous. In *Geological Factors and
Evolution of Plants*, ed. B. H. Tiffney, pp. 169–222. New Haven: Yale
University Press.

Reyment, R. A. (1973). Cretaceous history of the South Atlantic Ocean. In
Implications of Continental Drift for the Earth Sciences, ed. D. H. Tarling &
S. K. Runcorn, pp. 805–14. London: Academic Press.

Reyment, R. A., Pengtson, P. & Tait, E. A. (1976). Cretaceous transgressions
in Nigeria and Sergipe-Alagoas (Brazil). *Anais Academia Brasileira Ciencias*,
48 (supplement), 253–64.

Robinson, P. L. (1973). Palaeoclimatology and continental drift. In
Implications of Continental Drift to the Earth Sciences, ed. D. H. Tarling &
S. K. Runcorn, pp. 449–76. London: Academic Press.

Ronov, A. B. & Balukhovskii, A. N. (1982). Climatic zones on continents and
the general trend of climatic changes during the Late Mesozoic and
Cenozoic. *Lithology and Mineral Resources* **5**, 508–21.

Rowley, D. B., Raymond, A., Parrish, J. T., Lottes, A. L., Scotese, C. R. &
Ziegler, A. M. (1985). Carboniferous paleogeographic, phytogeographic, and
paleoclimatic reconstructions. *International Journal of Coal Geology*, **5**,
7–42.

Savin, S. M. (1977). The history of the Earth's surface temperature during
the past 100 million years. *Annual Review of Earth and Planetary Sciences*, **5**,
319–56.

Savin, S. M., Douglas, R. G. & Stehli, F. G. (1975). Tertiary marine
paleotemperatures. *Geological Society of America Bulletin*, **86**, 1499–510.

Sclater, J. G., Hellinger, S. & Tapscott, C. (1977). The paleobathymetry of
the Atlantic Ocean from the Jurassic to the present. *Journal of Geology*, **85**,
509–52.

Scotese, C. R., Bambach, R. K., Barton, C., Van der Voo, R. & Ziegler,
A. M. (1979). Paleozoic base maps. *Journal of Geology*, **87**, 217–77.

Shackleton, N. & Boersma, A. (1981). The climate of the Eocene ocean.
Journal of the Geological Society of London, **138**, 153–7.

Simpson, G. G. (1946). Tertiary land bridges. *Transactions of the New York Academy of Sciences*, **8**, 255–8.

Spaeth, C., Hoefs, J. & Vetter, U. (1971). Some aspects of the isotopic composition of belemnites and related paleotemperatures. *Geological Society of America Bulletin*, **82**, 2139–50.

Spicer, R. A. & Parrish, J. T. (1986). Paleobotanical evidence for cool North Polar climates in the middle Cretaceous (Albian-Cenomanian). *Geology*, **14**, 703–6.

Sues, H.-D. & Taquet, P. (1979). A pachycephalosaurid dinosaur from Madagascar and a Laurasia–Gondwanaland connection in the Cretaceous. *Nature*, **279**, 633–5.

Talwani, M. & Eldholm, O. (1977). Evolution of the Norwegian-Greenland Sea. *Geological Society of America Bulletin*, **88**, 969–99.

Talwani, M. & Udintsev, G. (1976). Tectonic synthesis. In *Initial Reports of the Deep Sea Drilling Project*, vol. 38, M. Talwani, G. Udintsev *et al.*, pp. 1213–42. Washington, DC: US Government Printing Office.

Talwani, M., Udintsev, G., *et al.* (1976). Sites 336 and 352. In *Initial Reports of the Deep Sea Drilling Project*, vol. 38, M. Talwani, G. Udintsev *et al.*, pp. 23–116. Washington, DC: US Government Printing Office.

Taquet, P. (1978). Niger et Gondwana. *Annales Societé Géologique du Nord*, **47**, 337–41.

Tedford, R. H. (1974). Marsupials and the new paleogeography. In *Paleogeographic Provinces and Provinciality*, ed. C. A. Ross, Society of Economic Paleontologists and Mineralogists Special Publication 21, pp. 109–26.

Thiede, J. (1977). Subsidence of aseismic ridges: evidence from sediments on Rio Grande Rise (southwest Atlantic Ocean). *American Association of Petroleum Geologists Bulletin*, **61**, 929–40.

Thiede, J. (1979). History of the North Atlantic Ocean: evolution of an asymmetric zonal paleo-environment in a latitudinal ocean basin. In *Deep Drilling Results in the Atlantic Ocean: Continental Margins and Paleoenvironment*, ed. M. Talwani, W. Hay & W. B. F. Ryan, Maurice Ewing Series 3, pp. 275–96. Washington, DC: American Geophysical Union.

Thiede, J. (1980). Palaeo-oceanography, margin stratigraphy and palaeophysiography of the Tertiary North Atlantic and Norwegian-Greenland Seas. *Philosophical Transactions of the Royal Society of London, ser. A*, **294**, 177–85.

Thomas, R. D. K. & Olson, E. C., eds. (1978). *A cold look at the warm-blooded dinosaurs*, Selected Symposia Series, no. 28. Washington, DC: American Association for the Advancement of Science.

Tomblin, J. F. (1975). The Lesser Antilles and Aves Ridge. In *The Ocean Basins and Margins*, vol. 3, *The Gulf of Mexico and the Caribbean*, ed. A. E. M. Nairn & F. G. Stehli, pp. 467–500. New York: Plenum Press.

Vail, P. R., Mitchum, R. M. & Thompson, S., III. (1977). Seismic stratigraphy and global changes of sea level. Part 4: Global cycles of relative changes of sea level. In *Seismic Stratigraphy – Applications to Hydrocarbon Exploration*, ed. C. E. Payton, American Association Petroleum Geologists Memoir 26, pp. 83–97.

Vinogradov, A. P. (ed.) (1967). *Atlas of the Lithological–Paleogeographical Maps of the U.S.S.R. IV. Paleogene, Neogene and Quaternary.* Moscow: Ministry of Geology, Akademia Nauk.

Vinogradov, A. P., ed. (1968). *Atlas of the Lithological–Paleogeographical Maps of the U.S.S.R. III. Triassic, Jurassic, Cretaceous.* Moscow: Ministry of Geology, Akademia Nauk.

West, R. M., Dawson, M. R. & Hutchison, J. H. (1977). Fossils from the Paleogene Eureka Sound Formation, N.W.T., Canada: occurrence, climatic and paleogeographic implications. In *Paleontology and Plate Tectonics*, ed. R. M. West, Milwaukee Public Museum Special Publications in Biology and Geology 2, pp. 77–93.

Wolfe, J. A. (1971). Tertiary climatic fluctuations and methods of analysis of Tertiary floras. *Palaeogeography, Palaeoclimatology, Palaeoecology*, **9**, 27–57.

Wolfe, J. A. (1979). Temperature parameters of humid to mesic forests of eastern Asia and relation to forests of other regions of the northern hemisphere and Australasia. *United States Geological Survey Professional Paper* **1106**.

Wolfe, J. A. & Hopkins, D. M. (1967). Climatic changes recorded in Tertiary land floras in northwestern North America. In *Tertiary Correlation and Climatic Changes in the Pacific*, ed. K. Hatai, 11th Pacific Science Congress, Symposium 25, pp. 67–76.

Ziegler, A. M., Scotese, C. R. & Barrett, S. F. (1983). Mesozoic and Cenozoic paleogeographic maps. In *Tidal Friction and the Earth's Rotation*, vol. II, ed. P. Brosche & J. Sündermann, pp. 240–52. Berlin: Springer-Verlag.

Ziegler, P. A. (1982). *Geologic Atlas of Western and Central Europe.* Amsterdam: Elsevier.

4

Mid-Cretaceous to Early Tertiary vegetation and climate: evidence from fossil leaves and woods

G.R.UPCHURCH, JR AND J.A.WOLFE

Climate and vegetation strongly influence the history of life. Climate exerts a major selective force on life cycle, morphology and physiology, the means by which organisms cope with the external environment. A partial result of climatic selection on numerous plant lineages from a given geographic region is vegetation, which has a characteristic structure determined by its component species. Vegetation, in turn, organizes the environment into a three-dimensional patchwork of microenvironments, each of which exerts its own unique selective regime on living organisms.

Adequate understanding of terrestrial biotic history during the rise of angiosperms requires knowledge of palaeoclimate and palaeovegetation. To date, however, detailed palaeobotanical reconstructions of climate and vegetation exist only for the Tertiary and Quaternary, and are lacking for the Cretaceous, a time of major angiosperm diversification. In this paper we summarize megafossil evidence of climate and vegetation during the diversification and rise to dominance of angiosperms (mid-Cretaceous to Eocene), concentrating on the role of angiosperms in the regional vegetation. Our results indicate a major change in terrestrial climate and vegetation during the rise of the angiosperms, which has important implications for the diversification of lineages and the evolution of adaptation.

Methods of analysis

High percentages of extinct genera and families characterize much of angiosperm history, precluding detailed interpretations of vegetation based on related extant taxa. Analyses of leaf and wood physiognomy offer a more precise method of inferring climatic and vegetational change, because, in extant vegetation, changes in physiognomy parallel overall changes in vegetational structure and hence climate (e.g. Richards, 1952; Wolfe, 1971). Features of value in determining vegetation and climate are detailed below.

Leaf size. Leaf size is high in understories of humid, evergreen, multistratal forests and decreases with decreasing temperature or precipitation. Large leaf size can also be caused by low light levels and moderate temperatures at high latitudes (Wolfe, 1985). Leaf size in fossil and modern vegetation is measured typically by a series of size classes (e.g. microphyll, notophyll, mesophyll), which, in turn, are used to construct the leaf size index (Wolfe, 1978; Wolfe & Upchurch, 1987*b*), a measure of average leaf size. The size classes used here are those of Raunkiaer (1934) as modified by Webb (1959).

Leaf margin. Serrate margins are typical of humid microthermal (mean annual temperature < 13 °C) vegetation and decrease in abundance with higher temperatures (Wolfe, 1979). In extant forests of eastern Asia, the percentage of entire-margined species correlates with mean annual temperature (*ca* 3%/1 degC), and the megathermal–mesothermal boundary (mean annual temperature = 20 °C) approximates 60%. In the Southern Hemisphere, 68% to 70% entire-margined species approximates to the megathermal–mesothermal boundary, and in mesothermal climates a decrease of 4% = 1 degC. Serrate margins also tend to correlate with thin (deciduous) leaves (Givnish, 1979).

Leaf texture. Thick, xeromorphic leaves are typically evergreen and predominate in megathermal and mesothermal climax vegetation, whereas thin, hygromorphic leaves are typically deciduous and predominate in microthermal climax vegetation or successional mesothermal vegetation. Some evergreen leaves are thin, making determination of relative evergreenness in fossil assemblages only approximate.

Drip-tips. Evergreen leaves in humid environments typically have highly attenuated apices, particularly in the understories of multistratal forests (Richards, 1952). The role of the drip-tip may be to accelerate draining of the moisture layer on the leaf surface, and thereby retard the growth of epiphytes. Deciduous leaves generally lack drip-tips, perhaps because of their shorter life span.

Leaf organization. Compound foliage tends to occur on plants of successional or disturbed vegetation. Such foliage is 'throw-away' and commonly associated with deciduousness (Givnish, 1979).

Leaf shapes. Narrow (stenophyllous) leaves tend to occur on stream-side plants (Richards, 1952). Broad, cordate leaves (typically palmately veined) occur on plants that are lianes, sprawling shrubs in successional vegetation, or light-gap colonizers (Givnish & Vermeij, 1976; Givnish, 1979). Lobed leaves occur typically in successional or understory species (Horn, 1971; Givnish, 1979).

Wood anatomy. Annual variation in precipitation and temperature is strongly reflected in the anatomy of secondary xylem. In climates with little seasonal variation in precipitation and temperature, wood usually lacks growth rings, whereas in climates with strong seasonal variation, woods have well-defined growth rings. In angiosperm woods, the diameter of vessel elements and their frequency in transverse section together provide an estimate of vulnerability of xylem to inactivation by air embolisms, which are caused by freezing or transpirational stress (Carlquist, 1975, 1977).

Failure to consider evolutionary factors can result in inaccurate determinations of vegetation and climate. For example, the megathermal–mesothermal boundary (mean annual temperature = 20 °C) in the Northern and Southern Hemispheres is expressed today by different percentages of entire-margined species in angiosperms (see above), which is probably a reflection of the different geologic and biotic histories of the two hemispheres (Bailey & Sinnott, 1915). For a second example, mid-Cretaceous floras show marked evolutionary changes in the foliar physiognomy of angiosperms and the overall ratio of angiosperms to gymnosperms (Crane, this volume, Chapter 5; cf. below). As a consequence extreme care must be used in making inferences of palaeotemperature and in extrapolating the structure of extant vegetational types to fossil plant assemblages.

Temperature estimates for the Cretaceous were made using a Southern Hemisphere (Australian) leaf-margin scale. A Southern Hemisphere scale was chosen for the following reasons: (1) The high percentages of gymnosperms and low diversity of deciduous angiosperms in Cretaceous floras is more similar to the modern situation in the Southern Hemisphere than to that for the Northern Hemisphere. (2) A Southern Hemisphere leaf-margin scale provides lower estimates of mean annual temperature than does a Northern Hemisphere leaf-margin scale, and hence gives a more conservative estimate of palaeotemparature at middle and higher palaeolatitudes.

Temperature estimates were made only during periods of slow vegetational evolution, and analogies to extant vegetation made only when the fossil vegetation resembled a modern analog in all physiognomic features.

The conclusions presented in this study are based largely on physiognomic data for North American megafloras; the data and methods of analysis are to be discussed in more detail elsewhere (Wolfe & Upchurch, 1987b). North America was examined in detail because it has numerous described megafloras that have good stratigraphic control; megafloras from other regions were surveyed to determine the generality of conclusions drawn from North America. Details of interpretation for the Cretaceous (e.g. precise mean annual temperatures) should be considered tentative because this interval has been poorly collected relative to the Tertiary, and because few leaf floras have species diagnoses based on detailed analyses of leaf architecture and cuticular anatomy. However, we think that the major vegetational patterns proposed in this paper are valid, as physiognomic analyses of recently revised floras give results congruent with analyses of published literature.

Vegetational history of the angiosperms

The history of the Earth's vegetation during the rise of angiosperms is divisible into three phases: (1) mid-Cretaceous (Aptian–Early Cenomanian), (2) Late Cretaceous (Middle Cenomanian–Maastrichtian) and (3) Early Tertiary. Each phase is characterized by distinct patterns of vegetational physiognomy and unique patterns of physiognomic change. Each phase is controlled by a combination of geologic and evolutionary factors.

The Aptian–Early Cenomanian comprises the early diversification and initial rise to dominance of angiosperms. Most physiognomic types of foliage characteristic of later phases gradually appear in the fossil record during this time. Leaf size is generally small at low and middle palaeolatitudes, and multistratal forests have strongly restricted distribution in time and space. Deciduous angiospermous vegetation with cold/dark season leaf fall appears at high palaeolatitudes by the Early Cenomanian. Angiosperms may dominate megafloral assemblages, but never comprise the bulk of the regional palynoflora. The order of appearance of foliar physiognomic types, along with changes in relative abundance and diversity, shows a general resemblance to ecological succession but occurs over a significantly longer evolutionary time scale (*ca* 20 million years). The Middle Cenomanian–Late Maastrichtian comprises a relatively stable physiognomic period. Individual climatic regions possess characteristic foliar types. Leaf size again is small at lower and middle palaeolatitudes, and forests typically have poorly developed stratification, apparently resulting from low precipitation. Fossil woods from palaeolatitudes below 45° N and 45° S show little evidence for annual variation in precipitation and temperature; fossil woods from higher palaeolatitudes show definite evidence for seasonality in precipitation and daylength. During this time angiosperms become the most abundant element in regional palynofloras from low and middle palaeolatitudes.

The Paleocene–Eocene comprises the expansion of humid multistratal forest at low and middle palaeolatitudes, due to more abundant precipitation. Evergreen rainforests became widespread in regions with megathermal temperatures. The vegetation of mesothermal regions from the Northern Hemisphere is anomalously deciduous during much of the Paleocene, and here extant families of deciduous angiosperms diversify. During the Eocene tropical dry-season deciduous vegetation first appears, and here families such as Leguminosae diversify. A dramatic decrease in global temperature occurs at the end of the Eocene, and with it the development of areally extensive cold-season deciduous vegetation at middle and high palaeolatitudes.

Aptian–Early Cenomanian

The initial rise of angiosperms is divisible into four parts, based largely on palynological and megafossil work from the Atlantic Coastal Plain of North America (Table 4.1). During the earliest part (Zone I of Brenner (1963) or Aptian–Early Albian) angiosperms were an unimportant early successional element of the regional vegetation. Their relative abundance is 1 % or less in regional palynofloras, and they are a sporadic element in megafloras from stream-margin facies (Hickey & Doyle, 1977). Approximately three-fourths of leaf species from this interval are entire margined; average leaf size is microphyllous, and only one species has a possible drip-tip. Most species are unlobed and pinnately veined, a condition commonly associated with late-successional vegetation (Figure 4.1); however, many are stenophyllous (highly elongate), a condition typical of extant streamside plants (Richards, 1952). Stenophylls are the

Table 4.1. *Changes in pollen abundance and foliar physiognomy for mid-Cretaceous angiosperms*

Age, European time-scale	Atlantic Coastal Plain pollen zonation	% Angiosperm pollen	% Entire-margined species	Leaf size index
Late Albian–Early Cenomanian	Subzone II–C and Zone III, Potomac Group			
	Platanoid facies		66	74
	Average	40	74	40
	Fort Harker, Dakota Fm		80	65
Middle–Late Albian	Subzone II–B			
	B-2	20	60	37
	B-1	10	65	38
Aptian–Early Albian	Zone I	< 1	76	13

All data are for the Potomac Group, except those for Fort Harker, which is from the Dakota Formation (Fm) of Kansas. Relative abundances of angiosperm pollen are based on data from Brenner (1963) and Upchurch & Doyle (1981), except for those of the Late Albian–Early Cenomanian, which are based on Upchurch & Doyle (1981) and Upchurch (unpublished results).

most diverse physiognomic type; others include large pinnately veined leaves typical of understory plants (*Ficophyllum*), bipinnately lobed leaves of possible semi-aquatic habit (*Vitiphyllum*), and reniform leaves with incipient palmate venation and a petiole/blade angle suggestive of an herbaceous or liana habit (*Proteaephyllum reniforme* and *Ficophyllum tenuinerve* in part) (Hickey & Doyle, 1977; Upchurch, unpublished results).

During the Middle–Late Albian (Subzone II-B of Brenner (1963)) angiosperms have a relative abundance of 20% or less in regional palynofloras, but can dominate local palynological and megafossil assemblages (Hickey & Doyle, 1977). Leaf megafossils occur in stream-margin facies and in overbank facies that often show evidence of environmental disturbance (Hickey & Doyle, 1977; Hickey, 1984a). In the Potomac Group there are somewhat lower percentages of entire-margined species in this interval than in earlier assemblages. In coeval assemblages from higher palaeolatitudes in the Western Interior, such as those from the Fall River Formation (Ward, 1899) and Cheyenne Sandstone (Berry, 1922), there can be lower percentages of entire-margined species than in the Potomac Group (e.g. the Fall River Formation), and some leaf genera common to both regions have more strongly developed teeth in the Western Interior (e.g. *Sapindopsis*). Average leaf size is larger than for Zone I, but species with drip-tips and probable vine habit are still of low diversity.

Physiognomic diversity for Middle–Late Albian angiosperm foliage is greater than that for Aptian–Early Albian angiosperms (Figure 4.1). All new foliar physiognomic types have features characteristic of extant early successional plants (cf. Givnish, 1979). Types characteristic of woody plants include pinnatifid and pinnately compound leaves (*Sapindopsis*), palmately lobed leaves with palinactinodromous primary venation (the 'platanoids'), and shallowly cordate leaves with serrate margin and palmate venation ('*Populus*' *potomacensis* and related 'trochodendroids'). Probable herbs include deeply cordate to peltate, orbicular leaves with palmate venation, a suite of features characteristic of scramblers and aquatics. All locally abundant species belong to one of these new physiognomic types; simple, pinnately veined leaves are rare in leaf assemblages.

Succeeding megafloras from the Late Albian (Subzone II-C of Brenner (1963) and equivalents) show a predominance of platanoid foliage and unlobed postulated derivatives (e.g. *Protophyllum*) in fluviatile sands, both from the Atlantic Coastal Plain and from the Soviet

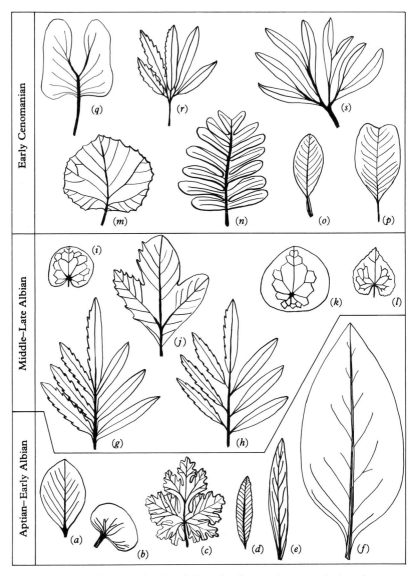

Figure 4.1. Origin of major foliar physiognomic types during the Aptian to Early Cenomanian. Each type illustrated for an individual zone persists into the overlying zones, although its systematic affinities can vary. Each taxon in parentheses represents the taxon used to illustrate a particular physiognomic type.

(*a*) Microphyllous, not elongate (*Celastrophyllum obovatum*); (*b*) incipiently palmate, semi-aquatic or vine (*Proteaephyllum reniforme*); (*c*) bipinnatifid (*Vitiphyllum multifidum*); (*d*) stenophyllous, serrate

Union (Vakhrameev, 1952; Hickey & Doyle, 1977). This abundance of platanoids may be due to their diversification and displacement of other groups of angiosperms during this time, as older and younger assemblages from sandy facies show higher physiognomic diversity. Average leaf size is large (Table 4.1), perhaps because platanoids are larger-leaved than most physiognomic types.

Early Cenomanian megafloras (Zone III of the Atlantic Coastal Plain and equivalents) are diverse and known from such widespread regions as Kazakhstan (Vakhrameev, 1952), Bohemia (Velenovský, 1882, 1884, 1886, 1887), the Atlantic Coastal Plain (Berry, 1916), the Western Interior (Lesquereux, 1892), and Alaska (Hollick, 1930; Spicer, Wolfe & Nichols, 1987). At lower middle palaeolatitudes in the Northern Hemisphere, angiosperms reach relative abundances of up to 40% in regional palynofloras (Table 4.1); however, angiosperms form the overwhelming majority of leaf megafossils. Reasons for the discrepancy between pollen and leaf abundances probably include: (1) high production of wind-dispersed pollen and spores by gymnosperms and ferns; (2) possible preference of gymnosperms for well-drained soils, where preservation of organic material is minimal (Retallack & Dilcher, 1981); and (3) the abundance and diversity of Laurales (Upchurch & Dilcher, 1984), whose pollen rarely survives in the geologic record (Muller, 1981, 1984). Angiosperm leaf megafossils occur predominantly in fluvial-deltaic rocks representing near-channel and overbank facies, which were deposited under freshwater to brackish conditions.

By the Early Cenomanian, the percentage of entire-margined angiosperm species declines markedly along a major latitudinal gradient, as in extant vegetation (Wolfe & Upchurch, 1987); however, these leaf-margin percentages show a lower latitudinal gradient than do those of

---

(Drewrys Bluff Leaf Type no. 1); (*e*) stenophyllous, entire-margined (*Rogersia*); (*f*) mesophyllous, unlobed (*Ficophyllum*); (*g*) pinnatifid (*Sapindopsis*); (*h*) pinnately compound (*Sapindopsis*); (*i*) cordate, orbicular (*Populophyllum reniforme*); (*j*) lobed, palinactinodromous primary venation (platanoid); (*k*) peltate, orbicular, aquatic (*Nelumbites*); (*l*) cordate, non-orbicular, serrate ('*Populus*' *potomacensis*); (*m*) unlobed, serrate, with highly branched basal secondary veins ('*Viburnum*'); (*n*) pinnatifid, more than one secondary vein per lobe ('*Liriodendron*' *snowii*); (*o*) simple, entire-margined, with winged petiole (*Citrophyllum*); (*p*) simple, entire-margined, emarginate apex (*Liriodendropsis*); (*q*) bilobed (*Liriophyllum*); (*r*) palinactinodromously compound (*Dewalquea*); (*s*) 'dichotomously' compound (*Halyserites*).

extant vegetation (*ca* 1% per degree palaeolatitude, $R^2 = 0.94$). At middle palaeolatitudes high percentages of entire-margined species and thick laminar texture indicate broadleaved evergreen vegetation, while, in the Arctic, low percentages of entire-margined species and thinner foliar texture indicate broadleaved deciduous vegetation (cf. Wolfe, 1979). At middle palaeolatitudes average leaf size is small (under notophyllous), while in the Arctic average leaf size is larger. Drip-tips are uncommon at all palaeolatitudes, and there are few species of probable vine habit. At lower–middle palaeolatitudes there are marked differences between individual leaf assemblages in leaf size and other physiognomic features. Sedimentary facies and salinity are two contributing factors, as the smaller-leaved assemblages come from overbank facies, and the smallest-leaved assemblage (Rose Creek, Dakota Formation of Nebraska) comes from brackish-water deposits (Retallack & Dilcher, 1981; Wolfe & Upchurch, 1987*b*). For larger-leaved assemblages, the factors controlling physiognomy cannot be related to differences in sedimentary environment. The largest-leaved assemblage comes from the Dakota Formation of Kansas, near Fort Harker, and differs from coeval sandstone and shale assemblages in having high percentages of species with entire margins, drip-tips, and probable vine habit (Wolfe & Upchurch, 1987*b*).

Early Cenomanian leaf assemblages are much more diverse physiognomically than Early Cretaceous assemblages (Figure 4.1), and late successional physiognomic types show higher species diversity and higher relative abundances. New leaf types characteristic of early successional plants include bilobed foliage (e.g. *Liriophyllum*), pinnatifid leaves with more than one primary vein entering each lobe ('*Liriodendron*', *Conospermites* in part), and palinactinodromously to 'dichotomously' compound leaves (*Dewalquea, Halyserites*). New leaf types characteristic of late successional plants include pinnately veined, unlobed leaves with strongly emarginate apices, a suite of features found today in canopy species of tropical forests (Richards, 1952), and simple unlobed leaves that show possible evidence for derivation from an ancestor with compound leaves (*Citrophyllum*) (Hickey, 1978). Unlobed, pinnately veined foliage becomes the dominant element in many leaf assemblages from both stream-margin and overbank facies. The highest physiognomic diversity and all major physiognomic innovations occur at middle palaeolatitudes. Arctic vegetation is physiognomically stereotyped and consists of a few, predominantly early successional, leaf types that are first seen earlier at middle palaeolatitudes (e.g. platanoids, trochodendroids).

In summary, the Aptian–Early Cenomanian records the gradual physiognomic assembly of the Late Cretaceous vegetation. The pattern of change in foliar physiognomy supports Doyle & Hickey's (1976) hypothesis that the angiosperms were initially early successional plants, and that by the Cenomanian they formed the canopy of many late successional forests. In general, the first types of foliage to dominate leaf assemblages have features characteristic of early successional species, while not until the Early Cenomanian are types of foliage characteristic of late successional vegetation a regular and diverse component of the megaflora. At middle palaeolatitudes, most leaf assemblages have small leaf size and few species with drip-tips or probable vine habit. In megathermal regions today, such features characterize vegetation of subhumid regions and sandy porous soils, which typically has an open canopy and a relatively short stature (e.g. Richards, 1952). However, a few assemblages from North America have high percentages of species with entire margins, large leaves, drip-tips and (in the case of Fort Harker) probable vine habit. These assemblages appear to represent megathermal, closed-canopy multistratal forest, but the factors controlling their distribution are unclear. One distinct possibility is that they document wetter and warmer periods of a duration of much less than two million years, the average length of a mid-Cretaceous pollen zone/subzone.

Temperature estimates for the Aptian–Early Cenomanian are uncertain because the vegetation was evolving rapidly and because there are no close modern analogs to the mid-Cretaceous vegetation. If Southern Hemisphere leaf-margin percentages are used (20 °C mean annual temperature = 68 % to 70 % entire-margined species), then mean annual temperature at lower–middle palaeolatitudes averaged 20 °C, with assemblages such as Fort Harker documenting possible intervals of higher temperatures (Wolfe & Upchurch, 1987*b*).

Middle Cenomanian–Late Maastrichtian

The Middle Cenomanian–Late Maastrichtian represents a stable period in the history of angiospermous vegetation. During this time, angiosperms rise to dominance in the palynological record of many regions, first at lower then at high palaeolatitudes. Few physiognomic types of foliage originate during this time, in contrast to the

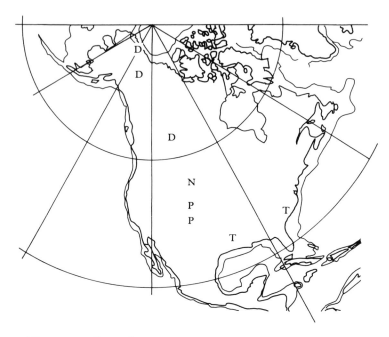

Figure 4.2. Generalized map of Late Cretaceous vegetation for
North America, based on the Late Maastrichtian. During cool
intervals the megathermal/mesothermal boundary was approxi-
mately 5° to 10° farther south. T, Tropical Forest; P, Paratropical
Forest; N, Notophyllous Broadleaved Evergreen Forest; D, Polar
Broadleaved Deciduous Forest.

Aptian–Early Cenomanian. Physiognomic differences between leaf
floras represent largely variations on a theme that can be explained by
differences in temperature and precipitation. Three major latitudinal
zones characterize vegetation of the Northern Hemisphere during this
time: (1) low-middle palaeolatitudes (*ca* 30° to 45° N), (2) high-middle
palaeolatitudes (*ca* 45° to 65° N) and (3) high palaeolatitudes (> 65° N)
(Figure 4.2).

Low-middle latitude vegetation. Low-middle latitude megafloras from
the eastern United States (Middle Cenomanian–Early Maastrichtian)
and southern Rocky Mountains (Campanian–Late Maastrichtian)
show many physiognomic similarities, despite their isolation by the
Western Interior Seaway. Floristically, eastern North America belonged to the Normapolles palynoprovince, which extended into west-

central Europe, while the southern Rocky Mountains belonged to the Colorado–Californian region (Batten, 1984). Angiosperms first rise to dominance in regional palynofloras during the Middle Cenomanian (Woodbridge member of the Raritan Formation) and remain dominant throughout the Late Cretaceous, except during postulated cool intervals (Wolfe, 1976). Proportions of entire-margined species usually exceed 70%, and most leaves are thick textured, indicating megathermal evergreen vegetation (Wolfe & Upchurch, 1987*b*). Leaf-margin percentages indicate paratropical temperatures (mean annual temperature 20 to 25 °C) for most intervals of the Late Cretaceous, with tropical temperatures (mean annual temperature > 25 °C) occurring in eastern North America during warm intervals. Leaf size in both coarse and fine-grained sedimentary facies is significantly smaller than that for extant megathermal rainforests, except during the Middle Cenomanian (Figure 4.3); as a rule, leaf size is smaller than notophyllous in fossil assemblages (Leaf Size Index < 40), while it is larger than notophyllous in extant megathermal rainforests (Leaf Size Index > 65, usually much higher). Species with drip-tips and probable vine habit are of low diversity. All of the fruit and seed assemblages analyzed for the Late Cretaceous of North America by Tiffney (1984) come from low-middle palaeolatitudes, and these have the small diaspore size indicative of seed germination in full sunlight. Together these features indicate open-canopy conditions caused by subhumid conditions. Subhumid conditions are indicated also by the prevalence of emarginate apices on the foliage, a condition that commonly occurs in extant tropical plants when growth of the developing leaf apex is prematurely terminated by conditions of high light and low humidity (Richards, 1952). Small leaf size, rarity of drip-tips, and small seed size also characterize megafloras from the Normapolles palyno-province of west-central Europe, located at low-middle palaeolatitudes, and small leaves characterize megafloras from southern Egypt, located within 5° of the palaeoequator (Wolfe & Upchurch, 1987*b*).

Simple, pinnately veined angiosperm leaves characterize leaf floras from low–middle palaeolatitudes. In general, the range of leaf types seen in the Middle Cenomanian–Late Maastrichtian of this region is similar to that seen in the Early Cenomanian, but with a lower relative abundance and diversity of groups such as the platanoids and inferred platanoid derivatives (cf. Hickey, 1984*b*).

The physiognomy of fossil wood assemblages from low-middle palaeolatitudes indicates that there was little seasonality to the climate. Distinct growth rings are absent in all angiosperms and are absent or weakly developed in gymnosperms, indicating little seasonal variation in rainfall (Creber & Chaloner, 1985; Wolfe & Upchurch, 1987*b*). This contrasts with the situation for modern subhumid climates, where there is typically a distinct seasonality in precipitation that is reflected in wood anatomy. The lack of seasonality to precipitation suggests that the closest living analog to the Late Cretaceous vegetation of low-middle palaeolatitudes may be the evergreen vegetation of sandy porous soils in the tropics, where there is abundant precipitation year-round, but low nutrients limit leaf size and vegetational stature in a manner similar to that of low rainfall (e.g. Givnish, 1979; Richards, 1952; Walter, 1973). Fossil angiosperm woods come from trunks up to 1 m in diameter, which indicates that at least some angiosperms were trees; these woods contain large volumes of parenchyma and wide, often solitary vessels that occur in low frequencies (Wheeler, Matten & Lee, 1987; Wolfe & Upchurch, 1987*b*). Both features indicate lack of freezing temperatures, as comparable extant woods belong typically to tropical taxa (cf. Chalk, 1955; Carlquist, 1975). The combination of high volumes of parenchyma and wide vessels is atypical for woods from modern subhumid climates (Carlquist, 1975, 1977), which are strongly seasonal, but is hypothesized to be adaptive under conditions of low seasonal variation in precipitation and no period of drought, both of which were present in this region during the Late Cretaceous (Wolfe & Upchurch, 1987*b*).

High-middle latitude vegetation. High-middle palaeolatitude megafloras are known primarily from the latter half of the Late Cretaceous (Senonian); little is known about Middle Cenomanian–Coniacian megafloras. Angiosperms rise to dominance in regional palynofloras from North America between the Turonian and Campanian (D. J. Nichols, unpublished) and, by the Santonian, angiosperms are the most abundant element in the megafossil record. Percentages of entire-margined species generally range from 40% to 65%, and the leaves are thick-textured, indicating mesothermal evergreen vegation (mean annual temperature = 13 to 20 °C). The rarity of drip-tips indicates the absence of rainforest vegetation, but average leaf size in most floras is slightly less than notophyllous and hence nearly as large as foliage in extant mesothermal evergreen vegetation from humid regions (cf. Wolfe, 1979). Leaf size, therefore, indicates more favorable moisture conditions for plant growth at high-middle palaeolatitudes than at low-middle palaeolatitudes. Fruit and seed assemblages are not known from Late Cretaceous mesothermal regions, but inferred vegetational structure would predict some large diaspores characteristic of plants with shade germination. Most mesothermal evergreen floras have physiognomic features comparable to those of extant Notophyllous Broadleaved Evergreen Forest (warm-month mean > 20 °C, cold-month mean > 1 °C), but the Patoot flora of Western Greenland represents Microphyllous Broadleaved Evergreen Forest, a vegetational type today restricted to low-latitude uplands (warm-month mean < 20 °C, mean annual temperature < 15 °C). Leaves with emarginate apices are less abundant and diverse at high-middle palaeolatitudes than at low-middle palaeolatitudes, supporting the interpretation of more abundant moisture.

Pinnately veined entire-margined species are present in Late Cretaceous mesothermal regions, but are less abundant and diverse than in megathermal regions. Other characteristic foliar types include serrate-margined leaves with simple craspedodromous secondary veins (e.g. '*Dryophyllum*'), palmately veined leaves with acrodromous lateral primary veins (e.g. '*Zizyphus*'), and serrate-margined leaves with actinodromous primary venation or highly branched basal secondary veins (e.g. '*Viburnum*'). Foliar types most characteristic of high-palaeolatitude vegetation, such as the platanoids and trochodendroids, have higher relative abundances and species diversity in mesothermal regions than in megathermal regions (cf. Hickey, 1984*b*).

Fossil wood assemblages from high-middle palaeolatitudes have important similarities and differences from those of low-middle palaeolatitudes. Angiosperm and gymnosperm woods described from the Early Maastrichtian of California (e.g. Page, 1979, 1980, 1981), located at palaeolatitude 45° N, and the late Senonian of South Africa (Mädel, 1960, 1962; Schultze-Motel, 1966), located at palaeolatitude 45° S, generally have no growth rings and therefore show little evidence of seasonality (Creber & Chaloner, 1985; Wolfe & Upchurch, 1987*b*). The angiosperm woods can have less parenchyma than those from low-middle palaeolatitudes (e.g. the South African woods) and somewhat smaller vessels; in addition, many of the taxonomic comparisons are with extant angiosperms that inhabit mesic climates. At somewhat higher palaeolatitudes (56 to 60° N) conifer woods have well-defined growth rings, indicating a definite seasonality of climate, caused by variation in precipitation or light during the growing season (Creber & Chaloner, 1985; Wolfe & Upchurch, 1987*b*).

High latitude vegetation. High-palaeolatitude megafloral assemblages are well known from North America and Siberia. The most complete megafloral sequence is from Alaska and ranges in age from Cenomanian to Campanian (Spicer *et al.*, 1987); Maastrichtian leaf assemblages are known from Central Alberta (Bell, 1949; Wolfe & Upchurch, 1986). Angiosperm leaves from these regions are associated with a variety of deciduous gymnosperms, and are characterized by low percentages of entire-margined species, thin leaf texture and large average leaf size. Such features indicate Polar Broadleaved Deciduous Forest (Wolfe, 1985), an extinct vegetational type, where temperatures ranged from mesothermal to microthermal and leaf fall was controlled by Arctic night. Large leaf size is consistent with inferred high precipitation, although much of the leaf size is probably due to long daylength and low intensity of sunlight.

Physiognomically, Polar Broadleaved Deciduous Forest of this interval is of low diversity. Characteristic leaf types include platanoids, trochodendroids, and '*Viburnum*'-type foliage. Pinnately veined, entire-margined foliage is a low-abundance element in high-latitude megafloras, and during the Late Cretaceous it declines in abundance and species diversity. Species with emarginate apices are known only from the Coniacian (Spicer *et al.*, 1987).

Woods from the Maastrichtian of central Alberta (Ramanujam & Stewart, 1969) show well-defined growth rings that can have significant latewood, indicating either a period of low moisture during the latter part of the growing season or growth during periods of winter twilight (Wolfe & Upchurch, 1987*b*). Woods from the Late Cenomanian of the North Slope, Alaska (Spicer & Parrish, 1986) have large and distinct growth rings with practically no latewood, indicating abundant precipitation throughout the growing season and the rapid onset of winter dormancy.

Physiognomic evolution. New physiognomic types for the Middle Ceno-manian to Late Maastrichtian are limited to monocotyledons and occur in broadleaved evergreen vegetation. Palms first appear in the Santonian (Coniacian?) of megathermal regions and spread to mesothermal regions by the Campanian (Bell, 1957; Daghlian, 1981; Hickey, 1984*b*). Zingiberales first appear during the Campanian (Friis, 1988). Both groups have large, often pinnately veined leaves that are borne typically on thick, parenchymatous, sparsely branched stems. Such physiognomy is strikingly similar to that of many cycadophytes, which undergo a marked reduction in species diversity and relative abundance during the Aptian–Early Cenomanian (Crane, this volume, Chapter 5). Such

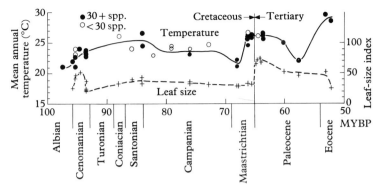

Figure 4.3. Percentage of entire-margined species and average leaf
size for Late Cretaceous assemblages from low-middle palaeolatitudes of North America. Leaf-margin percentages are standardized
for palaeolatitude, using inferred latitudinal temperature gradients.

similarities in physiognomy probably reflect similarities in vegetative
ecology, and thus indicate a lag between the decline of cycadophytes
during the mid-Cretaceous and the subsequent rise of angiospermous
analogs.

In summary, land temperatures during the Late Cretaceous were
higher than at present, especially at middle and high palaeolatitudes, and
latitudinal temperature gradients were approximately one-half to two-
thirds those of the present day (*ca* 0.5 degC per degree latitude) (see also
Parrish, this volume, Chapter 3). The latitudinal temperature gradient
inferred for the Campanian is approximately 0.3 degC, based on a
decline of 1.3 % of entire-margined species per degree palaeolatitude
(Wolfe & Upchurch, 1987*b*). Calculation of a latitudinal temperature
gradient for the Late Maastrichtian is more problematic because there
are no known floras below 46° N; if Early Paleocene floras from the Gulf
Coast are included, the estimated gradient is 0.3 degC (Wolfe &
Upchurch, 1987*b*). When standardized for palaeolatitude, leaf-margin
percentages for eastern North America and the Western Interior indicate
fluctuations in mean annual temperature during the Late Cretaceous
(Figure 4.3), with temperature maxima during the Santonian and Late
Maastrichtian. These temperature fluctuations appear to be at least
hemisphere-wide in extent, as the inferred Santonian maximum is
corroborated by physiognomic, floristic, and oxygen isotope data from
outcropping rocks in the Soviet Union (Vakhrameev, 1978), and the

inferred Late Maastrichtian maximum is corroborated by oxygen isotope data from deep-sea cores in the north and south Atlantic Ocean (Boersma, 1984).

Latitudinal variation in leaf size indicates a major latitudinal gradient in moisture during the Late Cretaceous (Wolfe & Upchurch, 1987b); in general, conditions were subhumid at lower palaeolatitudes and ameliorated towards the poles. Physiognomic data from fossil woods indicate that there was little seasonality to the climate at palaeolatitudes below 45° N and 45° S.

Early Tertiary

The events at the Cretaceous–Tertiary (K–T) boundary had a major impact on terrestrial climates and vegetational structure. Palynological and foliar remains together provide evidence for mass-kill of the vegetation at the K–T boundary, followed by a period of vegetational recovery that resembled modern secondary succession in many of its characteristics, but occurred over a longer time-scale. Precipitation levels greatly increased at low and middle northern palaeolatitudes, and these initiated the widespread development of multistratal rainforests in megathermal climates. Vegetation at high-middle palaeolatitudes was deciduous during much of the Paleocene, and this was apparently due to the extinction of broadleaved evergreen taxa, rather than to a major decline in mean annual temperature. Dry-season deciduous forests developed in southeastern North America during the Eocene, indicating the origin of winter drought in megathermal regions.

The K–T transition is well-represented in the Western Interior of North America, in continuous stratigraphic sections extending from New Mexico to Saskatchewan. Palynological studies of these sections provide strong evidence for mass-kill of the vegetation at the K–T boundary and recolonization of the landscape by ferns (e.g. Nichols et al., 1986; Tschudy et al., 1984). The mass-kill event occurs at the level of a distinctive boundary clay that contains high levels of iridium, shocked quartz, and other evidence of a bolide impact (e.g. Bohor et al., 1984; Orth et al., 1981). Detailed analysis of leaf megafossils and dispersed cuticles at the K–T boundary in the Raton Basin of New Mexico and Colorado corroborates palynological interpretations of mass-kill and recolonization; in addition, it indicates that vegetational recovery physiognomically mimicked secondary succession in extant megathermal vegetation but occurred over a longer time-scale (Wolfe & Upchurch, 1987a). No evidence for a decrease in temperature can be detected, but a marked increase in precipitation is indicated by an

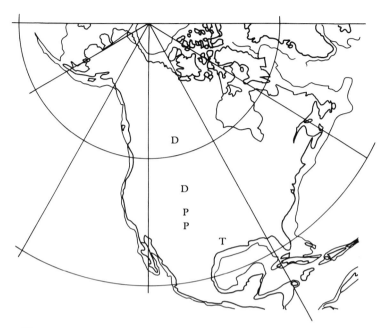

Figure 4.4. Early Paleocene vegetational map for North America. Note the absence of notophyllous broadleaved evergreen vegetation and the close proximity of Paratropical Rainforest and Broadleaved Deciduous Forest. T, Tropical Rainforest; P, Paratropical Rainforest; D, Broadleaved Deciduous Forest.

increase in leaf size, higher percentages of species with drip-tips, and higher percentages of species with hairless foliage. By the end of the early Paleocene multistratal Paratropical Rainforest was well developed, but with a much lower species diversity than that of modern analogs.

Throughout middle palaeolatitudes in North America during the Early Paleocene leaf megafloras are larger leaved than their Late Cretaceous counterparts (Fig. 4.3) and have a higher percentage of species with drip-tips and probable vine habit. This indicates an areally extensive increase in precipitation at the K–T boundary (Wolfe & Upchurch, 1986, 1987*b*). The few known leaf assemblages of possible Paleocene age from near the palaeoequator in North Africa also have a larger leaf size than their Cretaceous counterparts (Wolfe & Upchurch, 1987*b*). Analysis of North American leaf-margin percentages also indicates that the vegetation underwent a major reorganization along a latitudinal gradient, best explained by a brief low-temperature excursion at the K–T boundary (Wolfe & Upchurch, 1986) (Figure 4.4). Vegetation from 48° N and regions south of this represents megathermal rainforest, while vegetation from 54° N and regions north of this represents broadleaved deciduous forest. Mesothermal broadleaved evergreen forest is not represented by any known leaf assemblage, and could have occupied a narrow latitudinal belt of only a few degrees. Today, even with the current high latitudinal temperature gradient (0.5 degC/1° latitude) along the East Asian coast, mesothermal broadleaved evergreen vegetation extends through 9° of latitude. In eastern North America today, the distribution of mesothermal evergreen vegetation is limited by intense Arctic cold fronts (Wolfe, 1979). However, such cold fronts could not be responsible for the situation during the Early Paleocene, because of the abundance and diversity in high-middle palaeolatitude faunas of large ectothermic vertebrates (e.g. crocodiles, champosaurs) (e.g. Bartels, 1980), animals that today can tolerate only mild freezing (e.g. Estes & Hutchinson, 1980). Taken together, these data are best explained by a major extinction of mesothermal broadleaved evergreen taxa at the K–T boundary in North America, and their replacement by broadleaved deciduous taxa during the Early Paleocene.

Deciduous vegetation characterizes much of the Paleocene of the Northern Rocky Mountains and Great Plains. Floristically, this vegetation is characterized by extant deciduous-leaved families such as Betulaceae and Juglandaceae, which undergo a major diversification during this time (e.g. Crane, this volume, Chapter 5; Manchester, 1987). By the Early Eocene, however, megathermal vegetation occupied its greatest latitudinal extent and expanded to 50° to 60° N and similarly S (Wolfe, 1985). Physiognomically, much of this vegetation represents rainforest; floristically, it contains a diversity of families that are important in extant megathermal rainforest, including Annonaceae, Lauraceae, Menispermaceae and Icacinaceae (e.g. Reid & Chandler, 1933; Wolfe, 1977). By the Middle Eocene, megathermal vegetation was reduced in latitudinal extent, and during the remainder of the Eocene its latitudinal boundaries changed with fluctuations in mean annual temperature (Wolfe, 1985).

Two new vegetational types attained wide distribution by the Middle Eocene (Figure 4.5). In southeastern North America and Eurasia megathermal vegetation is best classified as semideciduous forest, and indicates the development of major regions of seasonally dry climates (Wolfe, 1985). Notable in this vegetation is the diversity of leaflets with the entire margins and cross-striated pulvini characteristic of Leguminosae, an important family in extant dry-season deciduous vegetation.

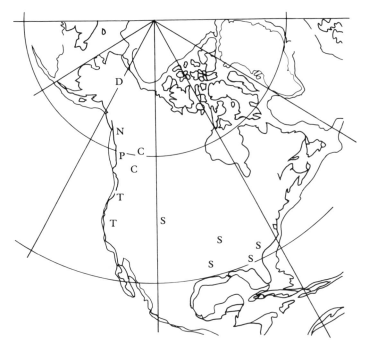

Figure 4.5. Late Middle Eocene vegetational map for North America (adapted from Wolfe, 1985). T, Tropical Rainforest; S, Tropical Semi-deciduous Forest; P, Paratropical Rainforest; N, Notophyllous Broadleaved Evergreen Forest; D, Polar Broadleaved Deciduous Forest; C, mixed coniferous forest.

Pollen of legumes is present in the Maastrichtian, but does not diversify until the Eocene (Muller, 1984). In volcanic uplands of high-middle palaeolatitudes of western North America, microthermal coniferous forests dominated by Pinaceae and Cupressaceae first appear (Wolfe, 1985). Broadleaved deciduous plants were a significant component of this vegetation, and major microthermal families such as Rosaceae and Aceraceae underwent an adaptive radiation (Wolfe, 1987).

In summary, the Paleocene–Eocene records the disappearance of certain Late Cretaceous vegetational types, coupled with the first appearance or expansion of vegetational types characteristic of the modern world. During the Early Paleocene, precipitation increased markedly to levels typical of many extant low- and middle-latitude climates, and subsequent to the Early Paleocene seasonal dryness became an important factor in megathermal vegetation. At the end of the Eocene a major temperature decline reduced the latitudinal extent of megathermal vegetation, and large regions of microthermal broadleaved deciduous forest developed for the first time during the history of the angiosperms (Wolfe, 1985).

Discussion

The proposed scenario of vegetational evolution has important implications for the history of angiosperm adaptation and diversification. First, physiognomic evidence suggests that during the Late Cretaceous megathermal forests were adapted primarily to conditions of aseasonal dryness, and that a major increase in precipitation at the K–T boundary provided the first areally extensive conditions under which true multistratal rainforests could develop. Because megathermal rainforests are fundamentally closed-canopy vegetation, such forests would have provided numerous new or greatly expanded niches for plant adaptation. These include:

Tall trees with slender, buttressed trunks. High precipitation would lead to close spacing of individual trees and, in turn, a closed canopy and low light levels. Selection would therefore favor tall trees that could grow above the competition.

The liana habit. An alternative strategy to evolving into a tall tree would be to use adjacent trees for structural support. In vegetation with short stature and open canopy, scrambling shrubs would be favored over lianas; development of a closed canopy and a tall forest would place a premium on rapid vertical growth.

The epiphytic habit. Epiphytes are rare in dry climates. The development of tall rainforests would increase selection for epiphytes because of low light levels on the forest floor; the abundance of precipitation would permit the accumulation of enough water to allow life on the stems of tall trees.

The understory habit. High-moisture/low-light conditions characterizing the understory would select for large leaf size and the ability to photosynthesize under suboptimal light conditions.

Shade germination. Low-light conditions of the forest floor would increase selective pressure for large nutrient reserves in seeds because of limited seedling photosynthesis. Selection for large seed size would decrease dispersability by wind currents, and hence increase the benefits of biotic dispersal.

Physiognomic analyses of Northern Hemisphere megafloral assemblages are all consistent with the origin of areally extensive, multistratal, closed-canopy vegetation. Foliar physiognomy indicates that lianas and understory plants are more abundant and diverse in the Early Tertiary than in the Late Cretaceous (Wolfe & Upchurch, unpublished results). Average diaspore size increases markedly from Late Cretaceous to Paleocene (Tiffney, 1984), as would be expected with an increase in shade germination and increase in biotic dispersal. The size of fossil tree trunks and the percentage of epiphytes also would be expected to increase, but insufficient data exist to test these predictions.

Many of the patterns of diversification seen in extant families during the Early Tertiary are readily explained by canopy closure coupled with an increase in forest height and canopy stratification. Extant families of lianas such as Icacinaceae, Menispermaceae and Vitaceae are abundant and diverse in the Early Eocene (Reid & Chandler, 1933; Wolfe, 1977), but are rare or absent in the Late Cretaceous (Muller, 1981, 1984). Extant families that form tall canopy trees in megathermal rainforests, such as Anacardiaceae, Lauraceae and Rutaceae, all are diverse components of Eocene floras (Reid & Chandler, 1933; Wolfe, 1977), but are rare or absent in the Late Cretaceous (Muller, 1981, 1984). These data suggest that the family-level modernization of angiosperm floras during the Early Tertiary was due, in part, to the expansion of multistratal vegetation over large geographic regions. Such modernization would have been assisted by the ecologic disruption at the K–T boundary, for which foliar data suggest extinction up to 75 % at the level of biological species (Wolfe & Upchurch, 1986), and for which pollen and leaf data indicate changes in abundance for many surviving lineages (Fleming, 1985; Tschudy *et al.*, 1984; Wolfe & Upchurch, 1987*b*).

The southerly expansion of broadleaved deciduous vegetation in the Northern Hemisphere during the Early Paleocene could very well have promoted the diversification of extant families of deciduous angiosperms. For example, Juglandaceae are a characteristic component of Paleocene palynofloras from the Northern Rocky Mountains and Great Plains; although present in low diversity and abundance during the Late Cretaceous, the family diversified rapidly during the Paleocene, and by the beginning of the Eocene most modern genera are present (Manchester, 1987; Nichols & Ott, 1978). For another example, Betulaceae are rare, low diversity elements in Late Cretaceous palynofloras and comprise just *Alnus* and possibly *Betula*; during the Paleocene the family becomes more abundant and diverse in the Northern Rocky Mountains, and new genera make their first appearance at this time (P. R. Crane, personal communication). Today, deciduous angiosperms are more diverse in the Northern Hemisphere than in the Southern Hemisphere (Bailey & Sinnott, 1915); this could be explained if areally extensive regions of broadleaved deciduous vegetation during the Paleocene were a strictly Northern Hemisphere phenomenon.

The general dryness of Cretaceous megathermal climates relative to those of the Tertiary and Recent also appears to have had important effects on early phases of angiosperm diversification, and can explain certain phenomena that appear to be anomalous by modern standards. For example, the ability to tolerate osmotic stress due to elevated salinities evolved early within the angiosperm adaptive radiation: it is definitely present by the Cenomanian (cf. Retallack & Dilcher, 1981) and could have evolved as early as the Aptian, on the basis of relatively high abundances of *Afropollis* pollen in Lower Cretaceous marine rocks from Sénégal (cf. Doyle, Jardiné & Doerenkamp, 1982). The early evolution of salinity tolerance in angiosperms is difficult to explain if extant primitive angiosperms are the sole ecological model for Early Cretaceous forms, because today primitive angiosperms occur predominantly in mesic upland environments characterized by abundant precipitation (Carlquist, 1975). One proposed solution to this problem is that early angiosperms were adapted almost exclusively to coastal habitats (Retallack & Dilcher, 1981); however, early angiosperms were also present in a variety of non-coastal habitats (Doyle *et al.*, 1982) and in coastal plain habitats that lack evidence of tidal influence (Hickey & Doyle, 1977; Upchurch & Doyle, 1981). An alternative explanation is that early angiosperms were adapted to the subhumid conditions that characterized Cretaceous climates, and that this facilitated evolution into environments of elevated salinity: plants adapted to conditions of high evaporation would need relatively little change to adapt to osmotic stress caused by elevated salinity. A dry climate also could explain some of the small leaf size characteristic of the earliest angiosperms, although evolutionary factors clearly played an important role as well.

Also explicable by the existence of areally extensive, subhumid climates during the Late Cretaceous is the rise of wind-pollinated angiosperms in megathermal regions. Today arborescent, wind-pollinated angiosperms characterize microthermal broadleaved deciduous vegetation (Wolfe, 1979), where seasonal leaf fall and close spacing of individuals permit efficient transfer of pollen by winds (Whitehead, 1969). Modern megathermal regions have few wind-pollinated species, apparently because dense canopy and high rainfall impede the efficient wind-transport of pollen. The first major group of wind-pollinated angiosperms to diversify during the Late Cretaceous is Normapolles (Doyle & Hickey, 1976), whose extant relatives include wind-pollinated families such as Juglandaceae and Betulaceae (Wolfe, 1974; Friis, 1983). Normapolles are a diverse and abundant component of low–middle palaeolatitude palynofloras from eastern North America and west-central Europe, regions with megathermal temperatures during the Late Cretaceous. Wind pollination would not be expected to evolve extensively in megathermal rainforest because a closed canopy and high precipitation would impede the transfer of pollen between conspecific individuals (Whitehead, 1969; Crepet, 1981). One proposed solution to this problem is deciduousness of foliage during a distinct dry season (Crepet, 1981); however, the lack of growth rings in nearly all woods indicates that there was little seasonality to precipitation in the Normapolles palynoprovince. An alternative hypothesis is that *aseasonal* dryness provided the conditions under which wind pollination could evolve, as an open canopy and low rainfall would permit the effective transfer of pollen between conspecific individuals. The strong resemblance seen between Late Cretaceous vegetation of the Normapolles palynoprovince and extant megathermal evergreen vegetation of sandy, porous soils is of particular interest, as a disproportionate number of wind-pollinated species from megathermal regions occur in this vegetational type today (Richards, 1952; D. Janzen, personal communication). If vegetational reconstructions of the Normapolles region are correct, then the origin of many extant wind-pollinated groups was in a region of aseasonally dry megathermal climate, and the present-day abundance of these groups in microthermal deciduous vegetation resulted from climatic change following the K–T boundary event.

The foliar physiognomic data anlayzed in this study provide no evidence for an equatorial belt of angiosperm-dominated rainforest during the Cretaceous. This raises the question: 'Did tropical rainforest with modern physiognomy exist prior to the Tertiary?' Unfortunately, data from low palaeolatitudes are too sparse to answer this question adequately. Known leaf megafloras from the Late Cretaceous of Egypt are within a few degrees of the palaeoequator, where a belt of high rainfall would be expected to develop; these floras show no evidence for abundant precipitation, suggesting an absence of tropical rainforest. However, fruit and seed floras described from the Campanian–Maastrichtian of Sénégal (Monteillet & Lappartient, 1981) and the Maastrichtian of Nigeria (Chesters, 1955) also are located within a few degrees of the palaeoequator, and these show the large diaspore size characteristic of plants from closed-canopy forests (Tiffney, 1984). Clearly more data on Late Cretaceous megafloras from low palaeolatitudes are needed to resolve this question, which has important implications for the origin of tropical rainforest, the origin of diversity, and the evolution of adaptation in tropical plants and animals.

We thank Peter Crane, Virginia Page, Judith Parrish, Elizabeth Wheeler and Scott Wing for their assistance and for helpful discussions of the ideas presented in this paper.

References

Bailey, I. W. & Sinnott, E. W. (1915). A botanical index of Cretaceous and Tertiary climates. *Science*, 41, 832–3.

Bartels, W. S. (1980). Early Cenozoic reptiles and birds from the Bighorn Basin, Wyoming. *University of Michigan Papers on Paleontology*, 24, 73–9.

Batten, D. J. (1984). Palynology, climate and the development of Late Cretaceous floral provinces in the Northern Hemisphere; a review. In *Fossils and Climate*, ed. P. J. Brenchley, pp. 127–70. New York: John Wiley and Sons.

Bell, W. E. (1949). Uppermost Cretaceous and Paleocene floras of western Alberta. *Geological Survey of Canada, Bulletin*, 13, 1–231.

Bell, W. E. (1957). Flora of the Upper Cretaceous Nanaimo Group of Vancouver Island, British Columbia. *Geological Survey of Canada Memoir*, 293, 1–84.

Berry, E. W. (1916). Systematic paleontology, Upper Cretaceous: Fossil plants. In *Upper Cretaceous*, ed. W. B. Clark, pp. 757–901. Baltimore: Maryland Geological Survey.

Berry, E. W. (1922). The flora of the Cheyenne Sandstone of Kansas. *US Geological Survey Professional Paper*, 127–I, 199–225.

Boersma, A. (1984). Campanian through Paleocene paleotemperature and carbon isotope sequence and the Cretaceous–Tertiary boundary in the Atlantic Ocean. In *Catastrophes and Earth History: The New Uniformitarianism*, ed. W. A. Berggren & J. A. Van Couvering, pp. 247–78. Princeton: Princeton University Press.

Bohor, B. F., Foord, E. E., Modreski, P. J. & Triplehorn, D. M. (1984). Mineralogic evidence for an impact event at the Cretaceous–Tertiary boundary. *Science*, 224, 867–9.

Brenner, G. J. (1963). The spores and pollen of the Potomac Group of Maryland. *Bulletin of the Maryland Department of Geology, Mines, and Water Resources,* 27, 1–215.

Carlquist, S. (1975). *Ecological Strategies of Xylem Evolution.* Berkeley: University of California Press.

Carlquist, S. (1977). Ecological factors in wood evolution: a floristic approach. *American Journal of Botany,* 64, 887–96.

Chalk, L. (1955). Ray volumes in hardwoods. *Tropical Woods,* 101, 1–10.

Chesters, K. I. M. (1955). Some plant remains from the Upper Cretaceous and Tertiary of West Africa. *Annals and Magazine of Natural History, Series* 12, 8, 498–504.

Creber, G. T. & Chaloner, W. G. (1985). Tree growth in the Mesozoic and Early Tertiary and the reconstruction of palaeoclimates. *Palaeogeography, Palaeoclimatology, Palaeoecology,* 52, 35–60.

Crepet, W. L. (1981). The status of certain families of the Amentiferae during the middle Eocene and some hypotheses regarding the evolution of wind pollination in dicotyledonous angiosperms. In *Paleobotany, Paleoecology, and Evolution,* vol. 2, ed. K. J. Niklas, pp. 103–28. New York: Praeger Publishers.

Daghlian, C. P. (1981). A review of the fossil record of monocotyledons. *The Botanical Review,* 47, 517–55.

Doyle, J. A., & Hickey, L. J. (1976). Pollen and leaves from the mid-Cretaceous Potomac Group and their bearing on early angiosperm evolution. In *Origin and Early Evolution of Angiosperms,* ed. C. B. Beck, pp. 139–206. New York: Columbia University Press.

Doyle, J. A., Jardiné, S. & Doerenkamp, A. (1982). *Afropollis,* a new genus of early angiosperm pollen, with notes on the Cretaceous palynostratigraphy of paleoenvironments of northern Gondwana. *Bulletin Centres Recherches Exploration–Production Elf-Aquitaine,* 6, 39–117.

Estes, R. & Hutchinson, J. H. (1980). Eocene lower vertebrates from Ellesmere Island, Canadian Arctic Archipelago. *Palaeogeography, Palaeoclimatology, Palaeoecology,* 30, 325–47.

Fleming, R. F. (1985). Palynological observations of the Cretaceous–Tertiary boundary in the Raton Formation, New Mexico. *Palynology,* 9, 242 (Abs.)

Friis, E. M. (1983). Upper Cretaceous (Senonian) floral structures of juglandalean affinity containing Normapolles pollen. *Review of Palaeobotany and Palynology,* 39, 161–88.

Friis, E. M. (1988), *Spirematospermum chandlerae* sp. nov., an extinct species of Zingiberaceae from the North American Cretaceous. *Tertiary Research,* 9, 7–12.

Givnish, T. (1979). On the adaptive significance of leaf form. In *Topics in Plant Population Biology,* ed. O. T. Solbrig, S. Jain, G. B. Johnson & P. Raven, pp. 375–407. New York: Columbia University Press.

Givnish, T. & Vermeij, G. (1976). Sizes and shapes of liane leaves. *American Naturalist,* 110, 743–76.

Hickey, L. J. (1978). Origin of the major features of angiospermous leaf architecture in the fossil record. *Courier Forschungsinstitut Senkenberg*, **30**, 27–34.

Hickey, L. J. (1984*a*). Road log and stops, Part 1. In *Cretaceous and Tertiary Stratigraphy, Paleontology, and Structure, Southwestern Maryland and Northeastern Virginia*, American Association of Stratigraphic Palynologists Field Trip Volume and Guidebook, ed. N. O. Frederiksen & K. Krafft, pp. 193–209. Dallas: American Association of Stratigraphic Palynologists Foundation.

Hickey, L. J. (1984*b*). Changes in the angiosperm flora across the Cretaceous–Tertiary boundary. In *Catastrophes in Earth History: The New Uniformitarianism*, ed. W. A. Berggren & J. A. Van Couvering, pp. 279–314. Princeton: Princeton University Press.

Hickey, L. J. & Doyle, J. A. (1977). Early Cretaceous fossil evidence for angiosperm evolution. *The Botanical Review*, **43**, 3–104.

Hollick, A. (1930). The Upper Cretaceous floras of Alaska. *US Geological Survey, Professional Paper*, **159**, 1–123.

Horn, H. (1971). *The Adaptive Geometry of Trees*, Monographs on Population Biology 3. Princeton: Princeton University Press.

Lesquereux, L. (1892). The flora of the Dakota Group. *US Geological Survey, Monograph*, **17**, 1–400.

Mädel, E. (1960). Monimiaceen-Hölzer aus den oberkretazischen Umzaba-Schichten von ost-Pondoland (S-Afrika). *Senkenbergiana lethaea*, **41**, 331–91.

Mädel, E. (1962). Die fossilen Euphorbiaceen-Hölzer mit besonderer Berücksichtigung neuer Funde aus der Oberkreide Süd-Afrikas. *Senckenbergiana lethaea*, **43**, 283–321.

Manchester, S. R. (1987). The fossil history of Juglandaceae. *Annals of the Missouri Botanical Garden, Monographs*, **21**, 1–132.

Monteillet, J. & Lappartient, J. R. (1981). Fruits et graines du Crétacé Supérieur des carrières de Paki (Sénégal). *Review of Palaeobotany and Palynology*, **34**, 331–44.

Muller, J. (1981). Fossil pollen records of extant angiosperms. *The Botanical Review*, **47**, 1–142.

Muller, J. (1984). Significance of fossil pollen for angiosperm history. *Annals of the Missouri Botanical Garden*, **71**, 419–43.

Nichols, D. J., Jarzen, D. M., Orth, C. J. & Oliver, P. Q. (1986). Palynological and iridium anomalies at the Cretaceous–Tertiary boundary, south-central Saskatchewan. *Science*, **231**, 714–17.

Nichols, D. J. & Ott, H. L. (1978). Biostratigraphy and evolution of the *Momipites–Caryapollenites* lineage in the early Tertiary in the Wind River Basin, Wyoming. *Palynology*, **2**, 93–112.

Orth, C. J., Gilmore, J. S., Knight, J. D., Pillmore, C. L., Tschudy, R. H. &

Fassett, J. E. (1981). An iridium abundance anomaly at the palynological Cretaceous–Tertiary boundary in northern New Mexico. *Science*, 214, 1341–3.

Page, V. M. (1979). Dicotyledonous wood from the Upper Cretaceous of central California. *Journal of the Arnold Arboretum*, 60, 323–49.

Page, V. M. (1980). Dicotyledonous wood from the Upper Cretaceous of central California. *Journal of the Arnold Arboretum*, 61, 723–48.

Page, V. M. (1981). Dicotyledonous wood from the Upper Cretaceous of central California. *Journal of the Arnold Arboretum*, 62, 437–55.

Ramanujam, C. G. K. & Stewart, W. N. (1969). Fossil woods of Taxodiaceae from the Edmonton Formation (Upper Cretaceous) of Alberta. *Canadian Journal of Botany*, 47, 115–24.

Raunkiaer, C. (1934). *Life Forms of Plants and Statistical Plant Geography*. Oxford: Clarendon Press.

Reid, E. M. & Chandler, M. E. J. (1933). *The London Clay Flora*. London: British Museum (Natural History).

Retallack, G. & Dilcher, D. L. (1981). A coastal hypothesis for the dispersal and rise to dominance of flowering plants. In *Paleobotany, Paleoecology, and Evolution*, vol. 2, ed. K. J. Niklas, pp. 27–77. New York: Praeger Publishers.

Richards, P. W. (1952). *The Tropical Rainforest: An Ecological Study*. Cambridge: Cambridge University Press.

Schultze-Motel, J. (1966). Gymnospermen-Hölzer aus den oberkretazischen Umzamba-Schichten von Ost-Pondoland (S-Afrika). *Senckenbergiana lethaea*, 47, 279–337.

Spicer, R. A. & Parrish, J. T. (1986). Fossil woods from northern Alaska and climate near the middle Cretaceous North Pole. *Geological Society of America, Abstracts of 1986 Annual Meeting* (in press).

Spicer, R. A., Wolfe, J. A. & Nichols, D. J. (1987). Alaskan Cretaceous–Tertiary floras and Arctic origins. *Paleobiology*, (in press).

Tiffney, B. H. (1984). Seed size, dispersal syndromes, and the rise of the angiosperms: evidence and hypothesis. *Annals of the Missouri Botanical Garden*, 71, 551–76.

Tschudy, R. H., Orth, C. J., Gilmore, J. S. & Knight, J. D. (1984). Disruption of the terrestrial plant ecosystem at the Cretaceous–Tertiary boundary, Western Interior. *Science*, 225, 1030–2.

Upchurch, G. R., Jr & Dilcher, D. L. (1984). A magnoliid leaf flora from the mid-Cretaceous Dakota Formation of Nebraska. *American Journal of Botany*, 71 (5), Part 2, 119 (Abs.).

Upchurch, G. R., Jr & Doyle, J. A. (1981). Paleoecology of the conifers *Frenelopsis* and *Pseudofrenelopsis* (Cheirolepidiaceae) from the Cretaceous Potomac Group of Maryland and Virginia. In *Geobotany II*, ed. R. C. Romans, pp. 167–202. New York: Plenum Press.

Vakhrameev, V. A. (1952). [Stratigraphy and fossil flora of Cretaceous deposits of Western Kazakhstan.] In Russian. *Regional'naya Stratigrafia SSSR*, **1**, 1–346.

Vakhrameev, V. A. (1978). The climates of the Northern Hemisphere in the Cretaceous in light of paleobotanical data. *Paleontological Journal*, **12**, 143–54.

Velenovský, J. (1882). Die Flora der böhmischen Kreideformation, Part 1. *Beiträge zur Paläontologie Österreich-Ungarns und des Orients*, **2**, 1–25.

Velenovský, J. (1884). Die Flora der böhmischen Kreideformation. Part 2. *Beiträge zur Paläontologie Österreich-Ungarns und des Orients*, **3**, 26–47.

Velenovský, J. (1886). Die Flora der böhmischen Kreideformation. Part 3. *Beiträge zür Paläontologie Österreich-Ungarns und des Orients*, **4**, 48–61.

Velenovský, J. (1887). Die Flora der böhmischen Kreideformation, Part 4. *Beiträge zur Paläontologie Österreich-Ungarns und des Orients*, **5**, 62–75.

Walter, H. (1973). *Vegetation of the Earth in Relation to Climate and the Eco-physiological Conditions.* Berlin: Springer-Verlag.

Ward, L. F. (1899). The Cretaceous formation of the Black Hills as indicated by the fossil plants. *US Geological Survey 19th Annual Report*, part 2, 523–712.

Webb, L. J. (1959). A physiognomic classification of Australian rain forests. *Journal of Ecology*, **47**, 551–70.

Wheeler, E., Matten, L. C. & Lee, M. R. (1987). Five dicotyledonous woods from the Late Cretaceous of Illinois. *Botanical Journal of the Linnean Society*, **95**, 77–100.

Whitehead, D. R. (1969). Wind pollination in the angiosperms: evolutionary and environmental considerations. *Evolution*, **23**, 28–35.

Wolfe, J. A. (1971). Tertiary climatic fluctuations and methods of analysis of Tertiary floras. *Palaeogeography, Palaeoclimatology, Palaeoecology*, **9**, 27–57.

Wolfe, J. A. (1974). Fossil forms of Amentiferae. *Brittonia*, **23**, 334–55.

Wolfe, J. A. (1976). Stratigraphic distribution of some pollen types from the Campanian and lower Maestrichtian rocks (Upper Cretaceous), of the middle Atlantic States. *US Geological Survey Professional Paper*, **977**, 1–18.

Wolfe, J. A. (1977). Paleogene floras from the Gulf of Alaska region. *US Geological Survey Professional Paper*, **997**, 1–108.

Wolfe, J. A. (1978). A paleobotanical interpretation of Tertiary climates in the Northern Hemisphere. *American Scientist* **66**, 694–703.

Wolfe, J. A. (1979). Temperature parameters of humid to mesic forests of eastern Asia and their relation to forests of other areas of the Northern Hemisphere and Australasia. *US Geological Survey Professional Paper*, **1106**, 1–37.

Wolfe, J. A. (1985). Distribution of major vegetational types during the Tertiary. In *The Carbon Cycle and Atmospheric CO_2: Natural Variations Archean to Present*, American Geophysical Union Monograph 32, ed. E. T. Sundquist & W. S. Broecker, pp. 357–76. Washington, DC: American Geophysical Union.

Wolfe, J. A. (1987). An overview of the origins of the modern vegetation and flora of the northern Rocky Mountains. *Annals of the Missouri Botanical Garden*, **74**, 785–803.

Wolfe, J. A. & Upchurch, G. R., Jr (1986). Vegetation, climatic and floral changes at the Cretaceous–Tertiary boundary. *Nature*, **324**, 148–52.

Wolfe, J. A. & Upchurch, G. R., Jr (1987*a*). Leaf assemblages across the Cretaceous–Tertiary boundary in the Raton Basin, New Mexico and Colorado. *Proceedings of the National Academy of Science, USA*, **84**, 4096–5100.

Wolfe, J. A. & Upchurch, G. R., Jr (1987*b*). North American nonmarine climates and vegetation during the Late Cretaceous. *Palaeogeography, Palaeoclimatology, Palaeoecology*, **61**, 33–77.

5

Vegetational consequences of the angiosperm diversification

P.R.CRANE

Palynological and megafossil evidence from many geographical areas indicates that the angiosperms first attained ecological importance during the mid-Cretaceous (Brenner, 1963; Doyle & Hickey, 1976; Hughes, 1976) and that this led to profound changes in terrestrial plant communities. This chapter briefly examines the transition from gymnosperm-dominated to angiosperm-dominated vegetation. First, it reviews a sequence of Early Cretaceous to Paleocene fossil floras, concentrating on megafossil data from middle palaeolatitudes in the Northern Hemisphere, and then discusses the major changes that occurred in the composition and structure of terrestrial vegetation between the earliest Cretaceous and the Paleocene. Particular attention is directed toward changes in the relative proportions of five major groups of plants: pteridophytes, cycadophytes, conifers, angiosperms and 'other' seed plants (e.g. *Czekanowskia, Ginkgo*).

Review of floras

Neocomian floras

During the Neocomian, as in the Middle and Late Jurassic, latitudinal climatic gradients and floristic provinces were much less marked than they are today, as a result of less dispersed land masses and a more equable global climate (Parrish, this volume, Chapter 3). In the Northern Hemisphere most authors recognize a northern, Siberian–Canadian, floristic province and a southern, Euro-Sinian (Indo-European) province (Figure 5.1(*a*), e.g. Vakhrameev, 1964; Batten, 1984; Vakhrameev *et al.*, 1978). The classic Middle Jurassic flora from Yorkshire (Harris, 1961*a*, 1964, 1969, 1979; Harris, Millington & Miller, 1974) has an abundance

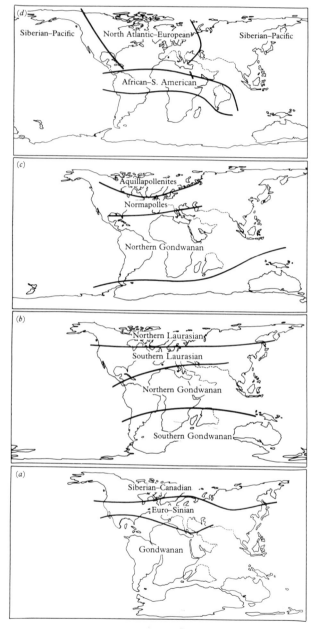

Figure 5.1. Floristic provinces in the Cretaceous and Early Tertiary.
(a) Neocomian (based on Vakhrameev et al., 1978; Batten, 1984). (b)
Cenomanian (based on Brenner, 1976; Batten, 1984). (c) Santonian–
Campanian (based on Batten, 1984). (d) Paleocene (based on
Krutzsch, 1967). (All base maps redrawn from Smith et al., 1981.)

of cycadophytes and is typical of the Euro-Sinian province, while Siberian–Canadian floras exhibit lower generic and specific diversity, less abundant cycadophytes, a higher proportion of Czekanowskiales, Ginkgoales, and conifers, and relatively few endemic taxa (Batten, 1984).

Pteridophytes typically account for 30% of the species in Jurassic and Neocomian floras (Figure 5.2(a)). Ferns were particularly abundant and diverse, and most were representatives of the extant families Dickson-iaceae, Dipteridaceae, Gleicheniaceae, Marattiaceae, Matoniaceae, Osmundaceae and Schizaeaceae. *Tempskya* (a tree fern of unknown systematic position; Ash & Read, 1976) and *Weichselia* (a xeromorphic plant, probably related to extant Matoniaceae; Alvin, 1968; Daber, 1968) are two of the most characteristic Early Cretaceous ferns. In forested situations, ferns probably dominated the ground cover, but a few may also have been epiphytes (e.g. *Todites princeps*, Osmundaceae; Harris, 1961a). Abundant megaspores of probable heterosporous water ferns (e.g. *Arcellites*; Batten, 1974) indicate that these plants were probably important colonizers of open water, and ferns also formed extensive stands in unwooded areas that would today be dominated by angiosperm herbs (Apert, 1973; Harris, 1981). Probable gleicheniaceous and schiz-aeaceous spores are especially abundant in Neocomian palynofloras, and many extant species in these families are characteristically plants of open habitats (R. G. Stolze, personal communication). In the Weald Clay of southern England the abundance of fusainized matoniaceous and gleicheniaceous leaves at some localities, combined with an almost complete absence of gymnosperm fossils, suggests a fern-savanna from which conifers and other gymnosperms were perhaps excluded by periodic fires (Harris, 1981).

Although lycopods and sphenopsids were less diverse than ferns in the Early Cretaceous, they were locally important in the herbaceous flora. The abundance of *Isoetes*-like and *Selaginella*-like megaspores (e.g. Kovach & Dilcher, 1985) and lycopod and sphenopsid megafossils (e.g. *Nathorstiana*; Mägdefrau, 1932; and *Equisetum*; Watson, 1983) suggests that lycopods and sphenopsids frequently dominated partially flooded areas and perhaps other open habitats.

Cycadophytes typically account for about 30% of the species in Jurassic and Neocomian megafloras (Figure 5.2(a)) and include at least three different kinds of plants with pinnate leaves: true cycads, Ben-nettitales (Cycadeoidales) and a heterogeneous group of seed ferns with pinnate foliage that is difficult to separate from that of true cycads. Few Mesozoic cycads are known in detail, but the data available (Harris,

1961*b*) suggest a reproductive biology similar to that of extant species. The seeds of *Beania* are among the largest known for Jurassic and early Cretaceous gymnosperms and, as in extant cycads, these seeds contain a thick megaspore membrane that probably correlates with a well-developed megagametophyte. At least some Mesozoic cycads were pachycaul (e.g. Smoot, Taylor & Delevoryas, 1985) but others were probably slender, highly branched (Harris, 1961*b*; Kimura & Sekido, 1975) and had significantly smaller leaves than do their extant relatives.

Despite the similarities between the leaves of the Bennettitales and Mesozoic cycads, the two groups differed substantially in most other features (Harris, 1976), and although few of these plants are well understood it is clear that extant cycads are a misleading guide to the biology and ecology of the Bennettitales (Crane, 1985). Studies of dispersed cuticles show that the Bennettitales were abundant in Neocomian vegetation (e.g. Oldham, 1976). In addition to pachycaul forms, such as *Cycadeoidea* (e.g. Crepet, 1974) and *Monanthesia* (Delevoryas, 1959), at least two other kinds of growth habit occur within the Bennettitales. The stems of *Zamites gigas* (Harris, 1969) and similar plants (e.g. *Ischnophyton*, Delevoryas & Hope, 1976) were probably sparsely branched, with a tuft of leaves and persistent leaf bases at the apex as in some palms. In contrast, the stems of *Wielandiella* (Nathorst, 1902; Harris, 1932) and *Ptilophyllum* plants were slender, frequently sympodially branched and bore leaves that abscissed cleanly from the stem (Harris, 1969, 1973). Even these slender bennettitalean stems were relatively thick in relation to the size of leaves that they bore, and among Recent angiosperm trees this is characteristic of plants that are ecologically intolerant of shade (White, 1983). Most bennettitalean leaves consist of an elongated main rachis bearing a segmented pinnately arranged lamina and thus conform to morphologies that are theoretically more advantageous for early successional rather than climax forest trees (Givnish, 1971). On average, bennettitalean leaves are an order of magnitude smaller than those of extant cycads (microphylls or small mesophylls) and even in otherwise mesomorphic vegetation they often exhibit marked xeromorphic tendencies, such as reflexed pinnae, well developed hairs or papillae on the abaxial surface, and sunken or occluded stomata (e.g. Harris, 1969). In their reproductive biology, most Bennettitales were probably insect pollinated (Crepet & Friis, this volume, Chapter 7) and the seeds were either dispersed by animals or perhaps released by fire (Harris, 1973). Small seed size is consistent with the idea that the Bennettitales colonized open habitats (Harris, 1973;

Harper, Lovell & Moore, 1970). Silicified seeds contain well-preserved dicotyledonous embryos (Wieland, 1906), but few gametophytes (Sharma, 1974). Combined with the lack of a thick maceration-resistant megaspore membrane (Harris, 1954) and the precocious differentiation of the integument (Crepet & Delevoryas, 1972) this may suggest that bennettitalean ovules, like those of angiosperms and Gnetales, were progenetic (Doyle, 1978) and passed rapidly through pollination, gametophyte development, fertilization and embryogenesis. Taken together, these features of reproductive biology, combined with stem and leaf morphology, support the suggestion that cycadophytes were the predominant shrubs of Mesozoic vegetation (Krassilov, 1981).

Current knowledge of the third group of cycadophytes, the pinnate-leaved seed ferns (e.g. *Ctenozamites*) is extremely poor. This group probably includes taxa of widely different relationships, and useful generalizations concerning their probable ecology and biology are not yet possible.

Conifers typically account for about 20% of the species in Neocomian and Jurassic megafloras (Figure 5.2(*a*)). The extant conifer families Araucariaceae, Cupressaceae, Cephalotaxaceae, Pinaceae, Podocar-paceae, and Taxaceae appear in the fossil record either during the Triassic or Jurassic and all are recorded from the Northern Hemisphere (Miller, 1982; Zhou, 1983). Several extinct conifer genera cannot be assigned reliably to extant families (Harris, 1976), while others are included in the extinct families Cheirolepidiaceae (also known as Hirmerellaceae, producers of *Classopollis* pollen) and Podozamitaceae (interpreted as cycadophytes in many older treatments of Mesozoic floras). The Cheirolepidiaceae were particularly important in Late Jurassic and Early Cretaceous vegetation and may have occupied a variety of habitats, particularly those that were dry or had saline ground-water (Alvin, 1983; Upchurch & Doyle, 1981).

The remaining category of 'other' seed plants includes taxa such as *Baiera*, *Czekanowskia*, Gnetales, *Ginkgo* and *Solenites* that together typically account for 10% to 15% of Neocomian and Jurassic mega-floras (Figure 5.2(*a*)). At least some Jurassic and Early Cretaceous *Ginkgo* seems to have been similar to extant *G. biloba*, although the seeds and pollen cones were smaller (Harris *et al.*, 1974). The Caytoniales were slender, branched, plants (Harris, 1971), with leaves that appear to be palmately compound, although the leaflets are attached apically in two pairs. *Caytonia* seeds were small and perhaps dispersed by animals, with the fleshy 'cupule' acting as the 'reward' for the dispersal vector.

Caytonia seeds have seed cuticles similar to those of the Bennettitales (Harris, 1954) and may also have been progenetic in their gametophyte and embryo development. The structure and biology of the Czekanowskiales is poorly understood. Their most distinctive vegetative feature is the presence of deciduous short shoots, and the abundance of Czekanowskiales at higher northern palaeolatitudes (Vakhrameev, 1978) suggests that this may be related to climatic seasonality. The Gnetales are very poorly represented in the megafossil record (Crane & Upchurch, unpublished) but are abundantly represented in palynofloras from tropical or arid areas (Brenner, 1976).

Barremian floras

Floristic provinces during the Barremian were similar to those of the Neocomian, and these Early Cretaceous floras are included together in the compilation of systematic data (Figure 5.2(*a*)). During the Barremian, or perhaps in the latest Neocomian, columellate semi-tectate pollen typical of flowering plants appears for the first time in palynofloras from southern England (Hughes, Drewry & Laing, 1979), Israel (Brenner, 1984), West Africa, Argentina and eastern North America (Hickey & Doyle, 1977). Although the abundance of angiosperm pollen in Barremian floras is typically less than 1 %, the systematic diversity may be considerable, and 15 kinds of presumed angiosperm pollen have been distinguished in the Barremian of southern England (Hughes *et al.*, 1979). In the Barremian of Gabon, sediments of the developing mid-Atlantic rift valley (Cocobeach sequence) that contain a diversity of monosulcate angiosperm pollen also have a high proportion of ephedroid grains relative to cheirolepidiaceous conifer pollen (Doyle, Jardiné & Doerenkamp, 1982), suggesting that early angiosperms initially colonized habitats in which these conifers were uncommon. Although there is geological evidence of aridity in the Barremian part of the Cocobeach sequence, angiosperm pollen occurs simultaneously in more equable areas (e.g. Israel, Sénégal), and thus there is no clear palaeontological support for the hypothesis (Stebbins, 1974) that angiosperms originated and diversified in seasonally arid environments (Doyle *et al.*, 1982).

Aptian and early Albian floras

During the mid-Cretaceous, Brenner (1976) recognized four floristic provinces on the basis of palynological assemblages: Northern Laurasia, Southern Laurasia, Northern Gondwana and Southern Gondwana

(Figure 5.1(*b*)). The Northern Laurasian province is broadly equivalent to the Jurassic–Neocomian Siberian–Canadian province, while the Southern Laurasian province corresponds approximately to the Euro-Sinian (Indo-European realm; Vakhrameev, 1975). Angiosperm leaves are not abundant in Aptian and early Albian floras. Where present they account for only about 5% of the species of leaf megafossils in floras of this age (Figure 5.2(*a*)). The percentage representation of cycadophytes and 'other' gymnosperm species in Aptian and early Albian floras decreases slightly with respect to Neocomian floras, while the percentage representation of pteridophytes and conifers shows a small increase.

Palynofloras from the Southern Laurasian and Northern Gondwanan provinces are more diverse than those from higher latitudes and contain angiosperm grains closely comparable to pollen of extant monocotyledons, the Chloranthaceae and Winteraceae (Walker, Brenner & Walker, 1983; Walker & Walker, 1984). Angiosperm pollen is most abundant and diverse in Northern Gondwana palynofloras, and in the Early Aptian of the Ivory Coast angiosperm pollen may account for over 10% of the pollen flora (Doyle *et al.*, 1982). Tricolpate pollen appears for the first time at around the Barremian–Aptian boundary in the Northern Gondwana province (Brenner, 1976; Doyle *et al.*, 1982), and this indicates the first appearance of the clade of non-magnoliid dicotyledons (subclasses Hamamelididae, Caryophyllidae, Dilleniidae, Rosidae, Asteridae) that accounts for over 70% of extant angiosperm species (Cronquist, 1981). During the Aptian, tricolpate pollen is common in Northern Gondwana, but there are few, if any, tricolpate grains recorded from the Aptian of Southern Laurasia.

Northern Gondwana palynofloras are typically low in pteridophyte spores and bisaccate conifer pollen (Brenner, 1976), whereas the reverse is true in the Southern Laurasian province, where the Schizaeaceae, Gleicheniaceae, Pinaceae and Podocarpaceae are all well represented. *Classopollis* is abundant in both provinces, but the most striking feature of Northern Gondwana palynofloras is the abundance (up to 40%) of *Ephedra*-like pollen (Figure 5.2(*b*)), which may also include forms with horn-like or elater-like projections (e.g. *Galeacornea*, *Elaterosporites*). *Ephedra*-like pollen is much less conspicuous in Southern Laurasia palynofloras. In Northern Gondwana, *Ephedra*-like pollen diversified almost simultaneously with angiosperm pollen, and, during the Aptian, fluctuations in abundance seem to follow similar fluctuations in angiosperm pollen content (Doyle *et al.*, 1982). This is consistent with the limited evidence available from megafossils that at least some early

Cretaceous Gnetales and angiosperms occupied similar habitats (Crane & Upchurch, unpublished).

During the Aptian and early Albian, palynofloras in the Northern Laurasian province typically lack angiosperm pollen, *Classopollis* and *Ephedra*-like pollen is rare, and, with the exception of spores of some Osmundaceae, Northern Laurasia palynofloras include only a relatively small component of pteridophyte spores. The proportion of bisaccate conifer pollen is correspondingly higher.

Very few Aptian to early Albian megafossil floras contain abundant angiosperms. In the Potomac Group, approximately 12 different taxa have been recognized among leaves of this age (Upchurch, 1984*a*). Aptian and early Albian angiosperm leaves have poorly organized venation (Wolfe, Doyle & Page, 1975), and those studied in detail show features characteristic of extant Magnoliidae, particularly Magnoliales, Chloranthaceae and Illiciales (Upchurch, 1984*a*, *b*). The leaves are generally small (Vakhrameev, 1981; Upchurch & Wolfe, this volume, Chapter 4) and in the Potomac Group occur consistently in moderately coarse to medium-fine sands and silts interpreted as near channel and levee deposits (Hickey & Doyle, 1977). At some Potomac Group localities, stems with attached leaves provide direct evidence of herbaceous angiosperms (Hickey & Doyle, 1977; Upchurch, 1984*a*; Upchurch, Hickey & Niklas, 1983), and sedimentological evidence suggests that, together with ferns and herbaceous gnetaleans, these were early successional colonizers of disturbed habitats (Crane & Upchurch, unpublished). Angiosperm megafossils are absent from Northern Laurasia megafloras, and, although cycadophytes are generally sparse (Vakhrameev, 1971), *Nilssonia* foliage may be common in some assemblages.

Late Albian floras

In the late Albian, floristic provinciality was very similar to that in the Aptian and early Albian (Figure 5.1(*b*)), but the geographic spread of flowering plants, at least in the Northern Hemisphere, was essentially completed (Samylina, 1968). By this time angiosperm megafossils and pollen are an important component of northern high latitude floras in areas from which they were previously absent (e.g. western Canada; Bell, 1956; Singh, 1975; and north slope of Alaska; R. A. Spicer, personal communication). Angiosperm leaves typically account for 40% to 50% of the taxa described in late Albian megafloras, and this expansion in the relative importance of angiosperms corresponds to a marked decline in the importance of cycadophytes, other seed plants, and pteridophytes in floras of this age (Figure 5.2(*a*); see also Vakhrameev, 1978). The conifers are the group least affected by the angiosperm diversification. Angiosperm pollen is significantly more abundant in late Albian than in early Albian palynofloras, and, in Northern Gondwana and Southern Laurasia, Aptian forms (e.g. *Clavatipollenites*) are joined by a variety of new angiosperm pollen types including monosulcate, tricolpate, tricolporate and periporate taxa (Muller, 1984). In the late Albian of the Potomac Group (Southern Laurasia), the relative abundance of tricolpate and tricolporate pollen grains may be as high as 70% at some localities (J. A. Doyle, personal communication). This increase in angiosperm pollen is accompanied, in the Potomac Group, by a slight decline in the abundance of schizaeaceous spores, *Classopollis* and *Exesipollenites* (probably pollen of Bennettitales; Harris, 1974), but there is a corresponding increase of *Araucariacites* and bisaccate conifer pollen (Brenner, 1963). At lower palaeolatitudes (Northern Gondwana), angiosperm pollen is more diverse, but palynofloras generally remained dominated by *Classopollis* and a variety of ephedroid grains (Figure 5.2(*b*); Muller, 1984). In Northern Laurasia, angiosperm pollen appears for the first time during the Albian, and monosulcate and tricolpate grains appear more or less synchronously (Singh, 1975).

Angiosperm leaves are much more diverse and abundant in the late Albian part of the Potomac Group (Southern Laurasia) than in the Aptian and early Albian, and it is estimated that about 36 different kinds are represented (Doyle & Hickey, 1976). Angiosperm leaves are locally dominant in some floras of this age and may account for almost all of the megafossils collected (e.g. West Brothers locality; Hickey & Doyle, 1977). Systematically angiosperm leaves, reproductive structures, and pollen from the late Albian part of the Potomac Group clearly demonstrate the presence of monocotyledons, magnoliids, hamamelidids and rosids (Doyle & Hickey, 1976; Hickey & Doyle, 1977; Upchurch, 1984*a*, *b*; Crane, Friis & Pedersen, 1986; Friis, Crane & Pedersen, 1986).

In addition to exhibiting increased systematic diversity, angiosperms had also attained considerable ecological amplitude by the end of the Early Cretaceous (Samylina, 1968). Late Albian leaves are generally larger, structurally more complex, and have more highly organized venation than those in the Aptian and early Albian. Evidence from functional morphology and sedimentology indicates that this increased variety of leaf morphologies reflects a major ecological radiation, during which angiosperms became established not only in early successional

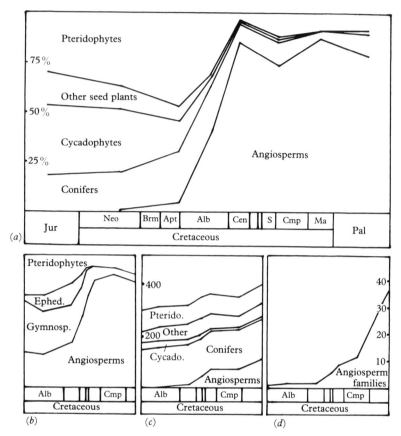

Figure 5.2. Estimates of the timing and magnitude of the angiosperm radiation. (a) Changes in mean percentage contribution of major plant groups to Jurassic (Jur), Cretaceous and Paleocene (Pal) floras. Calculated from mean species numbers in Jurassic–Paleocene floras listed in Appendix. Pteridophytes (Pterido.) include ferns and lycopods; cycadophytes (Cycado.) include cycads, Bennettitales and pinnate-leaved seed ferns. Identical major patterns of change are obtained if percentages are calculated from genera, or if mean species or mean genus numbers are used without conversion into percentages. Neo, Neocomian; Brm, Barremian; Apt, Aptian; Alb, Albian; Cmp, Campanian; Cen, Cenomanian; S, Santonian; Ma, Maastrichtian. (b) Percentage contribution of 'species' of major plant groups to Cretaceous palynofloras. The percentage of ephedroid (Ephed.) species is based on palynomorphs of the *Galeacornea* group. (Based on microfloral diversity curves for South America and Africa (Muller, 1984, Fig. 2).) (c) Estimated absolute species diversity of major plant groups between the Jurassic and the

habitats, but also as full aquatics and riparian trees (Doyle & Hickey, 1976; Hickey & Doyle, 1977). *Sapindopsis* and platanoid leaves that are particularly abundant at several Potomac Group localities and elsewhere in the Northern Hemisphere are interpreted as being plants of early successional angiosperm thickets. In the Potomac Group, conifers and Bennettitales are thought to have remained dominant in backswamp areas throughout the Albian (Hickey & Doyle, 1977).

Cenomanian floras

In the Cenomanian, floristic provinciality is very similar to that in the Aptian and Albian (Figure 5.1(*b*)). Angiosperms consistently dominate megafossil floras on a global scale for the first time, although locally, as in floras of this age from Czechoslovakia, gymnosperms may remain diverse (Velenovský & Viniklář, 1926, 1927, 1929, 1931; Pacltová, 1977). Angiosperm leaves typically account for about 75 % of the taxa described from Cenomanian megafloras, and their systematic representation remains at this level throughout the Late Cretaceous and into the Early Tertiary. The percentage representation of conifer taxa in Cenomanian floras is about half of that in Neocomian floras (approximately 15 %), but cycadophytes, pteridophytes and 'other' seed plants exhibit much more dramatic declines and generally only account for about 2 %, 6 % and 1 % of the megafossil taxa per flora, respectively (Figure 5.2(*a*)).

At low palaeolatitudes (Northern Gondwana), angiosperm pollen accounts typically for about 50 % of the total number of the dispersed palynomorph species (Muller, 1984), although some palynofloras may still be dominated by ephedroid and *Classopollis* pollen (Figure 5.2(*b*)). At middle and high palaeolatitudes, the abundance of angiosperm pollen varies considerably, but, even at megafossil localities that are dominated by angiosperm leaves, the relative abundance of angiosperm pollen may be only about 25 % (Penny, 1969). Other palynofloras are dominated by angiosperms, the remaining palynomorphs consisting largely of saccate pollen of probable Pinaceae and Podocarpaceae, and inaperturate pollen of probable Taxodiaceae. The angiosperm component of Cenomanian palynofloras is more diverse than in the Albian, and in particular there is a considerable variety of tricolpate, tricolporoidate and tricolporate

Paleocene. (Based on Niklas *et al.*, 1985, Fig. 3.) (*d*) First appearances of extant angiosperm families during the Cretaceous and the Paleocene, on the basis of fossil pollen (data from Muller, 1981, Table II). Families first recorded at the boundary between two stages were treated as appearing within the younger stage.

taxa. Triporate pollen grains appear for the first time in the Middle Cenomanian of Europe and North America (Southern Laurasia; Laing, 1975; Pacltová, 1978; Doyle & Hickey, 1976) and begin to diversify rapidly toward the end of this stage. Triporate pollen such as *Complexiopollis* are some of the earliest representatives of the Normapolles group, later members of which are known to be related to extant Juglandales (Friis, 1983), and perhaps other 'Amentiferae' such as Betulaceae and Myricaceae. Although some Normapolles grains may have been dispersed by insects (Batten, 1984), the size and lack of sculpture on most early representatives is consistent with wind pollination.

The diversity of angiosperm leaves in Cenomanian floras far exceeds that known from Albian floras. Magnoliid taxa are widespread (e.g. the *Myrtophyllum* and *Cocculophyllum* groups of Kvaček (1983)), but typically even these Cenomanian magnoliids have more highly organized venation than do their Aptian counterparts (G. R. Upchurch, personal communication). Conclusions based on leaves are therefore consistent with both floral and palynological data, which indicate that magnoliids had undergone a major radiation by the end of the Cenomanian. In addition, many Cenomanian leaf floras are dominated by a wide variety of platanoid taxa and related hamamelidids (e.g. the *Platanus–Credneria* and *Aralia–Debeya* groups of Kvaček (1983)). In the Early Cenomanian of central Kansas most of the leaves from the Dakota Sandstone Flora that have been assigned to *Aralia, Betulites, Platanus, Sassafras* and *Viburnum* (cf. Berry, 1916) belong to this highly variable platanoid–hamamelidid group. Most of the 'Dakota Sandstone' represents probable channel and levee sands (Retallack & Dilcher, 1981), and the abundance of platanoid leaves in such environments recalls the streamside habit of extant *Platanus*.

Turonian–Campanian floras

During the Turonian–Campanian, the angiosperm component of most eastern North American and European palynofloras is dominated by oblate triporate grains, frequently with complex compound apertures (Normapolles pollen). This Normapolles province (Figure 5.1(*c*)) corresponds approximately to the mid-Cretaceous southern Laurasian province and is apparently delimited to the north and south by climatic boundaries (see Upchurch & Wolfe, this volume, Chapter 4). Although Normapolles pollen does occasionally occur outside its characteristic range (e.g. California, China, India), the major abundance and diversity is longitudinally delimited by epicontinental seas (Batten, 1984). Within the Normapolles province, the eastern North American and European sub-provinces had different compositions, with the greatest variety of Normapolles forms being recorded from Europe (Batten, 1981). To the north of the Normapolles province the *Aquilapollenites* province (Figure 5.1(c)) is approximately equivalent to the mid-Cretaceous Northern Laurasian province, and has diverse palynofloras characterized by taxa such as *Aquilapollenites*, *Triprojectus*, *Wodehouseia* and *Proteacidites* (Norris, Jarzen & Awai-Thorne, 1975; Batten, 1984). Although *Aquilapollenites* shows some similarities to pollen of Santalales (Jarzen, 1977), the systematic relationships of these grains are poorly understood.

During the Turonian–Campanian, as in the Cenomanian, angiosperms account for approximately 70% of the species recorded in megafossil floras (Figure 5.2(a)). Pteridophytes and conifers each account for about 10%, while cycadophytes are usually very poorly represented (but see below).

In low palaeolatitude palynofloras (Northern Gondwana) angiosperms account typically for about 54% of the total palynomorph species in the Turonian, rising to 70% and 85% in the Coniacian and Campanian, respectively (Figure 5.2(b); Muller, 1984). By the Campanian, the considerable mid-Cretaceous diversity of *Ephedra*-like pollen had almost completely disappeared. In the Normapolles province, palynofloras are usually dominated by angiosperms, but at higher palaeolatitudes (*Aquilapollenites* province) angiosperms were less abundant relative to gymnosperms (Jarzen & Norris, 1975). In the Oldman Formation (southern Alberta, Campanian) the relative abundance of angiosperms is typically about 30% to 40% of the pollen flora, while gymnosperms account for 15% to 20% and ferns about 30% (Jarzen, 1982b). In terms of the diversity of palynomorph 'species', angiosperms account for only 43% of the species recognized. Pteridophyte and bryophyte spores account for 38%, while gymnosperm pollen accounts for 24% (Jarzen 1982b). Megaspores are diverse and abundant in some floras (e.g. Knobloch, 1984), suggesting that both lycopods and water ferns may have been important herbs at this time.

Cycadophytes are very poorly represented in most Turonian–Campanian floras, but occur sporadically (Berry, 1929) and persisted in greater numbers at high palaeolatitudes (Krassilov, 1975, 1983). In the Nanaimo flora (Vancouver Island; Bell, 1957), there are three genera of cycadophytes representing cycads, pteridosperms, and Bennettitales, but no numerical data are available to estimate their abundance in the vegetation. However, in the Mgachi flora (North Sakhalin; Krassilov,

1978), where cycads also occur, they account for less than 2% of the megafossil collection. The seed fern *Sagenopteris* (leaf of the *Caytonia* plant) also accounts for less than 2% of the plants collected, and both groups were probably only a minor component of the vegetation. The Mgachi flora is dominated by conifers (70% recorded megafossils) and indicates a conifer forest with an understory of angiosperms and ferns (Krassilov, 1978).

During the Turonian–Campanian interval, the number of angiosperms clearly related to extant groups at the ordinal level or below increases, and, by the Campanian, Muller (1981) recognized 13 orders and 12 families of extant angiosperms. As in the Cenomanian, it is the Magnoliidae, Hamamelididae and Rosidae that are best represented. In the middle and late Cretaceous, leaves related to the platanoid complex (*Debeya*, e.g. Němejc & Kvaček, 1975; *Dewalquea*, e.g. Knowlton, 1918) show features suggestive of a relationship to extant Fagaceae (Rüffle, 1977). Angiosperm reproductive remains of this age (Knobloch & Mai, 1984; Friis, 1985) can usually be assigned to a specific extant family or group of families. Normapolles grains (*Plicapollis*, *Trudopollis*) recorded from fossil flowers clearly demonstrate that at least some Normapolles were related to extant Juglandales (Friis, 1983). Palms also became important for the first time during the Turonian–Campanian interval, particularly at low palaeolatitudes and palm leaves and stems of probable Coniacian age are recorded from eastern North America (Daghlian, 1981).

Maastrichtian floras

During the Maastrichtian, floristic provinciality and the diversity and abundance of major plant groups is very similar to that of the Campanian (Figures 5.1(*c*), 5.2(*a*)). Angiosperms, however, appear to undergo a major modernization, with an approximate doubling of the number of extant families and orders recognizable on the basis of dispersed pollen (Figure 5.2(*d*); Muller, 1981). There is palynological evidence of all of the dicotyledonous subclasses except the Asteridae by the close of the Maastrichtian (Muller, 1981), palm pollen is common and diverse, and triporate pollen closely similar to that of extant Betulaceae, Juglandaceae, and Myricaceae is increasingly abundant.

In both the megafossil and palynological record, the Taxodiaceae are the most abundant and widespread conifers at higher palaeolatitudes, whereas at lower palaeolatitudes the Araucariaceae and Podocarpaceae are more common. With the exception of high latitude areas and conifer swamps, conifers rarely dominate the angiosperms during the Maastrichtian. Of the 12 localities in the Fox Hills, Lower Medicine Bow and Lance floras assessed quantitatively by Dorf (1942), only two are dominated by conifers. In middle and low palaeolatitude areas of the Northern Hemisphere corresponding to the Northern Gondwana and Normapolles provinces, megafossil floras contain a preponderance of leaves of probable Euphorbiaceae, Illiciales, Laurales and palms (Wolfe & Upchurch, unpublished; see also Bande, Prakash & Ambwani, 1981). In the *Aquilapollenites* province, the megafossil floras are dominated by several characteristic Paleocene groups, including the *Nyssidium/ Joffrea* lineage related to extant *Cercidiphyllum* (Brown, 1939; Crane, 1984), the *Nordenskioldia* lineage, probably related to extant *Trochodendron* and *Tetracentron*, as well as probable Trochodendrales, Platanaceae, Juglandaceae and Betulaceae (Shoemaker, 1966).

Paleocene floras

The large number of Paleocene megafloras from the Northern Hemisphere are generally rather uniform, particularly at middle to high palaeolatitudes (cf. Brown, 1962; Krassilov, 1976). Palynologically, however, in the area approximately corresponding to the mid-Cretaceous Northern and Southern Laurasian provinces and the Late Cretaceous Normapolles and *Aquilapollenites* provinces, Krutzsch (1967) has recognized a North Atlantic European province, including Europe, eastern North America and Greenland, and a Siberian–Pacific province including western North America, Asia and Australasia (Figure 5.1(*d*)). In terms of the relative abundance and diversity of major plant groups, Paleocene floras resemble those of the Maastrichtian (Figure 5.2(*a*)).

Further, apparently rapid modernization of the angiosperm flora occurs during the Paleocene, and the number of extant families recognized from dispersed pollen doubles from the Maastrichtian level (Figure 5.2(*d*); Muller, 1981). Triporate pollen is frequently very abundant in Paleocene palynofloras, particularly at middle and high palaeolatitudes (e.g. Zaklinskaya, 1967; Kedves, 1982), and contemporaneous megafossils indicate the presence of extinct Betulaceae and Juglandaceae (Crane, 1982; Manchester, 1987). Compared to those from the Upper Cretaceous, Paleocene palynofloras in these areas are relatively impoverished (Leffingwell, 1970; Tschudy, 1970), and estimates of extinction across the Cretaceous–Tertiary boundary based on megafossils vary from 40 % to 75 % (Hickey, 1981; Wolfe & Upchurch, unpublished). In areas approximately corresponding to the earlier Southern Laurasian and Normapolles provinces, Paleocene floras are typically dominated by palms and entire-margined leaves of the Euphorbiaceae and Laurales, many of which were apparently evergreen (Wolfe & Upchurch, unpublished). Serrate-margined leaves are probably assignable to the Juglandaceae and Tiliaceae, and, toward the end of the Paleocene, leaflets of the Leguminosae are common. Conifers are generally rare and herbs are represented by a variety of ferns.

At higher palaeolatitudes, the characteristic Paleocene conifers are probably deciduous Taxodiaceae (*Glyptostrobus*, *Metasequoia*, *Parataxodium*), and the typical ferns include *Lygodium*, *Onoclea* and *Dennstaedtia* (Brown, 1962; Hickey, 1977). The last records of characteristic Neocomian cycadophytes and ferns occur in high latitude Paleocene floras from Alaska (Hollick, 1936). The commoner angiosperms include Betulaceae and Platanaceae. Reproductive structure of the extinct species in the *Cercidiphyllum*/*Nyssidium*/*Joffrea* lineage and Platanaceae demonstrates that these plants may have been important opportunistic colonizers in the Paleocene vegetation (Crane & Stockey, 1985; Manchester, 1986), and this is strongly supported by the occurrence of large numbers of even-aged seedlings of both taxa over large areas (Stockey & Crane, 1983). In view of the apparent scarcity of angiosperm herbs at this time, it is possible that opportunistic shrubs fulfilled a more important role in the colonization of bare ground than in many present-day plant communities. Lianes may also have been effective early colonizers, as in disturbed tropical vegetation today (R. A. Spicer, personal communication), and this may account for the abundance of families such as the Menispermaceae and Vitaceae in many Early Tertiary floras (e.g. Reid & Chandler, 1933; Chandler, 1961; but see also Upchurch & Wolfe, this volume, Chapter 4; Collinson & Hooker, this volume, Chapter 10). The ubiquity of megaspores in Paleocene floras (e.g. Melchior & Hall, 1983) also suggests that lycopods may have been important contemporaneous herbs.

Vegetational consequences

Community composition

Analyses of the mean number of species of major plant groups in Jurassic, Cretaceous and Paleocene floras show clearly that, as the angiosperms increase, the mean number of cycadophytes, other seed plant and pteridophyte species abruptly declines (Figure 5.2(*a*)). Cretaceous megafossil sequences in Alaska (Smiley, 1972) and New Zealand (McQueen, 1956) that were not included in the data compilation show an almost identical pattern. The decrease in the mean number of conifer species per flora is small, although there are marked qualitative changes in the conifers present. The Cheirolepidiaceae are of much less importance in Late Cretaceous floras, while in the palynological record, at least at middle palaeolatitudes, there is a trend toward increasing abundance of pollen of probable Araucariaceae, Pinaceae, Podocarpaceae and Taxodiaceae (Doyle, 1977).

Analyses of species ranges in mid-Cretaceous floras from eastern North America show high levels of local extinction and systematic replacement for all major groups (Dorf, 1952; see also Vakhrameev, 1971). Pteridophytes and gymnosperms both play a more minor role in Late Cretaceous floras than in Early Cretaceous floras. This pattern is not reflected in recent estimates of global plant diversity through the Cretaceous (Niklas, Tiffney & Knoll, 1985), which show only minor declines in pteridophytes and cycadophytes, and a slow increase in angiosperm diversity (Figure 5.2(c)). Although the persistence of some typical Neocomian gymnosperms and pteridophytes in Paleocene high latitude floras will tend to preserve high levels of global diversity for these groups (cf. Krassilov, 1983), the estimate of angiosperm diversity is anomalously low (compare Figures 5.2(a),(c)). Analyses of mid-Cretaceous megafossil floras indicate that angiosperms were diverse and attained major importance in a variety of plant communities during the Cenomanian. Cenomanian floras from a variety of depositional settings are dominated by angiosperm leaves (Retallack & Dilcher, 1981; and see Appendix, p. 129), and, within the limitations common to most plant fossil assemblages, this does not appear to be an effect of biased palaeoenvironmental sampling. In the palynological record, however, angiosperm dominance is less clear. On the basis of analyses of palynomorph 'species' diversity in different low palaeolatitude areas, Muller (1984) concluded that the angiosperm 'ecological breakthrough' was not completed until the Turonian, and analyses of the relative abundance of pollen and spores in Cenomanian palynofloras from middle palaeolatitudes, show that angiosperms account typically for only about 40% to 60% of the dispersed palynomorphs. Underrepresentation of angiosperms in the palynoflora may suggest that uplands and well-drained areas remained dominated by conifers in the early Late Creceous, but could also result from the preponderance of insect-pollinated taxa in the earlier phases of the flowering plant radiation (Friis & Crepet, this volume, Chapter 6; Crepet & Friis, this volume, Chapter 7) and the probable lauralean affinities of many mid-Cretaceous magnoliid angiosperms

(G. R. Upchurch, personal communication). Pollen of most extant Laurales has a very thin sporopollenin exine, which is rarely preserved in the fossil record. The seemingly anomalous low diversity of angiosperm pollen in the Cenomanian may reflect palynological uniformity among early angiosperms, and in particular the practical difficulties of discriminating between the small tricolpate and tricolporoidate pollen grains that were probably produced by a variety of early representatives of the Hamamelididae and Rosidae (Doyle & Hickey, 1976; Crane *et al.*, 1986).

Although the marked declines in local importance of cycadophytes and pteridophytes that occurred as a result of the angiosperm radiation could be taken as evidence of competition between angiosperms and other tracheophytes, potential explanations for the success of flowering plants (e.g. Regal, 1977) are difficult to test with the data currently available. All of the major families of Jurassic and Early Cretaceous ferns become much less important in Late Cretaceous floras, and in particular there is a marked decline in the abundance of schizaeaceous spores (Brenner, 1963). Under certain circumstances Recent ferns and lycopods may be highly effective colonizers of bare ground (Spicer *et al.*, 1985), and families such as the Schizaeaceae and Gleicheniaceae, which probably dominated open environments at middle palaeolatitudes during the Early Cretaceous, may have been quickly affected by the appearance of opportunistic herbaceous or shrubby seed plants. Initially, middle palaeolatitude fern-savannas may have been replaced by angiosperm-fern-gnetalean communities, until the Gnetales themselves almost became extinct. Today the extant representatives of the characteristic Mesozoic fern families play a minor vegetational role compared to that of the 'higher' filicalean ferns (e.g. Polypodiaceae *sensu lato*) that diversified more recently, both figuratively and literally, in the shadow of the angiosperms. The correlation between extant ferns and damp shaded habitats may therefore have as much to do with the evolution of flowering plants as the biological limitations of pteridophytic reproduction (see also Upchurch & Wolfe, this volume, Chapter 4).

The most dramatic effect of the angiosperm radiation was on the cycadophytes (Figure 5.2(*a*)). The Bennettitales and pinnate-leaved seed ferns eventually became extinct, while the cycads became less diverse and more restricted in distribution. At middle palaeolatitudes these changes were more or less complete by the end of the Cenomanian (for an alternative view, see Krassilov, 1983). The demise of the Bennettitales is of particular interest because, like the Gnetales, many aspects of their reproductive biology were apparently similar to that of angiosperms, and this may have accentuated the competition between these groups. Angiosperms may not, therefore, have moved unchallenged into their role as the predominant early successional seed plants (cf. Tiffney, 1984), and at least some of the biological similarities of the angiosperms, Bennettitales and Gnetales may have been inherited from their common ancestor (Crane, 1985).

Community structure

Against the background of the major compositional and geographic changes in mid-Cretaceous vegetation there are some indications of an ecological pattern to the angiosperm radiation. Conclusions based on the Potomac Group need to be tested in other areas, but indicate that angiosperms first became established as early successional herbs or shrubs in disturbed habitats (Doyle & Hickey, 1976; Crane & Upchurch, unpublished). In particular, sedimentological data suggest that early angiosperms may have been important stream-side weeds in areas recently disturbed by erosion and deposition. During the late Albian, in the Potomac Group, angiosperms apparently diversified ecologically to occupy stream-side and aquatic habitats, the forest understory and early successional thickets (Doyle & Hickey, 1976; Hickey & Doyle, 1977). The average seed volume of around 1 mm^3 in Late Cretaceous angiosperms may have limited establishment to open conditions in which seedling photosynthesis could proceed unhindered (Tiffney, 1984). However, the stage at which angiosperms attained the status of dominant, canopy-forming large trees is more difficult to establish (see also Upchurch & Wolfe, this volume, Chapter 4). Among the largest angiosperm woods so far recorded from the Cretaceous are *Hythia* and *Cantia* from the Early Aptian and Early Albian, respectively (Stopes, 1915); however, there is controversy over the age of these specimens (Hughes, 1976), and most Late Cretaceous angiosperm woods are typically less than 10 cm in diameter (e.g. Page, 1979) in contrast to the large stumps of Mesozoic conifers (e.g. Bannan & Fry, 1957).

If the paucity of large fragments of fossil angiosperm wood reflects a scarcity of large arborescent dicotyledons (Upchurch & Wolfe, this volume, Chapter 4) it could support the view that Late Cretaceous angiosperms were ecologically subservient to gymnosperms and occupied only 'marginal or open habitats in the gymnosperm dominated vegetation' (Tiffney, 1984; see also Krassilov, 1983). However, even given the palaeoenvironmental bias toward lowland vegetation of most plant fossil assemblages, this hypothesis is difficult to reconcile with the

Figure 5.3. Diagram summarizing the distribution of major plant groups between major life-form categories for Neocomian, Aptian, Cenomanian, Campanian and Recent mid-latitude vegetation. (a), araucarian conifer; (b), taxodiaceous conifer; (c), cycadophyte shrub; (d), *Cycadeoidea* type bennettitalean; (e), herbaceous lycopod; (f), fern; (g), angiosperm shrub; (h), angiosperm herb; (i) gnetalean herb or small shrub; (j), angiosperm tree.

diversity and abundance of angiosperm megafossils, and the relative paucity of gymnosperms, in Late Cretaceous fossil floras (cf. Vakhrameev, 1981). A possible alternative scenario for mid-palaeolatitude vegetation is that small angiosperm trees or shrubs were sufficiently effective colonizers to exclude gymnosperms from most lowland habitats except perhaps taxodiaceous swamp forests. The result would be considerable areas dominated by shrubs or small trees (cf. the Mesozoic chaparral of Krassilov, 1973*a*), in which large arborescent angiosperms were uncommon and from which conifers such as Araucariaceae or Taxodiaceae were largely excluded or confined to occasional emergents. Other conifers (e.g. *Cephalotaxus, Torreya*) may themselves have been small trees or shrubs (Krassilov, 1973*b*). The hypothesized absence of large angiosperm trees during the mid-Cretaceous may lead to a misinterpretation of the effects of differential decay of angiosperm and conifer wood, and requires detailed examination, but it is apparently consistent with the observation, based on dinosaur remains, that during the Campanian (Dinosaur Provincial Park, Alberta) 'most of the primary productivity consumed by large herbivores must have been located between 0.5 and 4.0 m off the ground' (Béland & Russell, 1978; see also Coe *et al.*, this volume, Chapter 9).

The probable major changes in the structure and composition of terrestrial plant communities during the Cretaceous are summarized in Figure 5.3. The available evidence indicates that angiosperms first became established as early successional weedy herbs or shrubs, colonizing predominantly open environments or, at middle palaeolatitudes, habitats previously occupied by ferns and lycopods. Angiosperm small trees and shrubs that diversified very early in the Late Cretaceous may also have been highly effective colonizers, and apparently expanded first to dominate environments previously occupied by cycadophytes. Assuming appropriate climatic conditions (Upchurch & Wolfe, this volume, Chapter 4), close spacing of widely distributed small trees or shrubs would have provided ideal conditions for the development of wind pollination in the mid-Cenomanian. Compared to cycadophytes and pteridophytes, the relative abundance of conifers was little affected by the angiosperm radiation, and conifers are the only group of gymnosperms to have maintained their dominance over large areas in the face of competition from angiosperms (Regal, 1977). Large canopy-forming angiosperm forest trees may not have become widespread until the latest Cretaceous or Early Tertiary. The radiation of the major groups of extant angiosperm herbs, which account for about half of the number of extant angiosperm species (Van Valen, 1973), did not take place until later in the Tertiary.

Community stability and change

The increasing modernization of the angiosperm flora from the mid-Cretaceous is well documented by palynological evidence (Muller, 1984), and accumulating information from megafossils suggests that the available curves (Figure 5.2(d)) may even underestimate the extent to which Cretaceous angiosperms can be placed in extant families and orders (cf. Knobloch & Mai, 1984; Friis, 1985; Friis *et al.*, 1986). In parallel with this increasing modernization, is it possible to trace the history of Recent plant communities in the fossil record? For many years plant ecologists interpreted climax plant communities as 'super organisms' with their own 'growth and development' and 'ontogeny and phylogeny' (Clements, 1949, p. 123). The palaeobotanical mani-festation of Clementsian ecology is the concept of the geoflora as 'a group of plants that has maintained itself with only minor changes in composition for several periods of earth history' (Chaney in Chaney & Axelrod, 1959, p. 12). In the latest Cretaceous and Early Tertiary of the Northern Hemisphere, two major geofloras have been recognized traditionally. North of about 40° Paleocene palaeolatitude, the 'Arcto-Tertiary Geoflora' consisted of 'the group of temperate plants which is recorded in the older Tertiary rocks of Greenland, Spitzbergen, Alaska and arctic Canada, and which has spread southward through Eurasia and North America during the Cenozoic' (Chaney in Chaney & Axelrod, 1959, p. 12). The Arcto-Tertiary Geoflora is more or less equivalent to the Paleocene Greenland, and Eocene–Oligocene Turgaian Provinces of Krystofovich (1955), and has been interpreted as the once widespread precursor of the now more restricted 'mixed mesophytic forest' of eastern Asia and eastern North America (Wang, 1961). The more southerly Neotropical, or Palaeotropical, Geoflora consisted mainly of tropical evergreen dicotyledons, notably the Lauraceae, and is broadly equivalent to the Paleocene Gelinden, and Eocene–Oligocene Poltavian Provinces of Krystofovich (1955), and the Mastixioidean flora of European Eocene and Oligocene brown coals (Kirchheimer, 1957).

Although the geoflora concept dominated a generation of floristic work on Early Tertiary floras (e.g. Chaney, 1940), improved systematic and stratigraphic interpretation suggest that it is a potentially misleading oversimplification. For example, Paleocene 'Arcto-Tertiary' floras are highly depauperate compared to Recent mixed mesophytic forest (Wolfe,

1977) and also include a large proportion of extinct genera (cf. Crane & Stockey, 1985). Similarly, the extensively studied 'Palaeotropical' London Clay flora contains a mixture of plants that have their nearest living relatives in tropical southeast Asia (e.g. *Nipa*), as well as the mixed mesophytic forest of east central China (e.g. *Platycarya*, c.f. also Chaney, 1949). Although the geoflora concept correctly emphasizes the systematic homogeneity of higher-latitude Northern Hemisphere vegetation during the earliest Tertiary, recent studies on individual angiosperm lineages suggest that the history of extant plant communities is too complex to be usefully simplified in this way. This is particularly true during the Tertiary where increasing continental dispersion (Smith, Hurley & Briden, 1981; McKenna, 1983) and climatic changes (Wolfe, 1975, 1978) eventually led to a major increase in both diversity and floristic provinciality. A more detailed understanding of the development of Recent plant communities seems more likely to emerge from studies of individual plant lineages in conjunction with independent assessments of palaeoenvironmental change.

I thank W. C. Burger, W. G. Chaloner, J. A. Doyle, E. M. Friis, K. R. Pedersen, R. G. Stolze, G. R. Upchurch and J. A. Wolfe for helpful suggestions and constructive criticism. Ms C. D. Hult assisted in the systematic compilations. I am also grateful for many discussions on Mesozoic plants with the late P. D. W. Barnard. This work was partially supported by National Science Foundation grant BSR 8314592.

Appendix: List of Jurassic to Paleocene megafloras from the Northern Hemisphere

This list shows the number of species assigned to five major plant groups (not inclusive of reproductive structures or stems). Form genera (e.g. *Dicotylophyllum*) are treated as genera and varieties are treated as distinct species. Pteridophytes include lycopods and ferns; cycadophytes include cycads, Bennettitales and seed ferns with pinnate leaves; other seed plants include taxa such as *Baiera*, *Czekanowskia* and *Ginkgo* that are neither clearly cycadophytes nor conifers. The name of the flora is followed by the reference from which the counts of species were made. Sources for age determinations are given in parentheses only if different from, or more specific than, the source of systematic data.

1, Boroldaisky, USSR (Dolundenko & Orlovskaya, 1976); Middle Jurassic. 2, Borolsaisky, USSR (Dolundenko & Orlovskaya, 1976); Upper Jurassic. 3, Issyktasky, USSR (Dolundenko & Orlovskaya, 1976); Middle Jurassic. 4,

Karabastausky, USSR (Dolundenko & Orlovskaya, 1976); Upper Jurassic. 5, Oregon, USA (Fontaine, in Ward, 1905); Upper Jurassic (J. A. Doyle, personal communication). 6, Oroville, USA (Ward, 1900); Upper Jurassic (J. A. Doyle, personal communication). 7, Yorkshire, UK (Harris, 1961a, 1964, 1969, 1979; Harris et al., 1974); Middle Jurassic. 8, Hazelton, Canada (Bell, 1956); Neocomian–Barremian (restricted to plants from localities in the Hazelton–Cedarville area). 9, Horsetown, USA (Berry, 1911); Berriasian–Barremian (Hughes, 1976). 10, Kootenay, Canada (Bell, 1956); Berriasian (Stott, 1975). 11, Nikanassin, Canada (Bell, 1956); Berriasian (Stott, 1975). 12, Wealden, Germany (Berry, 1911); Neocomian (Hughes, 1976). 13, Wealden, Saxony (Berry, 1911); Neocomian (Hughes, 1976). 14, Wealden, U.K. (Berry, 1911); Neocomian (Hughes, 1976). 15, Upper Knoxville, USA (Berry, 1911); Neocomian (Hughes, 1976). 16, Almargem, Portugal (Teixeira, 1948); Aptian (Hughes, 1976). 17, Bullhead, Canada (Bell, 1956); Barremian–Aptian (Stott, 1975). 18, Cercal, Portugal (Teixeira, 1948); Upper Aptian–Lower Albian (Rey, 1972). 19, Fuson, USA (Berry, 1911); Aptian–Albian (McGookey et al., 1972). 20, Kome, Greenland (Heer, 1893); Barremian–Aptian (K. R. Pedersen, personal communication). 21, Kootenai, USA (La Pasha & Miller, 1984); Aptian. 22, Lower Blairmore, Canada (Bell, 1956); Aptian–Lower Albian (Stelk, 1975; Stott, 1975). 23, Luscar, Canada (Bell, 1956); Aptian. 24, Potomac Group, Zone I, USA (based on lists given by Berry (1911) for the Baltimore, Dutch Gap, Fredericksburg, Potomac Run and Trent's Reach localities); Aptian–Lower Albian (Brenner, 1963; Doyle & Hickey, 1976). 25, Buarcos, Portugal (Teixeira, 1948); Albian (Hughes, 1976). 26, Cheyenne, USA (Berry, 1921b); Albian (Ward, 1983). 27, Kingsvale, Canada (Bell, 1956); Albian. 28, Nazaré, Portugal (Teixeira, 1948); Albian (Hughes, 1976). 29, Pasayten, Canada (Bell, 1956); Albian. 30, Potomac Group, Zone II, USA (based on lists given by Berry (1911) for the Deep Bottom, Mount Vernon, Stump Neck, Welhams, Whitewater and Whitehouse Bluff localities); Middle–Upper Albian (Brenner, 1963; Doyle & Hickey, 1976). 31, Taverede, Portugal (Teixeira, 1948); uppermost Albian–Cenomanian (Hughes, 1976). 32, Upper Blairmore, Canada (Bell, 1956); Albian (Stott, 1975). 33, Bingen, USA (Berry, 1917); Cenomanian (?coeval with Raritan). 34, Dakota, USA (Berry, 1916); Cenomanian (Hickey & Doyle, 1977). 35, Dunvegan, Canada (Bell, 1963); Cenomanian (Stott, 1975). 36, Kaltag, Alaska, USA (Hollick, 1930); Cenomanian (R. A. Spicer, personal communication). 37, Melozi, Alaska, USA (Hollick, 1930); Cenomanian (R. A. Spicer, personal communication). 38, Nulato, Alaska, USA (Hollick, 1930); Cenomanian (R. A. Spicer, personal communication). 39, Raritan, USA (based on the lists given by Newberry (1895) for the Woodbridge and Sayreville localities); Cenomanian (Christopher, 1979). 40, Tuscaloosa, USA (Berry, 1919); Cenomanian (Hazel et al., 1977). 41, Woodbine, USA (Berry, 1921a); Cenomanian (Hedlund, 1966). 42, Atane, Greenland (Heer, 1893); Turonian–Santonian (K. R. Pedersen, personal communication). 43, Black Creek,

USA (Berry, 1914); Santonian–Campanian (Hazel *et al.*, 1977). 44, Eutaw, USA (Berry, 1925); Coniacian–Santonian (Hazel *et al.*, 1977). 45, Frontier, USA (Knowlton, 1918); Cenomanian–Turonian (Nichols & Jacobson, 1982. 46, Fruitland & Kirtland, USA (Knowlton, 1917*a*); Campanian (McGookey *et al.*, 1972). 47, Judith River, USA (Knowlton, in Stanton & Hatcher, 1905); Campanian (McGookey *et al.*, 1972). 48, Magothy, USA (Berry. 1916); Coniacian–Santonian (Christopher, 1979). 49, Milk River, Canada (Bell, 1963); Santonian (Stelk, 1975). 50, Nanaimo, Canada (Bell, 1957); Coniacian–Campanian. 51, Patoot, Greenland (Heer, 1893); Santonian–Campanian (K. R. Pedersen, personal communication). 52, Dawson Arkose, USA (Knowlton, 1930); Maastrichtian (?coeval with Raton Formation). 53, Lance, USA (Dorf, 1942); Maastrichtian (McGookey *et al.*, 1972). 54, Laramie, USA (Knowlton, 1922); Maastrichtian (McGookey *et al.*, 1972). 55, Lower Edmonton, Canada (Bell, 1949); Maastrichtian. 56, Lower Medicine Bow, USA (Dorf, 1942); Maastrichtian (McGookey *et al.*, 1972). 57, Raton, USA (Knowlton, 1917*b*); Maastrichtian (McGookey *et al.*, 1972). 58, St Mary River, Canada (Bell, 1949); Maastrichtian (McGookey *et al.*, 1972). 59, Upper Edmonton, Canada (Bell, 1949); Maastrichtian. 60, Vermejo, USA (Knowlton, 1917*b*); Maastrichtian (McGookey *et al.*, 1972). 61, Whitemud, Canada (Berry, 1935); Maastrichtian (Bell, 1949). 62, Atanikerdluk, Greenland (based on lists given by Heer (1893) for Atanikerdluk A and B floras); Paleocene (Pedersen, 1976). 63, Fort Union, USA (Brown, 1962); Paleocene. 64, Genesee, Canada (Chandrasekharam, 1974); Paleocene. 65, Menat, France (Laurent, 1912); Paleocene (Russell, in Kedves, 1982). 66, Middle Ravenscrag, Canada (Berry, 1935); Paleocene (Jarzen, 1982*a*). 67, Northwest Greenland (Koch, 1963); Paleocene. 68, Paskapoo, Canada (Bell, 1949); Paleocene. 69, Spitzbergen (Manum, 1962); Paleocene. 70, Upper Ravenscrag, Canada (Berry, 1935); Paleocene (Jarzen, 1982*a*).

Tabulation of data from the floras listed above

	Pteridophytes	Other seed plants	Cycadophytes	Conifers	Angiosperms
Jurassic					
1. Boroldaisky	19	9	9	7	0
2. Borolsaisky	1	2	6	6	0
3. Issyktasky	7	6	5	2	0
4. Karabastausky	10	8	17	17	0
5. Oregon	13	8	26	11	0
6. Oroville	9	2	11	2	0
7. Yorkshire	55	29	60	26	0

Tabulation of data (cont.)

	Pteridophytes	Other seed plants	Cycadophytes	Conifers	Angiosperms
Neocomian–Barremian					
8. Hazelton	9	3	6	2	0
9. Horsetown	6	3	6	7	3
10. Kootenay	9	6	10	4	0
11. Nikanassin	4	3	1	4	0
12. Wealden, Germany	4	1	5	3	0
13. Wealden, Saxony	15	2	7	3	0
14. Wealden, UK	17	2	18	13	0
15. Upper Knoxville	13	3	12	4	0
Aptian–Lower Albian					
16. Almargem	4	1	1	4	0
17. Bullhead	8	4	8	7	0
18. Cercal	7	1	0	4	4
19. Fuson	8	0	1	9	3
20. Kome	48	4	10	17	0
21. Kootenai	16	3	1	4	0
22. Lower Blairmore	15	5	9	10	1
23. Luscar	17	4	10	13	0
24. Potomac, Zone I	33	3	13	20	6
Middle–Upper Albian					
25. Buarcos	11	0	1	6	4
26. Cheyenne	4	0	0	3	11
27. Kingsvale	8	1	1	6	7
28. Nazaré	1	0	0	1	2
29. Pasayten	8	1	2	1	6
30. Potomac, Zone II	9	0	3	12	21
31. Taverede	5	0	0	5	8
32. Upper Blairmore	9	0	1	8	13
Cenomanian					
33. Bingen	2	0	0	6	20
34. Dakota	7	0	3	19	437
35. Dunvegan	10	2	2	8	42
36. Kaltag	6	9	5	17	75
37. Melozi	0	0	0	8	46
38. Nulato	1	0	1	6	8
39. Raritan	8	2	0	11	62
40. Tuscaloosa	9	0	0	14	118
41. Woodbine	0	0	0	2	37

Tabulation of data (*cont.*)

	Pteridophytes	Other seed plants	Cycadophytes	Conifers	Angiosperms
Turonian–Campanian					
42. Atane	34	5	8	20	97
43. Black Creek	2	0	0	12	58
44. Eutaw	0	0	0	7	37
45. Frontier	8	0	0	0	17
46. Fruitland and Kirtland	4	0	0	0	29
47. Judith River	1	0	0	6	14
48. Magothy	13	3	1	32	208
49. Milk River	6	0	0	6	15
50. Nanaimo	14	1	5	5	62
51. Patoot	21	0	0	15	76
Maastrichtian					
52. Dawson Arkose	5	0	0	0	65
53. Lance	6	1	0	2	54
54. Laramie	14	0	0	3	107
55. Lower Edmonton	1	1	1	5	4
56. Lower Medicine Bow	4	0	0	1	52
57. Raton	7	0	0	0	130
58. St Mary River	2	0	0	2	6
59. Upper Edmonton	1	0	0	1	6
60. Vermejo	14	0	0	7	77
61. Whitemud	1	1	0	2	19
Paleocene					
62. Atanikerdluk	9	1	0	17	143
63. Fort Union	20	1	2	7	120
64. Genesee	3	0	0	4	8
65. Menat	4	0	0	4	38
66. Middle Ravenscrag	2	0	0	11	25
67. Northwest Greenland	2	1	0	1	32
68. Paskapoo	6	1	0	5	23
69. Spitzbergen	6	6	0	14	54
70. Upper Ravenscrag	1	1	0	4	23

References

Alvin, K. L. (1968). The spore-bearing organs of the Cretaceous fern *Weichselia* Stiehler. *Botanical Journal of the Linnean Society*, **61**, 87–92.

Alvin, K. L. (1983). Reconstruction of a Lower Cretaceous conifer. *Botanical Journal of the Linnean Society*, **86**, 169–76.

Apert, O. (1973). Die Pteridophyten aus dem Oberen Jura des Manamara in Südwest-Madagascar. *Schweizerische Paläontologische Abhandlungen*, **94**, 1–62.

Ash, S. R. & Read, C. B. (1976). North American species of *Tempskya* and their stratigraphic significance. *United States Geological Survey, Professional Paper*, **874**, 1–42.

Bande, M. B., Prakash, U. & Ambwani, K. (1981). A fossil palm fruit *Hyphaeneocarpon indicum* gen. et sp. nov. from the Deccan Intertrappean Beds, India. *The Palaeobotanist*, **30**, 303–9.

Bannan, M. W. & Fry, W. L. (1957). Three Cretaceous woods from the Canadian Arctic. *Canadian Journal of Botany*, **35**, 327–37.

Batten, D. J. (1974). Wealden palaeoecology from the distribution of plant fossils. *Proceedings of the Geologists' Association*, **85**, 433–58.

Batten, D. J. (1981). Stratigraphic, palaeogeographic and evolutionary significance of Late Cretaceous and early Tertiary Normapolles pollen. *Review of Palaeobotany and Palynology*, **35**, 125–37.

Batten, D. J. (1984). Palynology, climate and the development of floral provinces in the northern hemisphere; a review. In *Fossils and Climate*, ed. P. Brenchley, pp. 127–64. London: Wiley and Sons.

Béland, P. & Russell, D. A. (1978). Paleoecology of Dinosaur Provincial Park (Cretaceous), Alberta, interpreted from the distribution of articulated vertebrate remains. *Canadian Journal of Earth Sciences*, **15**, 1012–24.

Bell, W. A. (1949). Uppermost Cretaceous and Paleocene floras of western Alberta. *Geological Survey of Canada Bulletin*, **13**, 1–231.

Bell, W. A. (1956). Lower Cretaceous floras of western Canada. *Geological Survey of Canada Memoir*, **285**, 1–331.

Bell, W. A. (1957). Flora of the Upper Cretaceous Nanaimo Group of Vancouver Island, British Columbia. *Geological Survey of Canada Memoir*, **293**, 1–84.

Bell, W. A. (1963). Upper Cretaceous floras of the Dunvegan, Bad Heart, and Milk River Formations of western Canada. *Geological Survey of Canada Bulletin*, **94**, 1–76.

Berry, E. W. (1911). The Lower Cretaceous Floras of the World, Correlation of the Potomac Formations, Pteridophyta, Cycadophytae, Gymnospermae, Monocotyledonae, Dicotyledonae. *Maryland Geological Survey, Lower Cretaceous*, pp. 99–165, 214–596.

Berry, E. W. (1914). The Upper Cretaceous and Eocene floras of South Carolina and Georgia. *United States Geological Survey, Professional Paper*, **84**, 1–200.

Berry, E. W. (1916). The Upper Cretaceous floras of the world, Thallophyta, Pteridophyta, Cycadophyta, Coniferophyta, Angiospermophyta. Text & plates. *Maryland Geological Survey, Upper Cretaceous*, pp. 183–313, 757–986.

Berry, E. W. (1917). Contributions to the Mesozoic flora of the Atlantic coastal plain. XII. Arkansas. *Bulletin of the Torrey Botanical Club*, 44, 167–90.

Berry, E. W. (1919). Upper Cretaceous floras of the eastern gulf region in Tennessee, Mississippi, Alabama and Georgia. *United States Geological Survey, Professional Paper*, 112, 1–141.

Berry, E. W. (1921*a*). The flora of the Woodbine Sand at Arthurs Bluff, Texas. *United States Geological Survey, Professional Paper*, 129G, 153–82.

Berry, E. W. (1921*b*). The flora of the Cheyenne Sandstone of Kansas. *United States Geological Survey, Professional Paper*, 129I, 199–226.

Berry, E. W. (1925). The flora of the Ripley Formation. *United States Geological Survey, Professional Paper*, 136, 1–94.

Berry, E. W. (1929). The flora of the Frontier Formation. *United States Geological Survey, Professional Paper*, 158I, 129–35.

Berry, E. W. (1935). A preliminary contribution to the floras of the Whitemud and Ravenscrag Formations. *Geological Survey of Canada Memoir*, 182, 1–107.

Brenner, G. J. (1963). The spores and pollen of the Potomac Group of Maryland. *State of Maryland Department of Geology, Mines and Water Resources Bulletin*, 27, 1–215.

Brenner, G. J. (1976). Middle Cretaceous floral provinces and early migrations of angiosperms. In *Origin and Early Evolution of Angiosperms*, ed. C. B. Beck, pp. 23–47. New York: Columbia University Press.

Brenner, G. J. (1984). Late Hauterivian angiosperm pollen from the Helez Formation, Israel. *Sixth International Palynological Conference, Calgary, Abstracts*, 1984, p. 15.

Brown, R. W. (1939). Fossil leaves, fruits and seeds of *Cercidiphyllum*. *Journal of Paleontology*, 13, 485–99.

Brown, R. W. (1962). Paleocene flora of the Rocky Mountains and Great Plains. *United States Geological Survey, Professional Paper*, 375, 1–119.

Chandler, M. E. J. (1961). *The Lower Tertiary Floras of Southern England*, vol. I *Palaeocene Floras. London Clay Flora (Supplement) and Atlas*. London: British Museum (Natural History).

Chandrasekharam, A. (1974). Megafossil flora from the Genesee locality, Alberta, Canada. *Palaeontographica B*, 147, 1–41.

Chaney, R. W. (1940). Tertiary forests and continental history. *Geological Society of America, Bulletin*, 51, 469–88.

Chaney, R. W. (1949). Early Tertiary ecotones in western North America. *Proceedings of the National Academy of Sciences, U.S.A.*, 35, 356–9.

Chaney, R. W. & Axelrod, D. I. (1959). Miocene floras of the Columbia Plateau. *Carnegie Institution of Washington, Contributions to Paleontology*, 617, 1–237.

Christopher, R. A. (1979). *Normapolles* and triporate pollen assemblages from the Raritan and Magothy Formations (Upper Cretaceous) of New Jersey. *Palynology*, **3**, 73–121.

Clements, F. E. (1949). *Dynamics of Vegetation*. New York: Wilson.

Crane, P. R. (1982). Betulaceous leaves and fruits from the British Upper Paleocene. *Botanical Journal of the Linnean Society*, **83**, 103–36.

Crane, P. R. (1984). A re-evaluation of *Cercidiphyllum*-like plant fossils from the British early Tertiary. *Botanical Journal of the Linnean Society*, **89**, 199–230.

Crane, P. R. (1985). Phylogenetic analysis of seed plants and the origin of angiosperms. *Annals of the Missouri Botanical Garden*, **72**, 716–93.

Crane, P. R., Friis, E. M. & Pedersen, K. R. (1986). Unisexual flowers from the Lower Cretaceous: fossil evidence on the early radiation of the dicotyledons. *Science*, **232**, 852–4.

Crane, P. R. & Stockey, R. A. (1985). Growth and reproductive biology of *Joffrea speirsii* gen. et sp. nov., a *Cercidiphyllum*-like plant from the late Paleocene of Alberta, Canada. *Canadian Journal of Botany*, **63**, 340–64.

Crepet, W. L. (1974). Investigations of North American cycadeoids: the reproductive biology of *Cycadeoidea*. *Palaeontographica B*, **148**, 144–69.

Crepet, W. L. & Delevoryas, T. (1972). Investigations of North American cycadeoids: early ovule ontogeny. *American Journal of Botany*, **59**, 209–15.

Cronquist, A. (1981). *An Integrated System of Classification of Flowering Plants*. New York: Columbia University Press.

Daber, R. (1968). A *Weichselia–Stiehleria*–Matoniaceae community within the Quedlinburg Estuary of Lower Cretaceous age. *Botanical Journal of the Linnean Society*, **61**, 75–85.

Daghlian, C. P. (1981). A review of the fossil record of monocotyledons. *The Botanical Review*, **47**, 517–55.

Delevoryas, T. (1959). Investigations of North American cycadeoids: *Monanthesia*. *American Journal of Botany*, **46**, 657–66.

Delevoryas, T. & Hope, R. C. (1976). More evidence for a slender growth habit in Mesozoic cycadophytes. *Review of Palaeobotany and Palynology*, **21**, 93–100.

Dolundenko, M. P. & Orlovskaya, E. R. (1976). Jurassic floras of the Karatau Range, southern Kazakhstan. *Palaeontology*, **19**, 627–40.

Dorf, E. (1942). Upper Cretaceous floras of the Rocky Mountain region. *Carnegie Institution of Washington, Contributions to Paleontology*, **508**, 1–168.

Dorf, E. (1952). Critical analysis of Cretaceous stratigraphy and paleobotany of Atlantic coastal plain. *Bulletin of the American Association of Petroleum Geologists*, **36**, 2161–84.

Doyle, J. A. (1977). Patterns of evolution in early angiosperms. In *Patterns of Evolution as Illustrated by the Fossil Record*, ed. A. Hallam, pp. 501–46. Amsterdam: Elsevier.

Doyle, J. A. (1978). Origin of angiosperms. *Annual Review of Ecology and Systematics*, **9**, 365–92.

Doyle, J. A. & Hickey, L. J. (1976). Pollen and leaves from the mid-Cretaceous Potomac Group and their bearing on early angiosperm evolution. In *Origin and Early Evolution of Angiosperms*, ed. C. B. Beck, pp. 139–206. New York: Columbia University Press.

Doyle, J. A., Jardiné, S. & Doerenkamp, A. (1982). *Afropollis*, a new genus of early angiosperm pollen, with notes on the Cretaceous palynostratigraphy of paleoenvironments of northern Gondwana. *Bulletin Centres Recherches Exploration–Production Elf–Aquitaine*, 6, 39–117.

Friis, E. M. (1983). Upper Cretaceous (Senonian) floral structures of juglandalean affinity containing *Normapolles* pollen. *Review of Palaeobotany and Palynology*, 39, 161–83.

Friis, E. M. (1985). Structure and function in Late Cretaceous angiosperm flowers. *Biologiske Skrifter Danske Videnskabernes Selskab*, 25, 1–37.

Friis, E. M., Crane, P. R. & Pedersen, K. R. (1986). Floral evidence for Lower Cretaceous chloranthoid angiosperms. *Nature*, 320, 163–4.

Givnish, T. J. (1971). On the adaptive significance of compound leaves, with particular reference to tropical trees. In *Tropical Trees as Living Systems*, ed. P. B. Tomlinson, pp. 351–80. Cambridge: Cambridge University Press.

Harper, J. L., Lovell, P. H. & Moore, K. G. (1970). The shapes and sizes of seeds. *Annual Review of Ecology and Systematics*, 1, 327–56.

Harris, T. M. (1932). The fossil flora of Scoresby Sound, east Greenland. Part 3. Caytoniales and Bennettitales. *Meddelelser om Grønland*, 85(5), 1–133.

Harris, T. M. (1954). Mesozoic seed cuticles. *Svenska Botaniska Tidskrift*, 48, 281–91.

Harris, T. M. (1961a). *The Yorkshire Jurassic Flora*, vol. I, *Thallophyta–Pteridophyta*. London: British Museum (Natural History).

Harris, T. M (1961b). The fossil cycads. *Palaeontology*, 4, 313–23.

Harris, T. M. (1964). *The Yorkshire Jurassic Flora*, vol. II, *Caytoniales, Cycadales & Pteridosperms*. London: British Museum (Natural History).

Harris, T. M. (1969). *The Yorkshire Jurassic Flora*, vol. III, *Bennettitales*. London: British Museum (Natural History).

Harris, T. M. (1971). The stem of *Caytonia*. *Geophytology*, 1, 23–9.

Harris, T. M. (1973). The strange Bennettitales. *Nineteenth Sir Albert Charles Seward Memorial Lecture, Birbal Sahni Institute of Palaeobotany*, 1970, pp. 1–11.

Harris, T. M. (1974). *Williamsoniella lignieri*: its pollen and the compression of spherical pollen grains. *Palaeontology*, 17, 125–48.

Harris, T. M. (1976). The Mesozoic gymnosperms. *Review of Palaeobotany and Palynology*, 21, 119–34.

Harris, T. M. (1979). *The Yorkshire Jurassic Flora*, vol. V, *Coniferales*. London: British Museum (Natural History).

Harris, T. M. (1981). Burnt ferns from the English Wealden. *Proceedings of the Geologists' Association*, 92, 47–58.

Harris, T. M., Millington, W. & Miller, J. (1974). *The Yorkshire Jurassic*

Flora, vol. IV, *Ginkgoales and Czekanowskiales*. London: British Museum (Natural History).

Hazel, J. E., Bybell, L. M., Christopher, R. A., Fredericksen, N. O., May, F. E., McLean, D. M., Poore, R. Z., Smith, C. C., Sohl, N. F., Valentine, P. C. & Whitmer, R. J. (1977). Biostratigraphy of the deep corehole (Clubhouse Crossroads corehole 1) near Charleston, South Carolina. *United States Geological Survey, Professional Paper*, **1028**, 71–89.

Hedlund, R. W. (1966). Palynology of the Red Branch Member (Woodbine Formation). *Bulletin of the Oklahoma Geological Survey*, **112**, 1–69.

Heer, O. (1893). Oversigt over Grønlands fossile Flora. *Meddelelser om Grønland*, **5**, 81–213.

Hickey, L. J. (1977). Stratigraphy and paleobotany of the Golden Valley Formation (Early Tertiary) of western North Dakota. *Geological Society of America Memoir*, **150**, 1–183.

Hickey, L. J. (1981). Land plant evidence compatible with gradual, not catastrophic, change at the end of the Cretaceous. *Nature*, **292**, 529–31.

Hickey, L. J. & Doyle, J. A. (1977). Early Cretaceous fossil evidence for angiosperm evolution. *The Botanical Review*, **43**, 3–104.

Hollick, A. (1930). The Upper Cretaceous floras of Alaska. *United States Geological Survey, Professional Paper*, **159**, 1–123.

Hollick, A. (1936). The Tertiary floras of Alaska. *United States Geological Survey, Professional Paper*, **182**, 1–185.

Hughes, N. F. (1976). *Palaeobiology of Angiosperm Origins*. Cambridge: Cambridge University Press.

Hughes, N. F., Drewry, G. E. & Laing, J. F. (1979). Barremian earliest angiosperm pollen. *Palaeontology*, **22**, 513–35.

Jarzen, D. M. (1977). *Aquilapollenites* and some santalalean genera: a botanical comparison. *Grana*, **16**, 29–39.

Jarzen, D. M. (1982*a*). Angiosperm pollen from the Ravenscrag Formation (Paleocene) southern Saskatchewan, Canada. *Pollen et Spores*, **14**, 119–55.

Jarzen, D. M. (1982*b*). Palynology of Dinosaur Provincial Park (Campanian), Alberta. *Syllogeus*, **38**, 1–69.

Jarzen, D. M. & Norris, G. (1975). Evolutionary significance and botanical relationships of Cretaceous angiosperm pollen in the western Canadian interior. *Geoscience and Man*, **11**, 47–60.

Kedves, M. (1982). Palynology of the Thanetian layers of Menat. *Palaeontographica B*, **182**, 87–166.

Kimura, T. & Sekido, S. (1975). *Nilssoniocladus* n. gen. (Nilssoniaceae n. fam.), newly found from the early Lower Cretaceous of Japan. *Palaeontographica B*, **153**, 111–8.

Kirchheimer, F. (1957). *Die Laubgewächse der Braunkohlenzeit*. Halle: Veb Wilhelm Knapp.

Knobloch, E. (1984). Megasporen aus der Kreide von Mitteleuropa. *Sbornik Geologických Věd, Paleontologie*, **26**, 157–95.

Knobloch, E. & Mai, H. D. (1984). Neue Gattungen nach Früchten und Samen aus dem Cenoman bis Maastricht (Kreide) von Mitteleuropa. *Feddes Repertorium*, **95**, 3–41.

Knowlton, F. H. (1917*a*). Flora of the Fruitland and Kirtland Formations. *United States Geological Survey, Professional Paper*, **98S**, 327–53.

Knowlton, F. H. (1917*b*). Fossil floras of the Vermejo and Raton Formations of Colorado and New Mexico. *United States Geological Survey, Professional Paper*, **101**, 223–435.

Knowlton, F. H. (1918). A fossil flora from the Frontier Formation of southwestern Wyoming. *United States Geological Survey, Professional Paper*, **108F**, 73–107.

Knowlton, F. H. (1922). The Laramie flora of the Denver Basin. *United States Geological Survey, Professional Paper*, **130**, 1–175.

Knowlton, F. H. (1930). The flora of the Denver and associated formations of Colorado. *United States Geological Survey, Professional Paper*, **155**, 1–142.

Koch, B. E. (1963). Fossil plants from the Lower Paleocene of the Agatdalen (Angmârtussut) area, central Nûgssuaq Peninsula, northwest Greenland. *Meddelelser om Grønland*, **172**, 1–120.

Kovach, W. I. & Dilcher, D. L. (1985). Morphology, ultrastructure, and paleoecology of *Paxillitriletes vittatus* sp. nov. from the mid-Cretaceous (Cenomanian) of Kansas. *Palynology*, **9**, 85–94.

Krassilov, V. A. (1973*a*). Mesozoic plants and the problem of angiosperm ancestry. *Lethaia*, **6**, 163–78.

Krassilov, V. A. (1973*b*). Climatic changes in eastern Asia as indicated by fossil floras. I. Early Cretaceous. *Palaeogeography, Palaeoclimatology, Palaeoecology*, **13**, 261–73.

Krassilov, V. A. (1975). Climatic changes in eastern Asia as indicated by fossil floras. II. Late Cretaceous and Danian. *Palaeogeography, Palaeoclimatology, Palaeoecology*, **17**, 157–72.

Krassilov, V. A. (1976).[*The Tsagayan flora of Amur region.*] In Russian – Moscow: Nauka.

Krassilov, V. A. (1978). Late Cretaceous gymnosperms from Sakhalin, U.S.S.R., and the terminal Cretaceous event. *Palaeontology*, **21**, 893–905.

Krassilov, V. A. (1981). Changes of Mesozoic vegetation and the extinction of dinosaurs. *Palaeogeography, Palaeoclimatology, Palaeoecology*, **34**, 207–24.

Krassilov, V. A. (1983). [Evolution of the flora in the Cretaceous period: is a Cenophytic era necessary?] In Russian. *Paleontologicheskiy Zhurnal*, 1983(3), pp. 93–6. (*Paleontological Journal*, **17**(3), 89–93.)

Krutzsch, W. (1967). Der Florenwechsel im Alttertiär Mitteleuropas auf Grund von Sporenpaläontologischen Untersuchungen. *Abhandlungen Zentrales Geologisches Institut Berlin*, **10**, 17–37.

Krystofovich, A. N. (1955). [Development of the phytogeographical regions of the northern hemisphere from the beginning of the Tertiary period.] In Russian. In *Problems of the Geology of Asia*, pp. 824–44. Moscow: Nauka.

Kvaček, Z. (1983). Cuticular studies in angiosperms of the Bohemian Cenomanian. *Acta Palaeontologica Polonica*, **28**, 159–70.

Laing, J. F. (1975). Mid-Cretaceous angiosperm pollen from southern England and northern France. *Palaeontology*, **18**, 775–808.

La Pasha, C. A. & Miller, C. N. (1984). Flora of the early Cretaceous Kootenai Formation in Montana; Palaeoecology. *Palaeontographica B*, **194**, 109–30.

Laurent, L. (1912). Flore fossile des schistes de Menat (Puy-de-Dôme). *Annales du Musée d'Histoire Naturelle de Marseille, Géologie*, **14**, 1–246.

Leffingwell, H. A. (1970). Palynology of the Lance (late Cretaceous) and Fort Union (Paleocene) Formations of the type Lance Area, Wyoming. *Geological Society of America Special Paper*, **127**, 1–64.

McGookey, D. P., Haun, J. D., Hale, L. A., Goodell, H. G., McCubbin, D. G., Wiemer, R. J. & Wulf, G. R. (1972). Cretaceous System. In *Geologic Atlas of the Rocky Mountain Region*, ed. W. W. Mallory, pp. 190–228. Denver: Rocky Mountain Association of Geologists.

McKenna, M. C. (1983). Holarctic land mass rearrangement, cosmic events and Cenozoic terrestrial organisms. *Annals of the Missouri Botanical Garden*, **70**, 459–89.

McQueen, D. R. (1956). Leaves of Middle and Upper Cretaceous Pteridophytes and Cycads from New Zealand. *Transactions of the Royal Society of New Zealand*, **83**, 673–85.

Mägdefrau, K. (1932). Über *Nathorstiana*, eine Isoetacee aus dem Neokom von Quedlinburg a. Harz. *Beihefte zum Botanischen Zentralblatt*, **49**, 706–18.

Manchester, S. R. (1986). Vegetative and reproductive morphology of an extinct plane tree (Platanaceae) from the Eocene of western North America. *Botanical Gazette*, **147**, 200–26.

Manchester, S. R. (1987). The fossil history of the Juglandaceae. *Annals of the Missouri Botanical Garden, Monographs*, **21**, 1–137.

Manum, S. (1962). Studies in the Tertiary flora of Spitzbergen, with notes on Tertiary floras of Ellesmere Island, Greenland and Iceland. *Norsk Polarinstitutt Skrifter*, **125**, 1–127.

Melchior, R. C. & Hall, J. W. (1983). Some megaspores and other small fossils from the Wannagan Creek site (Paleocene) North Dakota. *Palynology*, **7**, 133–45.

Miller, C. N. (1982). Current status of Paleozoic and Mesozoic conifers. *Review of Palaeobotany and Palynology*, **37**, 99–114.

Muller, J. (1981). Fossil pollen records of extant angiosperms. *The Botanical Review*, **47**, 1–142.

Muller, J. (1984). Significance of fossil pollen for angiosperm history. *Annals of the Missouri Botanical Garden*, **71**, 419–43.

Nathorst, A. G. (1902). Paläobotanische Mitteilungen. *Kongliga Svenska Vetenskapsakademiens Handlingar*, **36**, 1–28.

Němejc, F. & Kvaček, Z. (1975). *Senonian plant macrofossils from the region of Zliv and Hluboká (near České Budejovice) in south Bohemia.* Prague: Universita Karlova Praha.

Newberry, J. S. (1895). The flora of the Amboy Clays. *United States Geological Survey, Monographs,* **26**, 1–260.

Nichols, D. J. & Jacobson, S. R. (1982). Palynostratigraphic framework for the Cretaceous (Albian–Maestrichtian) of the overthrust belt of Utah and Wyoming. *Palynology,* **6**, 119–47.

Niklas, K. J., Tiffney, B. H. & Knoll, A. H. (1985). Patterns in vascular land plant diversification: a factor analysis at the species level. In *Phanerozoic Diversity Patterns: Profiles in Macroevolution,* ed. J. W. Valentine, pp. 97–128. Princeton: Princeton University Press.

Norris, G., Jarzen, D. M. & Awai-Thorne, B. V. (1975). Evolution of the Cretaceous terrestrial palynoflora in western Canada. In *The Cretaceous System in the Western Interior of North America,* ed. W. G. E. Caldwell, pp. 333–64. Waterloo: Geological Society of Canada.

Oldham, T. C. B. (1976). Flora of the Wealden plant debris beds of England. *Palaeontology,* **19**, 437–502.

Pacltová, B. (1977). Cretaceous angiosperms of Bohemia–Central Europe. *The Botanical Review,* **43**, 128–42.

Pacltová, B. (1978). Significance of palynology for the biostratigraphic division of the Cretaceous of Bohemia. *Paleontologická Konference* 1977, *Universita Karlova Praha,* vol. 7, pp. 93–111.

Page, V. (1979). Dicotyledonous wood from the Upper Cretaceous of Central California. *Journal of the Arnold Arboretum,* **60**, 323–49.

Pedersen, K. R. (1976). Fossil floras of Greenland. In *Geology of Greenland,* ed. A. Escher & W. S. Watt, pp. 519–35. Copenhagen: Geological Survey of Greenland.

Penny, J. S. (1969). Late Cretaceous and early Tertiary palynology. In *Aspects of Palynology,* ed. R. H. Tschudy & R. A. Scott, pp. 331–76. New York: Wiley.

Regal, P. J. (1977). Ecology and evolution of flowering plant dominance. *Science,* **196**, 622–9.

Reid, E. M. & Chandler, M. E. J. (1933). *The London Clay Flora.* London: British Museum (Natural History).

Retallack, G. & Dilcher, D. L. (1981). A coastal hypothesis for the dispersal and rise to dominance of flowering plants. In *Paleobotany, Paleoecology and Evolution,* vol. 2, ed. K. J. Niklas, pp. 27–77. New York: Praeger.

Rey, C. (1972). Recherches géologiques sur le Crétacé inférieur de l'Estremadura (Portugal). *Memorias Servicos Geológicos Portugal, Lisbon,* **21**, 1–477.

Rüffle, L. (1977). Entwicklungsgeschichtliche und ökologische Aspekte zur Oberkreide-Flora. *Zeitschrift für Geologische Wissenschaften, Berlin,* **5**, 269–303.

Samylina, V. A. (1968). Early Cretaceous angiosperms of the Soviet Union based on leaf and fruit remains. *Botanical Journal of the Linnean Society*, **61**, 207–18.

Sharma, B. D. (1974). Ovule ontogeny in *Williamsonia* Carr. *Palaeontographica B*, **148**, 137–43.

Shoemaker, R. E. (1966). Fossil leaves of the Hell Creek and Tullock Formations of eastern Montana. *Palaeontographica B*, **119**, 54–75.

Singh, C. (1975). Stratigraphic significance of early angiosperm pollen in the mid-Cretaceous strata of Alberta. In *The Cretaceous System in the Western Interior of North America*, ed. W. G. E. Caldwell, pp. 365–90. Waterloo: Geological Society of Canada.

Smiley, C. J. (1972). Plant megafossil sequences, North Slope Cretaceous. *Geoscience and Man*, **4**, 91–9.

Smith, A. G., Hurley, A. M. & Briden, J. C. (1981). *Phanerozoic Palaeocontinental World Maps*. Cambridge: Cambridge University Press.

Smoot, E. L., Taylor, T. N. & Delevoryas, T. (1985). Structurally preserved fossil plants from Antarctica. I. *Antarcticycas*, gen. nov., a Triassic cycad stem from the Beardmore Glacier area. *American Journal of Botany*, **72**, 1410–23.

Spicer, R. A., Burnham, R. J., Grant, P. & Glicken, H. (1985). *Pityrogramma calomelanos*, the primary colonizer of Volcán Chinchonal, Chiapas, Mexico. *American Fern Journal*, **75**, 1–5.

Stanton, T. W. & Hatcher, J. B. (1905). Geology and Paleontology of the Judith River Beds. *United States Geological Survey, Bulletins*, **257**, 1–174.

Stebbins, G. L. (1974). *Flowering plants: Evolution above the Species Level*. Cambridge, Massachusetts: Belknap Press.

Stelk, C. R. (1975). Basement control of Cretaceous sand sequences in western Canada. In *The Cretaceous System in the Western Interior of North America*, ed. W. G. E. Caldwell, pp. 427–40. Waterloo: Geological Society of Canada.

Stockey, R. A. & Crane, P. R. (1983). *In situ Cercidiphyllum*-like seedlings from the Paleocene of Alberta, Canada. *American Journal of Botany*, **70**, 1564–8.

Stopes, M. C. (1915). *Catalogue of Cretaceous Plants in the British Museum (Natural History)*, Part 2. London: British Museum (Natural History).

Stott, D. F. (1975). The Cretaceous system in northeastern British Columbia. In *The Cretaceous System in the Western Interior of North America*, ed. W. G. E. Caldwell, pp. 441–68. Waterloo: Geological Society of Canada.

Teixeira, C. (1948). *Flora Mesozóica Portuguesa*. Lisbon: Serviços Geológicos de Portugal.

Tiffney, B. H. (1984). Seed size, dispersal syndromes and the rise of the angiosperms: evidence and hypothesis. *Annals of the Missouri Botanical Garden*, **71**, 551–76.

Tschudy, R. H. (1970) Palynology of the Cretaceous–Tertiary boundary in the northern Rocky Mountain and Mississippi Embayment Regions. *Geological Society of America, Special Paper*, 27, 65–111.

Upchurch, G. R. (1984*a*). Cuticular evolution in early Cretaceous angiosperms from the Potomac Group of Virginia and Maryland. *Annals of the Missouri Botanical Garden*, 71, 522–50.

Upchurch, G. R. (1984*b*). Cuticular anatomy of angiosperm leaves from the Lower Cretaceous Potomac Group. I. Zone I leaves. *American Journal of Botany*, 71, 192–202.

Upchurch, G. R. & Doyle, J. A. (1981). Paleoecology of the conifers *Frenelopsis* and *Pseudofrenelopsis* (Cheirolepidiaceae) from the Cretaceous Potomac Group of Maryland and Virginia. In *Geobotany II*, ed. R. Romans, pp. 167–202. New York: Plenum.

Upchurch, G. R., Hickey, L. J. & Niklas, K. J. (1983). Leaves with chloranthoid characters from the Lower Cretaceous Potomac Group of Virginia. *American Journal of Botany*, 70, (5, part 2), 81 (abstract).

Vakhrameev, V. A. (1964). [Jurassic and early Cretaceous floras of Eurasia and the paleofloristic provinces of this period.] In Russian. *Academy of Sciences USSR, Geological Institute, Transactions*, 102, 1–261.

Vakhrameev, V. A. (1971). Development of the Early Cretaceous flora in Siberia. *Geophytology*, 1, 75–83.

Vakhrameev, V. A. (1975). [Main features of global phytogeography in Jurassic and early Cretaceous.] In Russian. *Paleontologicheskiy Zhurnal*, 1975(2), pp. 123–32. (*Paleontological Journal*, 9(2), 247–55.)

Vakhrameev, V. A. (1978). [The climates of the northern hemisphere in the Cretaceous in the light of paleobotanical data.] In Russian. *Paleontologicheskiy Zhurnal*, 1978(2), pp. 3–17. (*Paleontological Journal*, 12(2), 143–54.)

Vakhrameev, V. A. (1981). [Ancient angiosperms and the evolution of the flora in the middle of the Cretaceous period.] In Russian. *Paleontologicheskiy Zhurnal*, 1981(2), pp. 3–14. (*Paleontological Journal*, 15(2), pp. 1–11.)

Vakhrameev, V. A., Dobruskina I. A., Meyen, S. V. & Zaklinskaja, E. D. (1978). *Paläozoische und Mesozoische Floren Eurasiens und die Phytogeographie dieser Zeit*. Jena: Gustav Fischer.

Van Valen, L. (1973). Body size and numbers of plants and animals. *Evolution*, 27, 27–35.

Velenovský, J. & Viniklář, L. (1926). Flora Cretacea Bohemiae. I. *Rozpravy Státního Geologického Ústavu Československé Republiky*, 1, 1–57.

Velenovský, J. & Viniklář, L. (1927). Flora Cretacea Bohemiae. II. *Rozpravy Státního Geologického Ústavu Československé Republiky*, 2, 1–54.

Velenovský, J. & Viniklář, L. (1929). Flora Cretacea Bohemiae, III. *Rozpravy Státního Geologického Ústavu Československé Republiky*, 3, 1–33.

Velenovský, J. & Viniklář, L. (1931). Flora Cretacea Bohemiae, IV. *Rozpravy Státního Geologického Ústavu Československé Republiky*, 5, 1–112.

Walker, J. W., Brenner, G. J. & Walker, A. G. (1983). Winteraceous pollen in the Lower Cretaceous of Israel: early evidence of a magnolialean angiosperm family. *Science*, 220, 1273–5.

Walker, J. W. & Walker, A. G. (1984). Ultrastructure of Lower Cretaceous angiosperm pollen and the origin and early evolution of flowering plants. *Annals of the Missouri Botanical Garden*, 71, 464–521.

Wang, C. W. (1961). *The Forests of China*, Maria Moors Cabot Foundation, Publication 5, Cambridge, Massachusetts.

Ward, L. F. (1900). Status of the Mesozoic floras of the United States. First Paper. The Older Mesozoic. *United States Geological Survey, Twentieth Annual Report*, 2, 211–930.

Ward, L. F. (1905). Status of the Mesozoic floras of the United States. Part I, Text, Part II, Plates. *United States Geological Survey, Monographs* 48, 1–616.

Ward, J. V. (1983). Lower Cretaceous angiospermic pollen from the Cheyenne and Kiowa Formations (Albian) of Kansas, U.S.A. Ph.D. thesis, Arizona State University.

Watson, J. (1983). Two Wealden species of *Equisetum* found *in situ*. *Acta Palaeontologica Polonica*, 28, 265–9.

White, P. S. (1983). Corner's rules in eastern deciduous trees: allometry and its implications for the adaptive architecture of trees. *Bulletin Torrey Botanical Club*, 110, 203–12.

Wieland, G. R. (1906). American fossil cycads. I. Structure. *Carnegie Institute of Washington, Publication*, 34(1), 1–296.

Wolfe, J. A. (1975). Some aspects of plant geography of the northern hemisphere during the late Cretaceous and Tertiary. *Annals of the Missouri Botanical Garden*, 62, 264–79.

Wolfe, J. A. (1977). Paleogene floras from the Gulf of Alaska region. *United States Geological Survey, Professional Paper*, 997, 1–108.

Wolfe, J. A. (1978). A paleobotanical interpretation of Tertiary climates in the Northern Hemisphere. *American Scientist*, 66, 694–703.

Wolfe, J. A., Doyle, J. A. & Page, V. M. (1975). The bases of angiosperm phylogeny: paleobotany. *Annals of the Missouri Botanical Garden*, 62, 801–24.

Zaklinskaya, E. D. (1967). Palynological studies on late Cretaceous-Paleogene floral history and stratigraphy. *Review of Palaeobotany and Palynology* 2, 141–6.

Zhou, Z. (1983). *Stalagma samara*, a new podocarpaceous conifer with monocolpate pollen from the Upper Triassic of Hunan, China. *Palaeontographica B*, 185, 56–78.

6

·

Time of appearance of floral features

E. M. FRIIS AND W. L. CREPET

The presence of flowers and fruits defines the angiospermous condition, and floral structure provides the majority of characters used in angiosperm systematics. The success of the flowering plants and even the origin of flowers themselves may well be related to co-evolution with insect pollinators (Grant, 1950; Baker & Hurd, 1968; Burger, 1981; Stebbins, 1981). Insect pollination, especially pollination involving faithful pollinators may have advantages such as promotion of out-crossing, gametophytic competition and being energetically efficient (Cruden, 1977; Mulcahy, 1979). Most diverse angiosperm taxa are associated with faithful pollinators (Stebbins, 1981), and the most complex vegetation communities have a high ratio of faithfully pollinated taxa to wind-pollinated taxa (Kevan & Baker, 1983). Pollinator fidelity is apparently an important factor in angiosperm diversification (e.g. Grant, 1949; Crepet, 1984). The frequent partitioning of pollinator resources among closely related sympatric species (Armbruster & Herzig, 1984) also supports this view.

Specific floral structures may also confer evolutionary advantages. Pollen-tube growth through carpellary tissue may have promoted the evolution of effective incompatibility mechanisms, which in turn may have given angiosperms the initial evolutionary impetus for diversification. Once evolved, the carpel provided the opportunity for competition at the gametophytic level (Mulcahy, 1979; Endress, 1982) and provided the basic structure that later became modified for a wide diversity of dispersal mechanisms.

Understanding the radiation of angiosperms has been impeded by the lack of palaeontological information germane to the evolution of floral structure. In view of the growing number of hypotheses relating

145

angiosperm success to various features of their reproductive biology, fossil evidence on timing and circumstances of the origins of various floral features is badly needed. Advances in taxonomy, especially cladistic methodology, make the potential reward of a fossil-based chronology of various floral characters even greater.

Studies of flowers from the Cretaceous (Dilcher & Crane, 1984; Friis, 1984, 1985*a*) and Tertiary (Crepet, 1978, 1984; Crepet & Daghlian, 1980, 1981) prove that vital floral data are retrievable from the fossil record. There are now enough data to synthesize information on the historical development of angiosperm reproductive structures and certain important events in the evolution of angiosperms are already evident from these data. In this chapter, we summarize the data and in Chapter 7 we consider some of the implications of this information.

Pre-Albian records

Early Cretaceous angiosperm reproductive organs are scarce and most are compressions or impressions of fruits. About 20 different reproductive organs have been ascribed to the angiosperms from this time interval, but several of these are problematic. Apart from a superficial resemblance to some modern fruits, angiospermous affinities have not been demonstrated conclusively for any of the pre-Albian fossils. *Carpolithus* sp. reported by Chandler (1958) from Valanginian rocks of France and assigned to the angiosperms, is based on a single specimen that is no longer available for study. The fossil is a cast of an external mold and yielded no internal structure. Slightly younger (Hauterivian or Barremian) rocks from Mongolia have yielded several enigmatic reproductive organs interpreted by Krassilov (1982) as angiosperms (*Erenia stenoptera, Gurvanella dictyoptera*) and angiosperm-like (*Cyperacites* sp., *Problematospermum* sp., *Typhaera fusiformis, Potamogeton*-like spikes, *Sparganium*-like heads). All the fossils grouped as angiosperm-like show no evidence of fruit or seed structure and their affinities are uncertain. The two genera assigned to the angiosperms are based on small flattened fossils with central bodies surrounded by two-lobed wings. In *Gurvanella* the wing is thin with conspicuous reticulate nervation, while in *Erenia* the wing has no distinct structure. In *Gurvanella* there is a distinct median ridge that could be interpreted as a septum of a bilocular fruit, but none of the fossils shows definitive angiospermous features. Flattened, bilaterally symmetrical and winged seeds with similar morphology also occur in the conifers and *Welwitschia* (Gnetales), and an affinity with the gymnosperms cannot be excluded entirely.

Nyssidium orientale and *Nyssidium* sp. described from the Barremian flora of Suchan, South Primory, USSR (Samylina, 1961), are based on elliptical impressions about 10 mm long and 5 to 6 mm broad. They are ornamented for most of their length by a number of longitudinal ridges that leave an unsculptured area at one end. The other end of the fossils is slightly pointed, with a transverse groove separating a small pointed protrusion at the very tip. The fossils were referred to the angiosperms and compared by Samylina (1961) with fruits of modern *Nyssa*. However, these differ distinctly from the fossils in having ridges extending the full length of the endocarps and by the presence of germination valves. Seeds with a thick wall and pronounced longitudinal ridges also occur in several gymnospermous groups and *N. orientale* shows some morphological resemblance to seeds of the bennettitalean *Vardekloeftia*. In the latter, the inner integument extends typically beyond the outer integument, forming a small pointed protrusion (Harris, 1932). The pointed end of *Nyssidium orientale* may thus represent the micropylar area; and the unsculptured area a basal attachment scar. A bennettitalean affinity may be supported by the exclusive association of *N. orientale* with bennettitalean and coniferous leaves (Samylina, 1961).

The only other claimed angiosperm reproductive organ from the Barremian is *Onoana californica* described from the marine Horsetown group, California (Chandler & Axelrod, 1961; Hughes, 1976). The species is based on a single specimen that was embedded and cut into two halves. The fossil is about 2 cm long and has a thick outer layer with prominent pits and a thinner, yellowish inner layer. It was originally described as a one-loculed fruit and provisionally referred to the Icacinaceae by Chandler & Axelrod (1961). However, the fossil shows no definite angiospermous features, and may well be a gymnospermous seed with a thick sclerotesta and a well-developed megaspore membrane. Seeds of cycads may be of comparable size and are characterized by having a well-developed megaspore membrane (Crane, this volume, Chapter 5).

Albian records

A few flowers and several unquestionable angiosperm fruits and seeds are known from rocks of Albian age in Asia and North America. Although these fossils are scarce and their preservation is usually too poor to reveal the original floral morphology, it is clear that flowers with free carpels (apocarpous) were common and diverse among the late Early Cretaceous angiosperms. Syncarpy (fusion of carpels) is suggested by the morphology of the ?Middle Albian *Araliaecarpum kolymensis* and has been observed in a small fruit from the Upper Albian Patapsco Formation. Albian apocarpous fruits are elongated follicles with apparent ventral deshiscence or nutlets. The carpels are few (three to eight) or numerous (more than 100). The follicles are loosely spaced or tightly clustered and the receptacle varies from dish-shaped or slightly swollen to elongate. The length of the individual fruitlets ranges from 1 to 40 mm. The number of seeds in each fruitlet varies from one to several. The seeds are small, apparently anatropous and exhibit some variation in wall structure. Although angiosperm seeds with two integuments (bitegmic) have been reported from this time interval (Vakhrameev & Krassilov, 1979), there are no clear indications of integument number in any of the Albian angiosperm seeds. Anthers and pollen have rarely been associated with these reproductive organs, and in most cases the sexuality of the Albian angiosperms cannot be established. However, the structure of *Hyrcantha karatcheensis* (see below) and the presence of pistillate (female) and staminate (male) platanoid flowers indicate that both unisexual and bisexual flowers were present by the Late Albian.

Dispersed stamens obtained from Upper Albian sediments are small, apparently tetrasporangiate, with elongated pollen sacs. Anthers are sessile on a short filament and the connectives are apically expanded. Most late Early Cretaceous angiosperm reproductive organs are of magnoliidean affinity, but platanoid and possibly rosidean flowers and fruits are also of some importance in the North American floras. This is consistent with results from the fossil pollen record (Muller, 1970, 1981, 1984; Doyle & Hickey, 1976). The Aptian–Albian time interval (Muller's Phase I of differentiation) is characterized by the early appearance of Magnoliidae and by a later differentiation of the tricolpate Hamamelididae.

Several Albian angiosperm fruits have been reported from the Soviet Union. The most fully understood are the fruits of *H. karatcheensis*, described from the Middle Albian of Kazakhstan (Vakhrameev, 1952; Krassilov, Shilin & Vakhrameev, 1983). The fruits are apocarpous and formed from three to five elliptical follicles borne on a flattened receptacle (Figure 6.1 (*h*)). The stigmas are apparently sessile and the follicles have split along the ventral margin. There are no seeds in the follicles, but eight or nine transverse grooves on the follicle wall may be interpreted

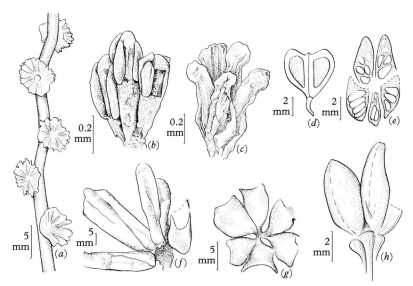

Figure 6.1. Early Cretaceous (Albian) flowers and fruits from North America ((*a*) to (*c*), (*f*) to (*g*)) and East Asia ((*d*) to (*e*), (*h*)). (*a*) Platanoid inflorescences ('*Sparganium' aspensis*); Wyoming. (*b*) Chloranthoid androecium; Maryland. (*c*) Pistillate platanoid flower; Maryland. (*d*) *Araliaecarpum kolymensis*; Kolyma Basin. (*e*) *Ranunculaecarpus quinquecarpellatus*; Kolyma Basin. (*f*) Apocarpous fruit; Virginia. (*g*) *Carpolithus conjugatus*; Virginia. (*h*) *Hyrcantha karatcheensis*; Kazakhstan. ((*d*) to (*e*) After Samylina, 1960; (*f*) to (*g*) after Dilcher, 1979; (*h*) after Krassilov *et al.*, 1983.)

as imprints of seeds. Some specimens show the remains of a perianth and an androecium, suggesting an hermaphroditic condition, with female and male parts in the same flower. According to Krassilov *et al.* (1983) the fruits share features with those of Ranuculaceae and Paeoniaceae. A related form, *Caspiocarpus paniculiger* from the same Middle Albian beds of Kazakhstan (Vakhrameev & Krassilov, 1979) has three to seven follicles apparently spirally arranged on an elongated receptacle. The follicles are split along the ventral margin and each contains two to four seeds, apparently developed from bitegmic and anatropous ovules (Vakhrameev & Krassilov, 1979; Krassilov, 1984). There are no remains of a perianth or of stamens so the organization of the original flower is obscure. Another apocarpous reproductive organ of supposed ranuncu-lalean affinity, *Ranunculaecarpus quinquecarpellatus* (Figure 6.1 (*e*)) from the Kolyma Basin, East Siberia, consists of five free carpels, each containing two to five small and thin-walled seeds (Samylina, 1960). The structure of the receptacle is unknown and no other floral parts are preserved with the fruits. *Araliaecarpum kolymensis* (Figure 6.1(*d*)), described from the same Albian beds of the Kolyma Basin, consists of small, flattened, two-lobed organs borne on a slender stalk (Samylina, 1960). Although the structure of the fossils is not completely understood, their morphology suggests that they were syncarpous, two-loculed fruits formed from two fused carpels, each apparently with a single seed. Three other reproductive organs from the Albian floras of the Kolyma Basin were assigned by Samylina (1960, 1968) to the angiosperms. *Caricopsis compacta* and *C. laxa* are small, elongated axes with loosely packed lateral organs, interpreted by Samylina (1960) as fruits and assigned to the Cyperaceae, but their structure and organization is obscure and their angiospermous affinity uncertain. *Kenella harrisiana* comprises molds and impressions of small ovate fossils with poorly developed longitudinal ridges and with a marginal fringe (Samylina, 1968). The fossils show no evidence of internal structure and their relationships are unclear. No angiosperm leaf fossils have been found in association with *Kenella* (Samylina, 1968) and a gymnospermous affinity cannot be excluded.

Angiosperm flowers, fruits and seeds have been recovered from various Albian localities in North America. Lower Albian rocks of Virginia have yielded two fruits formed from apocarpous gynoecia (Dilcher, 1979). *Carpolithus conjugatus* (Figure 6.1(*g*)) consists of three broad follicles that have dehisced along one margin (Fontaine, 1889; Dilcher, 1979), and a further, unnamed, fossil consists of from six to eight elongated follicles borne on a slightly swollen receptacle (Figure 6.1(*f*)). According to Crane & Dilcher (1984) '*Williamsonia*' *recentior* from the Albian Blairmore Group of Alberta is an apocarpous angiosperm fruit consisting of 150 to 200 tightly packed follicles, apparently spirally arranged on a swollen receptacle. The fossils resemble the Cenomanian fruit *Lesqueria elocata* (see p. 152) also assigned to the Magnoliidae by Crane & Dilcher (1984). Upper Albian rocks from the Potomac group in Maryland have yielded a diverse assemblage of small angiosperm flowers, fruits, seeds and dispersed anthers representing more than 20 different taxa (Friis, Crane & Pedersen, 1986; Crane, Friis & Pedersen, 1986). The plant fossils are lignitized or more rarely charcoalified, three-dimensionally preserved or slightly compressed. Apocarpous taxa are represented by several dispersed follicles and nutlets of probable magnoliidean affinity as well as small inflorescences of hamamelididean and rosidean affinities. A distinctive three-parted androecial structure is the first floral evidence of the early establishment of chloranthoid angiosperms (Friis *et al.*, 1986). The fossil consists of three cylindrical, fleshy stamens laterally fused at their bases. The connectives are slightly expanded apically and the stamens are tetrasporangiate (Figure 6.1(*b*)). Pollen grains found inside the stamens are about 10 μm long, coarsely reticulate, apparently three- or four-colpate, and covered by a thick coating resembling pollenkitt.

Small unisexual platanoid flowers are common in the same fossil flora (Crane *et al.*, 1986). The inflorescences are tightly packed sessile heads borne on an elongated axis. The pistillate (female) flowers are composed of five free carpels surrounded by a number of membranous parts, possibly representing staminodes (abortive stamens), perianth and bracts (Figure 6.1(*c*)). The staminate (male) flowers consist of five stamens surrounded by a number of membranous perianth parts and bracts. The anthers are basifixed on short filaments and the connectives are apically expanded. Pollen grains found inside the anthers are about 8 to 10 μm long, tricolpate and reticulate, similar to dispersed pollen grains described from the Potomac group as *Tricolpites minutus* (Brenner, 1963; Doyle, 1969). Other platanoid inflorescences (Figure 6.1(*a*)) resembling the Potomac fossils were described from the upper Albian Aspen Shale of Wyoming as *Sparganium aspensis* (Brown, 1933).

Cenomanian records

Cenomanian angiosperm reproductive organs are far more abundant and diverse than those of the Albian. Rich angiosperm assemblages with compressions or impressions of fruits and occasional flowers have been described from North America and Europe. Apocarpous gynoecia of magnoliidean affinity continued to be present in a significant number in the Cenomanian floras, but platanoid flowers and other small, apparently unisexual flowers crowded in catkin-like inflorescences were also of some importance. Particularly significant is the appearance of new floral types with a cyclic arrangement of floral parts and flowers with syncarpous gynoecia. The differentiation of the perianth into a distinct calyx and corolla was apparently established in both apocarpous and syncarpous flowers. Apocarpous reproductive organs are all preserved in the fruiting stage. Several show distinct spirally arranged carpels, and in some fossils a spiral arrangement of other floral parts is indicated by scars of abscissed parts at the bases of the receptacles. The length of the individual follicles ranges from 10 to 90 mm, and the length of the entire fruits from 20 to 140 mm. Flowers with cyclically arranged parts were mostly pentamerous (Figure 6.2(e)), but tetramerous and hexamerous forms were also present. Cenomanian syncarpous fruits were five-loculed capsules, except for a few taxa that were apparently trimerous. In the Early Cenomanian flowers, the ovaries were apparently superior, while flowers with inferior ovaries were established by the Middle Cenomanian.

The fossil pollen record shows increasing diversity of form during the Cenomanian (Muller's Phase II of differentiation) with further diversification of Aptian–Albian pollen types and the establishment of new tricolporate, triporate and periporate forms. Cenomanian angiosperm pollen is generally small with simple sculpture (Muller, 1970, 1981).

The Dakota Formation of the Western Interior of North America has rich fossil floras with well-preserved angiosperm reproductive organs (Dilcher, 1979; Retallack & Dilcher, 1981; Basinger & Dilcher, 1984; Crane & Dilcher, 1984; Dilcher & Crane, 1984). The plant-bearing strata of the Dakota Formation are generally accepted as being of Early Cenomanian or possibly late Late Albian age. Several multi-carpellate fruits composed of densely clustered follicles have been described and, of these, *Archaeanthus linnenbergeri* best reveals the original floral morphology (Dilcher & Crane, 1984). The fruits are composed of numerous spirally arranged follicles borne on a long stout receptacle. The follicles are stalked and dehisced along the ventral margin. The stigmatic area is apparently decurrent and extends for the full length of the carpel. Each carpel produced about 100 ovules, but apparently only 20 developed into mature seeds. Immediately below the follicles, the receptacle shows a discrete zone of various-sized scars, indicating that the fruit developed from a hermaphroditic flower with a perianth differentiated into a distinct calyx and corolla. The calyx probably consisted of three robust sepals and the corolla of six to nine more delicate petals. The androecium apparently consisted of numerous spirally arranged stamens. The structure of *A. linnenbergeri* indicates a close relationship with some modern members of the Magnoliidae, especially within the Magnoliales (Dilcher & Crane, 1984). The fruits of *Lesqueria elocata* are more compact than those of *A. linnenbergeri*, with the follicles tightly packed on a short cone-shaped receptacle. The structure of the perianth and that of the androecium are less clear. The fruits are attached to an elongated, cylindrical axis that bears numerous spirally arranged laminar appendages below the gynoecium. Crane & Dilcher (1984) interpreted the cylindrical axis as part of the receptacle and the laminar appendages as the remains of floral parts. Similar fossil fruits have been reported from several Cenomanian localities in Kansas, Texas and New

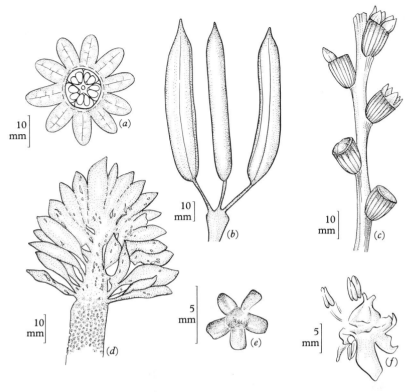

Figure 6.2. Late Cretaceous (Cenomanian) flowers and fruits from Czechoslovakia ((*a*) to (*c*)) and North America ((*d*) to (*f*)). (*a*) *Kalinaia decatepala*. (*b*) *Triplicarpus purkynei*. (*c*) '*Leptospermum*' *macrocarpum*. (*d*) Magnolialean multicarpellate fruit; Kansas. (*e*) Five-lobed calyx (*Calycites parvus*); New Jersey. (*f*) Pentamerous, bisexual flowers; Kansas. ((*a*) to (*b*) After Velenovský & Viniklář, 1926; (*c*) after Velenovský, 1889; (*d*) after Crane & Dilcher, 1984; (*e*) after Dilcher, 1979; (*f*) after Basinger & Dilcher, 1984.)

Jersey (Figure 6.2(*d*)) (Dilcher, 1979; Crane & Dilcher, 1984; Dilcher & Crane, 1984). An additional reproductive organ from the Dakota Formation is *Prisca reynoldsii*, described by Retallack & Dilcher (1981) as an apocarpous fruit formed from numerous loosely packed follicles spirally arranged on a long, slender receptacle. However, the interpretation of this fossil as a flower rather than as an inflorescence is equivocal. The material requires re-examination aimed particularly at clarifying the homologies and establishing more firmly the details of the seeds.

In addition to apocarpous fruits, the Early Cenomanian of North

America has also yielded a number of fossil floral structures with syncarpous ovaries. The most complete are flowers and fruits from the Dakota Formation of Rose Creek, Nebraska (Basinger & Dilcher, 1984). The flowers are bisexual and actinomorphic, with the floral parts in whorls of five (Figure 6.2(f)). The perianth is hypogynous and differentiated into calyx and corolla. The sepals are free and persistent, and the petals (usually abscissed from the fossils) are also free. There is apparently a single whorl of stamens opposite the petals, with anthers that dehisced by longitudinal slits and produced numerous, minute tricolporate pollen grains. The ovary is formed from five fused carpels and the fruits are apparently loculicidal capsules. There are five free, short styles. A slightly swollen area at the base of the ovary has been interpreted as a nectary (Basinger & Dilcher, 1984).

Several angiosperm reproductive organs have also been reported from rich Middle Cenomanian floras in Czechoslovakia (Velenovský, 1889; Bayer, 1914; Velenovský & Viniklář, 1926, 1927, 1929, 1931; Pacltová, 1977). Apocarpous fruits are less common in these floras but are represented by platanoid heads and *Triplicarpus purkynei* (Figure 6.2(*b*)). The identification of *T. purkynei* is based on a single fruit composed of three large (90 mm × 15 mm) stalked follicles borne on a dish-shaped receptacle. It was compared to modern members of the Magnoliidae (Velenovský & Viniklář, 1926). The considerable size of the follicles is remarkable because most Cretaceous angiosperm fruits and seeds are small, with larger fruits and seeds not becoming common until the end of the Cretaceous.

Several syncarpous, capsular fruits have been described from the Middle Cenomanian of Czechoslovakia. They comprise three- and five-loculed types formed from superior or inferior ovaries. *Kalinaia decatepala* (syn: *Rhizocarpus decapetalus*) (Figure 6.2(*a*)) includes five-loculed fruits, about 15 mm in diameter, with a persistent, epigynous ten-lobed perianth (Bayer, 1914; Velenovský & Viniklář, 1926). *Asterocelastrus cretaceus* is a smaller (8 to 10 mm in diameter), five-loculed fruit with a persistent and epigynous calyx (Velenovský & Viniklář, 1926). In both *Kalinaia* and *Asterocelastrus*, the locules apparently contained two seeds borne on the central axis. The fossil fruits described as *Leptospermum macrocarpum* (Figure 6.2(*c*)) are three-loculed with a persistent epigynous and five-lobed calyx (Velenovský, 1889). Two fruits (*Rutaecarpus quadrilobus, Ceratocarpus fendrychii*) are three-loculed and apparently developed from flowers with superior ovaries (Velenovský & Viniklář, 1926, 1931).

Several small catkins were also reported from the Cenomanian floras of Czechoslovakia, assigned to the Myricaceae. Few details of their floral structure are known and their organization is uncertain.

Turonian–Campanian records

Few angiosperm flowers and fruits are known from the early part of this time interval (Turonian and Coniacian). The Santonian–Campanian record of angiosperm reproductive organs is, on the other hand, extensive, and has provided evidence of considerable morphological and organizational diversity. Several rich angiosperm assemblages including many flowers, fruits and seeds have been recovered from Asia, Europe and North America.

Apocarpy was retained in a smaller group of flowers, while taxa with syncarpous gynoecia diversified considerably during the Late Cretaceous and comprise most of the fossil flowers from the Santonian–Campanian. The flowers are generally radially symmetrical (actinomorphic), but the occurrence of zingiberaceous seeds in the Campanian provides indirect evidence that bilateral symmetrical (zygomorphic) flowers may have developed by this time. The cyclic flowers include taxa with two whorls of perianth parts differentiated into a calyx and corolla (heterochlamydeous) and taxa with a single whorl of perianth parts (monochlamydeous). Further reproductive specialization is suggested by the first appearance of sympetalous flowers, fusion of stylar tissue to form a single style, and the development of various nectar-producing tissues. Stamens are in one or two whorls and the anthers are usually tetrasporangiate with longitudinal dehiscence, but more specialized forms were also present. One has bisporangiate anthers dehiscing by valves and another has partly fused stamens that form a laminar structure. Pollen grains have been recovered from several Santonian–Campanian flowers. Those from heterochlamydeous flowers are minute, about 6 to 10 μm in length, tricolpate or tricolporate often with a coating, resembling pollenkitt of modern flowers, that apparently held the pollen grains together, an important adaptation in insect-dispersed pollen. In the monochlamydeous flowers, the pollen grains are larger, about 14 to 20 μm in diameter, triporate, with almost smooth surfaces and apparently without any coating.

The syncarpous gynoecia of Santonian–Campanian angiosperms show a considerable diversity in structure. In Cenomanian angiosperms, the syncarpous gynoecia were formed from three or five carpels, while the

Santonian–Campanian record reveals a dominance of taxa with gynoecia formed from two or three carpels. Capsular fruits were still common among Santonian–Campanian forms, but several other fruit types, including nuts and drupes, had appeared by this time. While most of the Santonian–Campanian flowers apparently produced numerous ovules per ovary, the monochlamydeous taxa developed only a single or few ovules per ovary. The seeds were commonly anatropous, but orthotropous, campylotropous and amphitropous seed types had also developed. Seed coat structure suggests that both unitegmic and bitegmic forms were present, and that they may have been equally important.

A remarkable feature of Santonian–Campanian angiosperms was the apparent dominance of those taxa having epigynous flowers and generally minute flowers, fruits and seeds. Except for an apocarpous fruit from Japan that is about 20 mm × 25 mm in diameter (Nishida, 1985), the flowers and fruits are usually about 1 to 4 mm long, and do not exceed 10 mm (Friis, 1984, 1985a).

The reproductive structures from the Turonian–Campanian include taxa comparable to modern members of the Magnoliidae, Hamamelididae, Caryophyllidae, Dilleniidae, Rosidae, and the Liliidae. A large number are of rosidean affinity and the Saxifragales are especially well represented. Fossils related to the Hamamelididae (Hamamelidales, Fagales and Myricales/Juglandales) are also important among angiosperms from this time interval. The assemblage revealed by floral structures is consistent with the fossil pollen record (Muller, 1970, 1981). The Turonian (Muller's Phase IIIa of differentiation) is characterized by the appearance of the Hamamelididae and Rosidae, and during the Coniacian–Campanian time interval (Muller's Phase IIIb of differentiation) there is a marked diversification of these two groups, with the establishment of the Fagales, Betulales and Juglandales.

Upper Cretaceous floras from Western Greenland that may be of Turonian–Campanian age (Pedersen, 1976) have yielded a few angiosperm reproductive organs (*Magnoliaestrobus gilmourii*, *Platanus* sp., *Leptospermites arcticus*; Seward & Conway, 1935). They are poorly preserved and their general organization does not indicate any major innovations in floral morphology.

The fossil flora from Scania, southern Sweden, has probably yielded the most diverse and best-preserved angiosperm assemblage known from the Santonian–Campanian (Friis & Skarby, 1981, 1982; Friis, 1983, 1984, 1985a, b). The fossils are mostly preserved as three-dimensional charcoalifications and include more than 100 different taxa of angio-

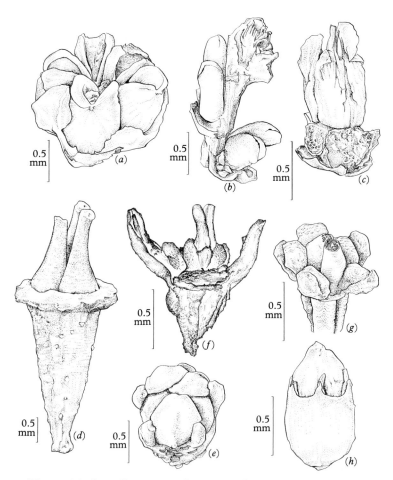

Figure 6.3. Late Cretaceous (Santonian–Campanian) flowers and fruits from Scania, Sweden. (*a*) Apocarpous fruit of five whorled carpels. (*b*) Apocarpous fruits of two carpels, borne in a spike. (*c*) Sympetalous flower (*Actinocalyx bohrii*). (*d*) to (*e*) Epigynous saxifragalean flower with three stout styles; mature flower (*d*) and flower bud (*e*). (*f*) *Scandianthus costatus* with ten-lobed nectary. (*g*) Epigynous saxifragalean flower with five-lobed nectary. (*h*) *Antiquocarya verruculosa* with inconspicuous perianth.

sperm reproductive structures. Most of the fossils are fruits and seeds, but flowers are also relatively common. Each taxon is usually represented by many specimens, often representing different developmental stages (Figures 6.3(*d*), (*e*)).

The Scanian material comprises relatively few apocarpous flowers and fruits. One form has five cyclically arranged follicles (Figure 6.3(a)), subtended by a persistent calyx of five partly fused sepals. The stigmas are globular and sessile, with minute, tricolpate pollen grains, about 6 μm long, adhering to the surface. The pollen has a thick coating that may represent stigmatic mucilage or pollenkitt, as is typical in plants with entomophilous flowers (Hesse, 1981, and personal communication). Other apocarpous floral remains consists of small bicarpellate fruits borne on a spike (Figure 6.3(b)). The fruitlets are follicles with a stigmatic area that is decurrent from the apical part of the carpels. Well-preserved pistillate and staminate platanoid inflorescences also occur among the Scania fossils. They are identical with the Lower Cretaceous platanoid fossils (see p. 151) in having carpels and stamens in fives, but the male flowers differ in having thick, conspicuous perianth parts and a distinct cap-like extension of the connective (Friis, 1985a). The Cretaceous platanoid fossils are distinguished from modern *Platanus* by the number of parts and lack of hairs and they probably represent an extinct lineage in the Platanaceae.

Syncarpy is the most common gynoecial condition among the Scanian fossils and most syncarpous flowers have a distinct calyx and corolla, while a smaller group has a single whorl of perianth parts. About one-third of the flowers have superior ovaries, only one taxon has a semi-inferior ovary, and all other flowers have inferior ovaries. The character combinations observed in the heterochlamydeous flowers suggest that they were pollinated by insects. The flowers of *Scandianthus costatus* and *S. major* are typical examples of the entomophilous flowers from the Santonian–Campanian of Scania (Friis & Skarby, 1982). They are small, about 1.0 to 2.5 mm long, actinomorphic and bisexual, with a shallow, dish-to-bowl-shaped perianth. The perianth and androecium are in whorls of five, while the gynoecium is two-carpellate. The perianth parts are free, and the calyx is persistent. The stamens are in two whorls with small, tetrasporangiate anthers that produced numerous minute tricolporate and almost smooth pollen grains coated with a thick pollenkitt-like substance. The ovary is inferior and unilocular with numerous small ovules. There are two stout styles, and the fruit is a capsule that opened apically between the styles. A distinct ten-lobed nectary disk is inserted between the androecium and gynoecium (Figure 6.3(f)). The surface of the nectary bears stomata-like openings that are mainly distributed on the apical and inner surfaces. Several similar, apparently related, flowers have also been recovered from the fossil flora of Scania. One has a distinct five-lobed nectary disk with nectariferous tissue also at the bases of the styles (Figure 6.3(g)). Another flower has three stout styles (Figure 6.3(d)).

While most of the heterochlamydeous flowers from Scania have free petals, the flowers of *Actinocalyx bohrii* (Figure 6.3(c)) provide the earliest known occurrence of sympetally (Friis, 1985b). The flowers are small, about 1 mm long and 0.6 mm in diameter. They are hypogynous, with parts in whorls of five except for the gynoecium, which is three-carpellate. The sepals are free and the petals are partly fused to form a tubular corolla that constricts to a narrow throat below the lobes. The stamens are in one whorl and pollen grains are minute (7 to 10 µm), tricolporate and almost smooth with a thick pollenkitt-like coating. The fruit is a three-locular capsule, with basal axile placentation.

The monochlamydeous flowers from Scania all show character combinations indicative of wind pollination. Three genera (*Manningia, Antiquocarya, Caryanthus*) have been recognized and all are small bisexual flowers with simple, epigynous perianth parts (Friis, 1983).

The perianth parts in *Antiquocarya* are very small (Figure 6.3(h)) or absent. The ovary is unilocular, with a single basal, orthotropous ovule, and is composed of two (*Caryanthus*) or three (*Manningia, Antiquocarya*) carpels. The perianth and androecium in *Manningia* are in whorls of five, but in whorls of six in *Antiquocarya*. In *Caryanthus* there are four sepals in two decussate pairs and an androecium of six stamens. Pollen in each of these flowers is triporate, oblate, almost smooth, and assignable to the Normapolles group.

A number of fossil floras with small angiosperm fruits and seeds and occasional flowers or fruits with persistent perianth parts are recorded from Santonian–Campanian rocks of Czechoslovakia and Germany (Vangerow, 1954; Knobloch, 1964, 1971; Knobloch & Mai, 1983, 1986). Although information on floral structure is not extensive, the fossils show concordance in general organization with the Scanian taxa. This is also true for angiosperm reproductive organs recovered from Turonian–Campanian rocks of east and southeast North America. One flora from the Turonian Tuscaloosa Formation comprises two undescribed flowers (material in the Smithsonian Institution and Field Museum), with semi-inferior to inferior ovaries. Rich Campanian angiosperm assemblages including flowers, fruits and seeds have been recovered from the Martha's Vineyard locality, Massachusetts (Tiffney, 1977, and personal communication), New Jersey (Tiffney, personal communication) and from the Neuse River locality, North Carolina (Friis, 1985a, 1987). A zingiberaceous seed referred to the extinct genus *Spirematospermum* has been described from the Neuse River locality (Friis, 1987), but most of the fossils are undescribed.

Other reports of Late Cretaceous reproductive remains include permineralized angiosperm fructifications from the Santonian of Hokkaido, Japan. One is a globular apocarpous fruit, about 20 to 25 mm in diameter, formed from more than 40 carpels on a dish-shaped receptacle (Nishida, 1985). The other, *Cretovarium japonicum*, comprises small three-loculed fruits, about 2 mm in diameter, apparently formed from superior ovaries (Stopes & Fujii, 1910). Further, two angiosperm reproductive structures have been described by Krassilov *et al.* (1983) from the Santonian–Campanian of Kazakhstan based on impression fossils. *Sarysua pomona* includes a number of incompletely preserved flowers and fruits with persistent calyces. They are apparently pentamerous with a five-loculed, superior ovary and five short styles. The structure of the other angiosperm reproductive organ from Kazakhstan, *Taldysaja medusa*, is unclear.

Maastrichtian records

Maastrichtian angiosperm reproductive organs are known from Europe, East Asia, India, Africa and Mexico. The material from India, Africa and Mexico constitutes the only Cretaceous evidence of floral morphology at low palaeolatitudes. The Maastrichtian fossils from Europe and Asia are sparse and do not differ from the Santonian–Campanian material in general organization and size (Krassilov, 1973; Knobloch, 1975; Jung, Schleich & Kästle, 1978; Knobloch & Mai, 1983), but those from the lower palaeolatitude localities record the first appearance of several reproductive features. Some of these may reflect differences in climate rather than in post-Campanian angiosperm evolution. There is, however, consistency between the taxonomic diversity of the reproductive organs and that demonstrated from the fossil pollen record (Muller, 1970, 1981).

While most Cretaceous angiosperm flowers and fruits have a dicotyledonous organization, several new trimerous floral types that are related to the monocotyledons appeared by the Maastrichtian, even though the fossil record of other organs suggests an earlier appearance for this group (Daghlian, 1981; Friis, 1987). Maastrichtian flowers are generally actinomorphic, but one taxon, *Raoanthus intertrappea*, exhibits slight zygomorphy, and zygomorphy is also indicated by the presence of *Musa*-like fruits (Zingiberales; Jain, 1963). Ovaries with secondary partitions of the locules were established in some flowers, and the earliest appearance of berries is documented. Most Maastrichtian fossil flowers are small, about 2.5 to 10 mm long, but several relatively large fruits and seeds, up to 100 mm long, had evolved before the end of the Cretaceous.

The fossil flowers and fruits from the Maastrichtian include taxa related to most of the major plant groups, except for the Asteridae, and fossils related to the Myrtanae and the Arecidae are especially well represented. This is consistent with the results obtained from the fossil pollen record (Muller, 1970, 1981). In this time interval (Muller's Phase IV of differentiation), angiosperm diversification reached a maximum, with a major radiation of the Arecaceae (palms).

The Intertrappean beds of Mohgoan-kalan in India have yielded a number of flowers and fruits preserved as petrifactions in chert. These beds were previously believed to be of Early Tertiary age (Paleocene or Eocene) based on their content of fossil plants (Sahni, 1934). The presence of dinosaur fragments in the Intertrappean beds, however, suggests that the beds are of Maastrichtian age (Sahni, Rana & Prasad, 1984). The fossil flowers of *Sahnianthus* and its allied fruits, *Enigmocarpon parijai*, form a conspicuous part of the petrified flora (Sahni, 1943; Shukla, 1944; Chitaley, 1977). The flowers are actinomorphic and hermaphroditic, with a superior ovary. They are apparently monochlamydeous, with eight connate sepals and with eight stamens. The ovary is six- to eight-loculed, with one stout style and a capitate stigma. The placentation is axile, with two rows of ovules in each locule. A nectary is present as a single scale in *S. parijai* (Shukla, 1944) or as two ovate stalked bodies in *S. dinectrianum* (Shukla, 1958). The fossils were compared to modern members of the Myrtales. According to Shukla (1944), the flowers were heterostylous, but variations in the length of styles and stamens are probably due to ontogenetic differences (Chitaley, 1955). Three other floral structures related to the Myrtales have been described from the Intertrappean beds. *Sahnipushpam glandulosum* is apparently pentamerous and monochlamydeous, with united sepals, and has a five-loculed ovary with secondary partitions. The stigma is discoid, covering most of the flower (Prakash, 1956; Chitaley, 1964; Prakash & Jain, 1964). *Chitaleypushpam* has seven connate sepals in one whorl and stamens fused to the base of the perianth (Paradkar, 1973). *Raoanthus*, like the other presumed myrtalean flowers, has one whorl of fused perianth parts, but it differs in being slightly zygomorphic. It has nine sepals and stamens and seven fused carpels (Chitaley & Patel, 1975). A few other dicotyledonous reproductive organs and several monocotyledonous flowers and fruits were described. *Deccananthus, Nipa,*

Tricoccites, *Palmocarpon* and *Viracarpon* are all trimerous and related to the Arecidae (Rode, 1933; Sahni, 1934; Prakash, 1954; Chitaley & Nambudiri, 1969; Chitaley & Kate, 1974).

An arecaceous infructescence resembling modern *Manicaria* has been reported from the Lower Maastrichtian Olmos Formation in Mexico (Weber, 1978). The infructescence is about 25 cm long and bears numerous three-valved fruits.

Several large fruits and seeds have been described from the late Senonian (Late Campanian or Maastrichtian) of Sénégal (Monteillet & Lappartient, 1981). They are apparently preserved as molds with no internal structures preserved. They show affinity to modern tropical plants of the Annonaceae, Caesalpinioideae, Meliaceae, Rubiaceae, Sterculiaceae and Arecaceae. Fossil seeds related to the Annonaceae have also been described from the Maastrichtian of Nigeria, together with seeds possibly related to the Icacinaceae and Passifloraceae (Chesters, 1955).

Paleocene–Oligocene records

Information on Early Tertiary angiosperm reproductive organs is extensive, based mainly on rich assemblages of fruit and seeds recovered from many parts of the world, but also on well-preserved flowers. During this time interval, specialization in many major features of reproductive biology reached a modern level. A significant innovation in Early Tertiary angiosperms is the establishment of many wind-pollinated unisexual floral types of hamamelididean/rosidean affinity. Unisexuality in insect-pollinated flowers was also established by the Early Tertiary. Adaptation to more specialized biotic pollinators is indicated by the development of strongly zygomorphic flowers and flowers with deep, funnel-shaped to tubular corollas, androecia with variously fused stamens (columnar and monadelphous types) and sterile ray flowers. Further specialization in dispersal mechanisms is indicated by strong diversification of fleshy fruit types and also by differentiation in dry fruit types, with the development of several distinct types of winged fruits and schizocarps. The size range of the fossil flowers, fruits and seeds is considerable.

The major evolutionary groups established by the end of the Cretaceous diversified further, and many modern angiosperm families and genera were established during the Early Tertiary. This also is consistent with the fossil pollen record (Muller, 1970, 1981). Muller's phase V of differentiation (Paleocene–Eocene) is characterized by the large number of pollen types that can be attributed convincingly to modern families and by an increase of pollen types referable to modern genera.

Late Paleocene flowers from western Tennessee exhibit a variety of innovations important in advanced insect pollination (Crepet & Taylor, 1986). Mimosoid brush inflorescences are well developed for lepidopteran or bee pollination (Figure 6.4(*a*)). They are more primitive than younger mimosoid blossoms (Crepet & Dilcher, 1977) and conform to the mimosoid floral archetype that has been suggested by Elias (1981) from studies of extant taxa. Tubular stigmas, less than 0.1 mm in diameter, are well preserved and packed with one kind of pollen (Figure 6.4(*b*)). These suggest pollination by a faithful pollinator with a single kind of pollen in the appropriate spot on its body (Crepet & Taylor, 1986).

Papilionoid blossoms from the same Paleocene locality (Figure 6.4(*c*)) constitute the first distinctly zygomorphic flower in the fossil record (Crepet & Taylor, 1985). Nevertheless, they are modern in aspect and even exhibit sculptured wing petals associated with bee pollination in extant papilionoids. Other probable papilionoid flowers from the same locality reveal that connate stamen filaments had also evolved by the Paleocene.

Inflorescences of an advanced, now bee-pollinated, tribe of the Euphorbiaceae (Hippomaneae) from the same deposits provide the first floral evidence of unisexuality in insect-pollinated flowers. Inflorescences of the same tribe are also known from an Eocene locality in western Tennessee (Crepet & Daghlian, 1981). Pseudanthia also appear in the Paleocene fossil inflorescences assignable to the tribe Euphorbieae of the Euphorbiaceae (Crepet & Taylor, unpublished). The pseudanthia are sessile on an elongate axis and each has a whorl of bracts subtending an apparently tricarpellate pistillate flower with fused style and a trilobed stigma (Figure 6.4(*d*)). There are at least three staminate florets in a pseudanthium and locules of the anthers are somewhat reflexed and widely separated on an expanded connective. The tricolporate pollen is typical of the Euphorbieae (Webster, 1975).

The Middle Eocene floras of Tennessee have also provided significant information on floral structure in Early Tertiary angiosperms (Crepet, 1978). Zygomorphic flowers are also present in these floras (e.g. Zingiberaceae), as are brush-flowered mimosoid inflorescences. The first flowers with relatively long funnel-shaped corollas and narrow corolla

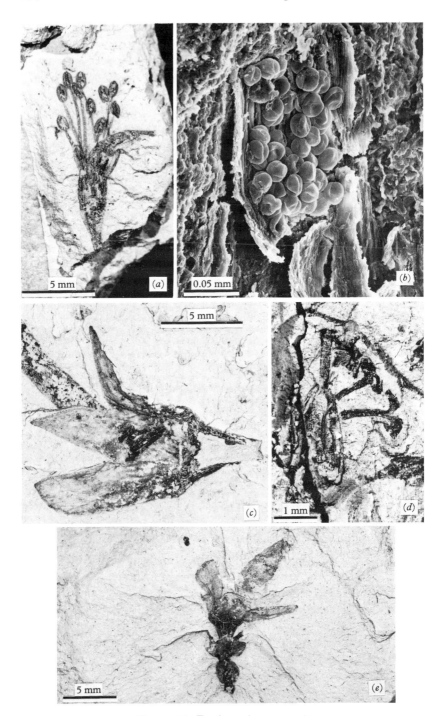

Figure 6.4. For legend see opposite.

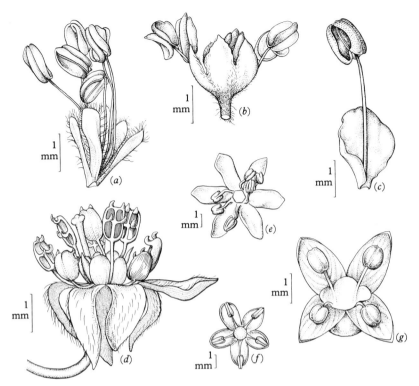

Figure 6.5. Early Tertiary flowers from the Baltic Amber. (*a*) *Quercus meyeriana*, staminate flower. (*b*) *Quercus taeniatopilosa*, staminate flower. (*c*) *Forskohleanthium nudum*, staminate flower. (*d*) *Cinnamomum prototypum*, with valvate anthers. (*e*) to (*g*) Isolated corollas with stamens attached; *Sambucus succinea* (*e*), *Berendtia primuloides* (*f*), *Myrsinopsis succinea* (*g*). ((*a*) to (*g*) after Conwentz, 1886.)

tubes are found in these deposits (Figure 6.4(*e*)). Flowers of the same age, with shorter funnel-shaped corollas and wider corolla tubes, are known from the Eocene Green River Formation (Crepet, 1984).

A further important source of information on Tertiary fossil flowers

---

Figure 6.4 (*a*) to (*d*) Paleocene flowers from southeastern North America. (*a*) *Protomimosoidea buchananensis*, the earliest evidence of Mimosoideae. (*b*) Scanning electron micrograph of the tubular stigma of *P. buchananensis* illustrating that it is full of tricolporate pollen (see Crepet & Taylor, 1986). (*c*) A papilionoid legume flower; the earliest fossil evidence of the Papilionoideae. (*d*) Pseudanthium of Euphorbieae. (*e*) Eocene flower with a narrow tubular corolla.

is the Baltic Amber, which in its original state of free-flowing resin
incorporated many small plant structures as well as insects. The Baltic
Amber is believed to have been formed over a considerable period in the
Early Tertiary. According to Larsson (1978), the amber resin production
in the Baltic area ceased by the beginning of the Oligocene, and the
richest amber deposits in the Baltic area, including the blue earth, are
of Late Eocene age (Piwocki *et al.*, 1985). The flowers included in the
amber are three-dimensionally preserved, often with delicate parts
intact. A detailed survey of the flora was given by Conwentz (1886). The
material is characterized by a high proportion of unisexual flowers with
a simple perianth in which the Fagaceae is especially well-represented
by many species of *Quercus, Castanea,* and *Fagus* (Figure 6.5(*a*) to (*c*)).
Trimerous lauraceous flowers are also a significant element in the amber
flora. There are several examples of valvate anther dehiscence and
Cinnamomum prototypum (Figure 6.5(*d*)) is probably the best known of
the amber flowers (Conwentz, 1886). The amber flora also contains a
considerable number of isolated corollas with more or less connate petals
(Figure 6.5(*e*) to (*g*)).

Chronological appearance of major floral features

Although information on fossil floral structure is relatively restricted, a
picture concordant with that provided from the study of dispersed fossil
pollen emerges of increasing morphological advancement. The major
trends in floral evolution are summarized in the following paragraphs
and in Figures 6.6 and 6.7, on the basis of the evidence from fossil
reproductive organs reviewed above. A full discussion of the evolutionary
and functional significance of the various characters is presented in
Chapter 7 (this volume).

Phyllotaxy and number of floral parts. The phyllotaxy of most Albian
floral structures is obscure, but a few forms show evidence of spiral
arrangement of parts. By the Early Cenomanian, the three major types
of phyllotaxy in the angiosperm flower were established: acyclic flowers
with all parts spirally arranged (*Lesqueria elocata*), hemicyclic flowers
with the perianth parts in whorls, or spiral pseudowhorls, and the
stamens and carpels spirally arranged (*Archaeanthus linnenbergeri*), cyclic
flowers with all floral parts arranged in whorls (Rose Creek flower;
Basinger & Dilcher, 1984). While acyclic and hemicyclic flowers were
apparently widespread in Early and Middle Cenomanian angiosperms,

Figure 6.6. Time of appearance of major floral types. (*a*) Small simple flowers with few floral parts (chloranthoid), may comprise both bisexual and unisexual forms. (*b*) Acyclic or hemicyclic flowers with numerous parts. (*c*) Small monochlamydeous, unisexual flowers. (*d*) Cyclic, heterochlamydeous and actinomorphic flowers. (*e*) Epigynous and heterochlamydeous flowers. (*f*) Sympetalous flowers. (*g*) Epigynous and monochlamydeous flowers. (*h*) Zygomorphic flowers. (*i*) Brush-type flowers. (*j*) Papilionoid flowers. (*k*) Deep funnel-shaped flowers. Solid lines are based on direct floral evidence, while dashed lines are based on other fossil evidence, mostly pollen.

their importance decreased as the cyclic flowers diversified. The dominance of cyclic flowers is well documented in Santonian–Campanian fossil floras, but the fossil pollen record suggests that it was probably already established by the Late Cenomanian.

Information on number of floral parts in the Albian reproductive structures is scarce. Carpel numbers range from a few (three to eight) to numerous (more than 100). Numbers of stamens are three or five, but are known only from two floral structures. Polymerous, acyclic and hemicyclic flowers, with an indefinite number of floral parts, were apparently important in the Cenomanian. Cyclic flowers are mostly with parts in fives, but flowers with parts in fours or in sixes were apparently also present by the Cenomanian. The Cenomanian floral material is scant, but it appears that the earliest cyclic flowers were isomerous, with an equal number of floral parts in all whorls. Heteromerous forms, with a smaller number of parts in the gynoecium, became established by the mid-Cenomanian. By the Santonian–Campanian heteromerous flower types were dominant and usually with the perianth and androecium in whorls of five, and a gynoecium of two or three carpels. By the Maastrichtian, true trimerous floral types were established and relatively common.

Perianth. Information on perianth parts in the Albian angiosperm flowers is scarce. The small platanoid flowers have a number of undifferentiated perianth parts not clearly arranged in whorls. By the Early Cenomanian, flower types with distinct calyx and corolla were well established. Several flower types with a single whorl of undifferentiated perianth parts are known from the Santonian–Campanian, and apparently naked flowers are also known from this time interval. Not all floral parts leave distinct scars when shed, and it may be extremely difficult to demonstrate whether the lack of floral parts in a fossil flower reflects the original structure of the flower or is due merely to the state of preservation.

Information on aestivation in the fossil flowers is generally scarce. In Santonian–Campanian flowers imbricate, open and decussate types have so far been observed. The earliest floral evidence of contorted aestivation is provided by small ebenaceous flowers from the Late Eocene of Australia (Christophel & Basinger, 1982).

Floral symmetry and fusion of parts. Albian and Cenomanian flowers were apparently all actinomorphic, with radial floral symmetry and free perianth parts. By the Santonian–Campanian bisymmetric forms with two planes of symmetry had developed. The occurrence of zygomorphic flowers with one plane of symmetry has been demonstrated from the Maastrichtian (*Raoanthus*), but indirect evidence from zingiberaceous seeds suggests that zygomorphy may have been established as early as the Campanian. In *Raoanthus* the monosymmetry is only weakly developed. Distinct zygomorphic differentiation of floral parts first occurs in Late Paleocene papilionoid flowers that have a keel, wing petals and a standard (Crepet & Taylor, 1985). The first sympetalous flowers are from the Santonian–Campanian and several sympetalous flowers have been recorded from the Maastrichtian. While Cretaceous sympetalous flowers were generally shallow, deep, as well as open, funnel-shaped forms were established by the Paleocene–Early Eocene.

Position of floral parts. The structure of the fossil fruits and flowers from the Albian and Early Cenomanian indicate that early angiosperm flowers were hypogynous. Epigyny was, however, well established as early as the mid-Cenomanian. The radiation of epigynous flowers apparently reached a maximum in the Santonian–Campanian, and epigyny is present in about two-thirds of all recorded floral structures of that age. The proportion of epigynous flowers decreased considerably during the Early Tertiary. In living angiosperms, epigyny is present in about one-fourth of all families (Grant, 1950).

Androecium. The two Albian androecia so far described comprise one form with three stamens fused at their base and one form with five stamens in a unisexual flower. By the Early Cenomanian, multistaminate, spirally arranged stamens and cyclic androecia were both present. In the Rose Creek flower of Basinger & Dilcher (1984), the stamens are arranged in a single whorl opposite the petals. This arrangement is generally considered as relatively advanced and may suggest that androecia with one whorl of stamens opposite the sepals were also present in the Cenomanian as well as androecia formed from two whorls of stamens. By the Santonian–Campanian, these forms were well-established. Cretaceous stamens usually have free filaments. The only exceptions are the dispersed chloranthoid androecia from the Albian of North America and the Santonian–Campanian of Sweden. A staminal column was described for the Late Paleocene *Sezannella* of France (Viguier, 1908) and the monadelphous androecium is known in Paleocene legumes. The anthers of Cenomanian to Campanian flowers were usually free from the perianth, and only in *Actinocalyx* from the Santonian–Campanian are the stamens fused to the corolla. Several Maastrichtian flowers from the Deccan Intertrappean beds show stamens fused to the petals and this is also relatively common among the Early Tertiary flowers from the Baltic Amber.

The number of pollen sacs may be difficult to establish for fossil material, as the septum in tetrasporangiate anthers often degenerates as the anthers mature. However, all the anthers known from Albian rocks are clearly tetrasporangiate. Clustering of pollen in four elongated groups, indicates that tetrasporangiate anthers were apparently also present in the Cenomanian flowers (Dilcher, 1979) and have been documented also in several fossil flowers from younger sediments. The earliest fossil evidence of bisporangiate forms is in hamamelidaceous

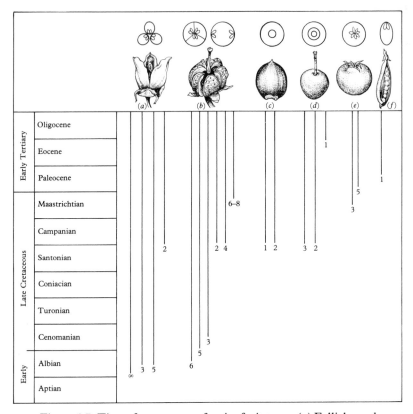

Figure 6.7. Time of appearance of major fruit types. (*a*) Follicles and nutlets from apocarpous ovaries. (*b*) Capsules. (*c*) Nuts. (*d*) Drupes. (*e*) Berries. (*f*) Pods. ((*b*) to (*e*) Syncarpous ovaries.) The figures indicate number of carpels in the ovary.

stamens from the Santonian–Campanian of Sweden. Dehiscence in Cretaceous anthers was by longitudinal slits or in the hamamelidaceous anthers from Sweden by two lateral valves. Lauraceous anthers dehiscing by two or more valves have been recorded from the Lower Tertiary Baltic Amber and from the Upper Tertiary rocks of Germany (Weyland, 1938). The earliest evidence of dehiscence by apical pores is in *Sezannella* from the Upper Paleocene (Viguier, 1908).

Gynoecium. The earliest angiosperm reproductive organs known were apocarpous, with free carpels. This was the predominant condition among the Albian and Early Cenomanian angiosperms. Syncarpy was established at least by the Late Albian and is represented by several different taxa by the Early Cenomanian. Syncarpous forms diversified strongly in the Late Cretaceous and are the most common type of gynoecium in Santonian–Campanian flowers. In extant angiosperms, 83% of all taxa have syncarpous gynoecia (Endress, 1982).

The earliest syncarpous ovaries from the Albian and Cenomanian were apparently all septate, and divided into a number of locules corresponding to the number of carpels. By the Santonian–Campanian, several distinct types of unilocular ovaries had developed. Secondary divisions of the locules first appeared in Maastrichtian angiosperms.

Apocarpous ovaries and fruits known from the fossil record show no evidence of distinct styles and the stigmatic area is usually inconspicuous. Syncarpous ovaries from the Cretaceous usually have distinct styles that are generally free. The fusion of stylar tissue to form a single style was established by the Santonian–Campanian, but not common until the Maastrichtian. The styles observed in the Cretaceous flowers are usually stout, while slender styles become more common in Early Tertiary material.

Fossil fruits known from Albian and Cenomanian rocks were apparently all dry and show no obvious modifications for dispersal. Fruits derived from apocarpous ovaries were follicles, or in the platanoid fossils, nutlets. Fruits derived from syncarpous ovaries were septate capsules. By the Santonian–Campanian several other dry fruit types such as nuts and samaras were established, and fleshy fruits appeared for the first time. The first evidence of berries is in the Late Maastrichtian. During the Early Tertiary, fleshy fruits became relatively common, and the size range of fruits increased considerably, indicating a wide range of dispersal mechanisms (Tiffney, 1984).

Very little is known about the seed structure in the Albian and Cenomanian angiosperms, but available information suggests that seeds were generally small and thin-walled, apparently anatropous or in a single taxon from the Middle Cenomanian (*Curvospermum marketense*; Knobloch & Mai, 1983) slightly campylotropous. By the Santonian–Campanian angiosperm seeds were varied in form and structure. Anatropous seeds were most abundant, but orthotropous, campylotropous and amphitropous seeds were also present. The structure of the seed-coat in these seeds varies considerably from very thin, membranous, and apparently formed from a single integument to thicker with several distinct tissues, and apparently formed from bitegmic ovules.

The presence of an aril has been documented for well-preserved seeds of *Spirematospermum* from Miocene deposits (Koch & Friedrich, 1971), but although similar seeds have also been recovered from Campanian rocks (Friis, 1988) it has not been possible to demonstrate the development of an aril in any of the Cretaceous material.

Nectaries. The earliest unequivocal evidence of nectar secretory tissue in angiosperm flowers is provided by the Santonian–Campanian flowers from Sweden. Several of the Swedish flowers possess distinct structures that are comparable to nectaries of modern flowers in their position and morphology. The nectaries are usually pronounced and form a five-, eight-, or ten-lobed disk inserted between the androecium and gynoecium, or they may be a swollen disk fused to the basal part of the gynoecium. In a small flower from the Early Campanian of Aachen, Germany, the nectary is developed as two-lobed structures at the bases of the stamens. The surface of the nectaries is rugulate or smooth, often with distinct stomata-like openings. The nectaries observed in the Santonian–Campanian flowers were apparently well exposed, while concealed nectaries are known from the Maastrichtian *Sahnianthus* and also inferred from several Paleocene and Eocene flowers with deep funnel-shaped corollas.

Conclusions

The study of Cretaceous and Tertiary floral structures demonstrates a general increase in morphological and organizational diversity of angiosperm reproductive organs through time, as summarized in the previous section and in Figures 6.6 and 6.7. Although the fossil record of flowers is incomplete, particularly from the earliest phases of angiosperm diversification, it is consistent with that of other organs such as pollen and leaves. It fails to provide an unequivocal basic archetype, but indicates that small and simple flower types related to the Chloranthaceae and Platanaceae appeared early in the history of the angiosperms and co-existed in the Albian with more elaborate floral structures of magnolialean affinity. Several evolutionary events that apparently represent major radiations of the angiosperm are recognized, including the appearance in the mid-Cretaceous of syncarpous ovaries, and of sympetalous corollas and zygomorphic floral symmetry late in the Cretaceous. These, as well as other floral features, are believed to have originated and diversified in response to pollination agents. Aspects of the evolution of these floral features and their interrelationship with insect pollinators are discussed in Chapter 7.

We thank W. G. Chaloner, P. R. Crane, M. J. Donoghue and J. A. Doyle for helpful suggestions and constructive criticism. Thanks are also due to Mary Jane Spring for preparing the illustrations. The work has been supported by a Niels Bohr Fellowship (E.M.F.), the NATO Senior Fellowship Scheme (E.M.F.), and by National Science Foundation Grant DEB810217 (W.L.C.).

References

Armbruster, W. S. & Herzig, A. L. (1984). Partitioning and sharing of pollinators by four sympatric species of *Dalechampia* (Euphorbiaceae) in Panama. *Annals of the Missouri Botanical Garden*, 71, 1–16.

Baker, H. G. & Hurd, P. D. (1968). Intrafloral ecology. *Annual Review of Entomology*, 13, 385–414.

Basinger, J. F. & Dilcher, D. L. (1984). Ancient bisexual flowers. *Science*, 224, 511–13.

Bayer, E. (1914). Fytopalaeontologicke přispěvky ku Poznání Českých křídových vrstev peruckych. *Archiv přirodovedecký vyzkum Čech*, 25, 1–66.

Brenner, G. J. (1963). The spores and pollen of the Potomac group of Maryland. *State of Maryland Department of Geology, Mines and Water Resources, Bulletin*, 27, 1–215.

Brown, R. W. (1933). Fossil plants from the Aspen Shale of southwestern Wyoming. *Proceedings of the United States National Museum*, 82(12), 1–10.

Burger, W. C. (1981). Why are there so many kinds of flowering plants? *BioScience*, 31, 572–81.

Chandler, M. E. J. (1958). Angiosperm fruits from the Lower Cretaceous of France and Lower Eocene (London Clay) of Germany. *Annals and Magazine of Natural History*, 13, 354–8.

Chandler, M. E. J. & Axelrod, D. I. (1961). An early Cretaceous (Hauterivian) angiosperm fruit from California. *American Journal of Science*, 259, 441–6.

Chesters, K. I. M. (1955). Some plant remains from the Upper Cretaceous and Tertiary of West Africa. *Annals and Magazine of Natural History*, 12, 498–504.

Chitaley, S. D. (1955). A further contribution to the knowledge of *Sahnianthus*. *Journal of the Indian Botanical Society*, 34, 121–9.

Chitaley, S. D. (1964). Further observations on *Sahnipushpam*. *Journal of the Indian Botanical Society*, 43, 69–74.

Chitaley, S. D. (1977). *Enigmocarpon parijai* and its allies. *Frontiers of Plant Sciences – Prof. P. Parija Felicitation Volume*, pp. 421–9.

Chitaley, S. D. & Kate, U. R. (1974). *Deccananthus savitrii*, a new petrified flower from the Deccan Intertrappean beds of India. *The Palaeobotanist*, 21, 317–20.

Chitaley, S. D. & Nambudiri, E. M. V. (1969). Anatomical studies of *Nypa* fruits from Deccan Intertrappean beds of India. I. *Recent Advances in Anatomy of Tropical Seed Plants*, 1969, pp. 235–48.

Chitaley, S. D. & Patel, M. Z. (1975). *Raoanthus intertrappea*, a new petrified flower from India. *Palaeontographica B*, **153**, 141–9.

Christophel, D. C. & Basinger, J. F. (1982). Earliest floral evidence for the Ebenaceae in Australia. *Nature*, **296**, 439–41.

Conwentz, H. (1886). *Die Flora des Bernsteins*, vol. 2 *Die Angiospermen des Bernsteins*. Leipzig: Wilhelm Engelmann.

Crane, P. R. & Dilcher, D. L. (1984). *Lesqueria*: an early angiosperm fruiting axis from the Mid-Cretaceous. *Annals of the Missouri Botanical Garden*, **71**, 384–402.

Crane, P. R., Friis, E. M. & Pedersen, K. R. (1986). Angiosperm flowers from the Lower Cretaceous: Fossil Evidence on the Early Radiation of the Dicotyledons. *Science*, **232**, 852–4.

Crepet, W. L. (1978). Investigations of angiosperms from the Eocene of North America: an aroid inflorescence. *Review of Palaeobotany and Palynology*, **25**, 241–52.

Crepet, W. L. (1984). Advanced (constant) insect pollination mechanisms: pattern of evolution and implications *vis à vis* angiosperm diversity. *Annals of the Missouri Botanical Garden*, **71**, 607–30.

Crepet, W. L. & Daghlian, C. P. (1980). Castaneoid inflorescences from the Middle Eocene of Tennessee and the diagnostic value of pollen (at the subfamily level) in the Fagaceae. *American Journal of Botany*, **67**, 739–57.

Crepet, W. L. & Daghlian, C. P. (1981). Euphorbioid inflorescences from the Middle Eocene Claiborne Formation. *American Journal of Botany*, **69**, 258–66.

Crepet, W. L. & Dilcher, D. L. (1977). Investigations of angiosperms from the Eocene of North America: a mimosoid inflorescence. *American Journal of Botany*, **64**, 714–25.

Crepet, W. L. & Taylor, D. W. (1985). The diversification of the Leguminosae: first fossil evidence of the Mimosoideae and Papilionoideae. *Science*, **228**, 1087–9.

Crepet, W. L. & Taylor, D. W. (1986). Primitive mimosoid flower from the Paleocene–Eocene and their systematic and evolutionary implications. *American Journal of Botany*, **73**, 548–63.

Cruden, R. W. (1977). Pollen–ovule ratios: a conservative indicator of breeding systems in flowering plants. *Evolution*, **31**, 32–46.

Daghlian, C. P. (1981). A review of the fossil record of monocotyledons. *The Botanical Review*, **47**, 517–55.

Dilcher, D. L. (1979). Early angiosperm reproduction: an introductory report. *Review of Palaeobotany and Palynology*, **27**, 291–328.

Dilcher, D. L. & Crane, P. R. (1984). *Archaeanthus*: An early angiosperm from the Cenomanian of the Western Interior of North America. *Annals of the Missouri Botanical Garden*, **71**, 351–83.

Doyle, J. A. (1969). Cretaceous angiosperm pollen of the Atlantic Coastal Plain and its evolutionary significance. *Journal of the Arnold Arboretum*, **30**, 1–35.

Doyle, J. A. & Hickey, L. J. (1976). Pollen and leaves from the Mid-Cretaceous Potomac Group and their bearing on early angiosperm evolution. In *Origin and Early Evolution of Angiosperms*, ed. C. B. Beck, pp. 139–206. New York: Columbia University Press.

Elias, T. S. (1981). Mimosoideae. In *Advances in Legume Systematics*, vol. 1 *Proceedings of the International Legume Conference, Kew, 1978*, ed. R. M. Polhill & P. H. Raven, pp. 143–52. Kew: Royal Botanic Gardens.

Endress, P. K. (1982). Syncarpy and alternative modes of escaping disadvantages of apocarpy in primitive angiosperms. *Taxon*, **31**, 48–52.

Fontaine, W. M. (1889). The Potomac or younger Mesozoic flora. *United States Geological Survey Monographs*, **15**, 1–377.

Friis, E. M. (1983). Upper Cretaceous (Senonian) floral structures of juglandalean affinity containing Normapolles pollen. *Review of Palaeobotany and Palynology*, **39**, 161–88.

Friis, E. M. (1984). Preliminary report of Upper Cretaceous angiosperm reproductive organs from Sweden and their level of organization. *Annals of the Missouri Botanical Garden*, **71**, 403–18.

Friis, E. M. (1985*a*). Structure and function in Late Cretaceous angiosperm flowers. *Biologiske Skifter Danske Videnskabernes Selskab*, **25**, 1–37.

Friis, E. M. (1985*b*). *Actinocalyx* gen. nov., sympetalous angiosperm flowers from the Upper Cretaceous of southern Sweden. *Review of Palaeobotany and Palynology*, **45**, 171–83.

Friis, E. M. (1988). *Spirematospermum chandlerae* sp. nov., an extinct species of Zingiberaceae from the North American Cretaceous. *Tertiary Research*, **9**, 7–12.

Friis, E. M., Crane, P. R. & Pedersen, K. R. (1986). Floral evidence for Cretaceous chloranthoid angiosperms. *Nature*, **320**, 163–4.

Friis, E. M. & Skarby, A. (1981). Structurally preserved angiosperm flowers from the Upper Cretaceous of southern Sweden. *Nature*, **291**, 485–6.

Friis, E. M. & Skarby, A. (1982). *Scandianthus* gen. nov., angiosperm flowers of saxifragalean affinity from the Upper Cretaceous of southern Sweden. *Annals of Botany*, **50**, 569–83.

Grant, V. (1949). Pollination systems as isolating mechanisms in angiosperms. *Evolution*, **3**, 82–97.

Grant, V. (1950). The protection of the ovules in flowering plants. *Evolution*, **4**, 179–201.

Harris, T. M. (1932). The fossil flora of Scoresby Sound East Greenland. Part 3. Caytoniales and Bennettitales. *Meddelelser om Grønland*, **85**(5), 1–133.

Hesse, M. (1981). The fine structure of the exine in relation to the stickiness of angiosperm pollen. *Review of Palaeobotany and Palynology*, **35**, 81–92.

Hughes, N. F. (1976). *Palaeobiology of Angiosperm Origins*. Cambridge: Cambridge University Press.

Jain, R. K. (1963). Studies in Musaceae. 1. *Musa cardiosperma* sp. nov., a fossil banana fruit from the Deccan Intertrappean Series. *The Palaeobotanist*, **12**, 45–8.

Jung, W., Schleich, H.-H. & Kästle, B. (1978). Eine neue, stratigraphisch gesicherte Fundstelle für Angiospermen-Früchte und -Samen in der oberen Gosau Tirols. *Mitteilungen der Bayerischen Staatssammlung für Paläontologie und historische Geologie*, **18**, 131–42.

Kevan, P. C. & Baker, H. G. (1983). Insects as flower visitors and pollinators. *Annual Review of Entomology*, **28**, 407–53.

Knobloch, E. (1964). Neue Pflanzenfunde aus dem südböhmischen Senon. *Jahrbuch der Staatlichen Museums für Mineralogie und Geologie zu Dresden*, 1964, pp. 133–201.

Knobloch, E. (1971). Fossile Früchte und Samen aus der Flyschzone der mährischen Karpaten. *Sborník Geologických Věd, Paleontologie*, **13**, 7–43.

Knobloch, E. (1975). Früchte und Samen aus der Gosauformation von Kössen in Österreich. *Věstník Ústředního ústavu geologického*, **50**, 83–91.

Knobloch, E. & Mai, D. H. (1983). Carbonized seeds and fruits from the Cretaceous of Bohemia and Moravia and their stratigraphical significance. *Knihovnička Zemního plynu a nafty*, **4**, 305–32.

Knobloch, E. & Mai, D. H. (1986). Monographie der Früchte und Samen aus der Kreide von Mitteleuropa. *Rozpravy Ústředního ústavu geologického*, **47**, 1–219.

Koch, B. E. & Friedrich, W. L. (1971). Früchte und Samen von *Spirematospermum* aus der miozänen Fasterholt-Flora in Dänemark. *Palaeontographica B*, **136**, 1–46.

Krassilov, V. A. (1973). Upper Cretaceous staminate heads with pollen grains. *Palaeontology*, **16**, 41–4.

Krassilov, V. A. (1982). Early Cretaceous flora of Mongolia. *Palaeontographica B*, **181**, 1–43.

Krassilov, V. A. (1984). New paleobotanical data on origin and early evolution of angiospermy. *Annals of the Missouri Botanical Garden*, **71**, 577–92.

Krassilov, V. A., Shilin, P. V. & Vakhrameev, V. A. (1983). Cretaceous flowers from Kazakhstan. *Review of Palaeobotany and Palynology*, **40**, 91–113.

Larsson, S. G. (1978). *Baltic Amber – A Palaeobiological Study*. Klampenborg: Scandinavian Science Press.

Monteillet, J. & Lappartient, J.-R. (1981). Fruits et graines du Crétacé supérieur des carrières de Paki (Sénégal). *Review of Palaeobotany and Palynology*, **34**, 331–44.

Mulcahy, D. L. (1979). The rise of the angiosperms. A genecological factor. *Science*, **206**, 20–3.

Muller, J. (1970). Palynological evidence on early differentiation of angiosperms. *Biological Review*, **45**, 417–50.

Muller, J. (1981). Fossil pollen records of extant angiosperms. *The Botanical Review*, **47**, 1–142.

Muller, J. (1984). Significance of fossil pollen in angiosperm history. *Annals of the Missouri Botanical Garden*, **71**, 419–43.

Nishida, H. (1985). A structurally preserved magnolialean fructification from the mid-Cretaceous of Japan. *Nature*, **318**, 58–9.

Pacltová, B. (1977). Cretaceous angiosperms of Bohemia–Central Europe. *The Botanical Review*, **43**, 128–42.

Paradkar, S. A. (1973). *Chitaleypushpam mohgaoense* gen. et sp. nov. from the Deccan Intertrappean beds of India. *The Palaeobotanist*, **20**, 334–8.

Pedersen, K. R. (1976). Fossil floras of Greenland. In *Geology of Greenland*, ed. A. Escher & W. S. Watt, pp. 519–35. Copenhagen: Geological Survey of Greenland.

Piwocki, M., Olkowicz-Paprocka, I., Kosmowska-Ceranowicz, B., Grabowska, I. & Odrzywolska-Bieńkowa, E. (1985). Stratygrafia trzeciorzędowych osadów bursztynonośnych okolic Chłapowa koło Pucka. *Prace Muzeum Ziemi*, **37**, 61–77.

Prakash, U. (1954). *Palmocarpon mohgaoense* sp. nov., a palm fruit from the Deccan Intertrappean Series, India. *The Palaeobotanist*, **3**, 91–6.

Prakash, U. (1956). On the structure and affinities of *Sahnipushpam glandulosum* sp. nov. from the Deccan Intertrappean series. *The Palaeobotanist*, **4**, 91–100.

Prakash, U. & Jain, R. K. (1964). Further observations on *Sahnipushpam Shukla*. *The Palaeobotanist*, **12**, 128–38.

Retallack, G. & Dilcher, D. L. (1981). Early angiosperm reproduction: *Prisca reynoldsii* gen. et sp. nov. from Mid-Cretaceous coastal deposits in Kansas, U.S.A. *Palaeontographica B*, **179**, 103–37.

Rode, K. P. (1933). A note on fossil angiospermous fruits from the Deccan Intertrappean beds of Central Provinces. *Current Science*, **2**, 171–2.

Sahni, A., Rana, R. S. & Prasad, G. V. R. (1984). SEM studies of thin egg shell fragments from the Intertrappeans (Cretaceous–Tertiary Transition) of Nagpur and Asifabad, Peninsular India. *Journal of the Palaeontological Society of India*, **29**, 26–33.

Sahni, B. (1934). The silicified flora of the Deccan Intertrappean Series. Part II. Gymnospermous and angiospermous fruits. *Proceedings of the 21st Indian Science Congress*, pp. 317–18. Calcutta: Society instituted in Bengal for inquiring into history and antiquities, the arts, sciences and literature in Asia.

Sahni, B. (1943). Indian silicified plants. II. *Enigmocarpon parijai*, a silicified fruit from the Deccan, with a review of the fossil history of the Lythraceae. *Proceedings of the Indian Academy of Science*, **17B**, 59–96.

Samylina, V. A. (1960). [Angiosperms from the lower Cretaceous of the Kolyma basin.] In Russian. *Botanicheskiy Zhurnal*, **45**, 335–52.

Samylina, V. A. (1961). [New data on the lower Cretaceous flora of the southern part of the maritime territory of the R.F.S.R.] In Russian. *Botanicheskiy Zhurnal*, **46**, 634–45.

Samylina, V. A. (1968). Early Cretaceous angiosperms of the Soviet Union based on leaf and fruit remains. *Botanical Journal of the Linnean Society*, **61**, 207–18.

Seward, A. C. & Conway, V. M. (1935). Fossil plants from Kingigtok and Kagdlunguak, West Greenland. *Meddelelser om Grønland*, **93**(5), 1–41.

Shukla, R. K. (1958). *Sahnianthus dinectrianum*, sp. nov., a new species of the petrified flower *Sahnianthus* from the Eocene beds of the Deccan. *Journal of the Palaeontological Society of India*, **3**, 114–18.

Shukla, V. B. (1944). On *Sahnianthus*, a new genus of petrified flowers from the Intertrappean Beds at Mohgaon Kalan in the Deccan and its relation with the fruit *Enigmocarpon parijai* Sahni from the same locality. *Proceedings of the National Academy of Science of India*, **14**, 1–39.

Stebbins, G. L. (1981). Why are there so many species of flowering plants? *BioScience*, **31**, 573–7.

Stopes, M. C. & Fujii, K. (1910). Studies on the structure and affinities of Cretaceous plants. *Philosophical Transactions of the Royal Society*, **201**, 1–90.

Tiffney, B. H. (1977). Dicotyledonous angiosperm flower from the Upper Cretaceous of Martha's Vineyard, Massachusetts. *Nature*, **265**, 136–7.

Tiffney, B. H. (1984). Seed size, dispersal syndromes and the rise of angiosperms: evidence and hypothesis. *Annals of the Missouri Botanical Garden*, **71**, 551–76.

Vakhrameev, V. A. (1952). [Stratigraphy and fossil flora of the Cretaceous deposits in the western Kazakhstan.] In Russian. *Regional Stratigraphy of the USSR*, **1**, 1–340.

Vakhrameev, V. A. & Krassilov, V. A. (1979). [Reproductive organs of flowering plants from the Albian of Kazakhstan.] In Russian. *Paleontologicheskiy Zhurnal* 1979(1), pp. 121–8.

Vangerow, E. F. (1954). Megasporen und andere pflanzliche Mikrofossilien aus der Aachener Kreide. *Palaeontographica B*, **96**, 24–38.

Velenovský, J. (1889). Květena Českého Cenomanu. *Rozpravy Královské České Společnosti Nauk VII*, **3**, 1–75.

Velenovský, J. & Viniklář, L. (1926). Flora Cretacea Bohemiae. I. *Rozpravy Státního Geologického Ústavu Československé Republiky*, **1**, 1–57.

Velenovský, J. & Viniklář, L. (1927). Flora Cretacea Bohemiae. II. *Rozpravy Státního Geologického Ústavu Československé Republiky*, **2**, 1–54.

Velenovský, J. & Viniklář, L. (1929). Flora Cretacea Bohemiae. III. *Rozpravy Státního Geologického Ústavu Československé Republiky*, **3**, 1–33.

Velenovský, J. & Viniklář, L. (1931). Flora Cretacea Bohemiae. IV. *Rozpravy Státního Geologického Ústavu Československé Republiky*, **5**, 1–112.

Viguier, R. (1908). Recherches sur le genre *Sezannella*. *Revue générale de Botanique*, **20**, 6–13.

Weber, R. (1978). Some aspects of the Upper Cretaceous angiosperm flora of Coahuila, Mexico. *Courier Forschungsinstitut Senckenberg*, **30**, 38–46.

Webster, G. L. (1975). Conspectus of a new classification of the Euphorbiaceae. *Taxon*, **24**, 593–601.

Weyland, H. (1938). Beiträge zur Kenntnis der Rheinischen Tertiärflora. III. Zweite Ergänzungen und Berichtigungen zur Flora der Blätterkohle und des Polierschiefers von Rott im Siebengebirge. *Palaeontographica B*, **83**, 123–71.

7

The evolution of insect pollination in angiosperms

W.L.CREPET AND E.M.FRIIS

The origin of the angiosperm flower may be interpreted as the result of a co-evolutionary relationship between insects and as yet unknown angiosperm ancestors (Grant, 1950; Baker & Hurd, 1968; Hickey & Doyle, 1977; Crepet, 1979, 1983). Evidence based on the fossil record of insects and the nature of various gymnosperm reproductive organs suggests that insects became increasingly specialized for feeding on plant reproductive structures during the Late Carboniferous–Mesozoic interval (Scott & Taylor, 1983; Crepet, 1979, 1984). Both the hermaphroditic condition in the Bennettitales and the angiosperm carpel are thought to have evolved from relationships where insect feeding on reproductive organs resulted in pollination (Grant, 1950; Crepet, 1979, 1983; however, for a different interpretation of carpel origin, see Zavada & Taylor, 1985). The renewed (see Arber & Parkin, 1907) appreciation of the relationship between the Bennettitales and angiosperms (see Crane, 1985; Doyle & Donoghue, this volume, Chapter 2), coupled with the Jurassic appearance of the hermaphroditic taxon *Williamsoniella*, makes it clear that the flowering plants evolved after the origin of insect pollination in at least one of their sister groups. Insect pollination may therefore be plausibly considered as primitive in the angiosperms.

The course of the subsequent relationship between the pollinating insects and the flowering plants and its possible significance in angiosperm success and diversification pattern are considered in this chapter.

Setting the scene – pollination in the Bennettitales

The hermaphroditic fructifications of the Bennettitales are functional analogs of certain magnoliidean angiosperm flowers (Figure 7.1 (*a*) to (*c*)).

181

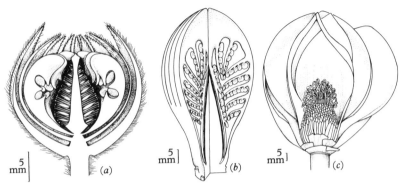

Figure 7.1. (a) *Williamsoniella*, a Jurassic hermaphroditic bennettitalean. (b) *Cycadeoidea*, a Late Jurassic–Early Cretaceous
hermaphroditic bennettitalean. (c) *Magnolia grandiflora*, a living
angiosperm, ((a) After Crane, 1985; (b) after Crepet, 1974.)

Together with the nature of the diverse contemporaneous anthophilous
insects (Diptera and Coleoptera), they provide an archetypal syndrome
of insect pollination. Bennettitaleans vary in fructification structure from
those that have open, small, exposed, radially symmetrical 'flowers'
(*Williamsoniella*; Figure 7.1 (a)) to those with moderate-sized flowers and
permanently closed microsporophylls that are surrounded by bracts
bearing a dense ramentum (*Cycadeoidea*; Figure 7.1(b)) to the possibly
ancestral type flowers of *Wielandiella* that have rounded ovulate receptacles and apparently separate microsporophylls with linear synangia
and abaxial dehiscence.

Fundamental bennettitalean features suggest that the group may have
been beetle pollinated. They are similar in gross morphology to modern
beetle-pollinated flowers (Figure 7.1 (c)); their pollen is often relatively
large and sometimes has a sculptured exine (*Williamsoniella*; Crepet,
unpublished); and the position of the cones of *Cycadeoidea*, a taxon that
was probably highly specialized for beetle pollination, mitigates against
the involvement of insects that could not enter tight spaces (Crepet,
1974; Thien, White & Yatsu, 1983). The nature of the ovulate receptacle,
one of the most uniformly consistent bennettitalean structures, is also
highly suggestive of beetle pollination. Tiny ovules are distributed
among sterile interseminal scales that have expanded heads (Crepet,
1974). The polygonal–triangular heads of the scales surround and armor
the ovules and the micropyles project above the surface of the interseminal scales as short minute tubes. This evident protection of the

ovules is consistent with pollination by insects with chewing mouth-parts and suggests that beetles were significant early (pre-angiospermous) insect pollinators (see Grant, 1950). Beetles were diverse in the Mesozoic (Handlirsch, 1906–8) with flower-visiting Elateridae and Nitidulidae present as early as the Triassic. Diptera were also well developed during the Mesozoic, with Tipulidae (crane flies) and Mycetophilidae (fungus gnats) rather diverse by the Jurassic (Rohdendorf, 1974). These are relatively unimportant pollinators today (Proctor & Yeo, 1973), but it is possible that they, along with the beetles, were visitors of *Williamsoniella*. Even the Hymenoptera (Symphyta, Xyelidae) were present and also cannot be ruled out as possible visitors (Burnham, 1978; Krassilov & Rasnitzen, 1982; and see discussion below).

Pollination in Barremian–Albian angiosperms

Assessments of insect pollination during this interval of angiosperm history must be based largely on inferences from pollen and leaf remains, pollination in extant families related to Cretaceous taxa, and the fossil record of insects.

The monosulcate palynoflora from the uppermost Lower Cretaceous suggests that the Magnoliidae were fairly diverse by the Aptian or earlier (e.g. Doyle *et al.*, 1977; Walker & Walker, 1984). Certain monosulcates have been compared with pollen of the Chloranthaceae (Walker & Walker, 1984) and Lactoridaceae (M. S. Zavada, personal communication), and tetrads of monosulcate-derived pollen similar to those of modern Winteraceae have been reported from the Aptian–Albian (Walker, Brenner & Walker, 1983). Contemporaneous leaves generally have loose, poorly organized venation, consistent with the modern Magnoliidae (Hickey & Doyle, 1977) and the pollen record. Upchurch (1984*a*, *b*) compared some of these leaf types with modern Illiciales and Chloranthaceae, in respect of cuticular features and tooth type.

The earliest recognizable angiosperm pollen is columellate–reticulate in wall structure (Doyle, 1977). Columellate–reticulate wall structure may be primitive in the angiosperms (Doyle, 1978), but granular wall structure is considered as primitive within the flowering plants by Walker & Skvarla (1975). Zavada (1984) has reported a correlation between reticulate wall structure and sporophytic self-incompatibility that may suggest such an incompatibility mechanism had already developed in the earliest recognizable angiosperms. Alternatively, this might have been the original condition in angiosperms, because granular-walled angiosperm pollen older than reticulate (semi-tectate) walled pollen has not been discovered. Granular-walled pollen may also be associated with incompatibility (gametophytic), but is not strong evidence of self-incompatibility. Givnish (1982) has pointed out a relationship between passive seed dispersal (presumed in early angiosperms, Tiffney, 1984; Friis & Crepet, this volume Chapter, 6) and self-incompatibility in extant angiosperms that supports the possibility that early angiosperms were self-incompatible.

Magnoliidae are often considered as generalized in their pollination syndromes and indeed they are pollinated by a range of insects. Pollination systems vary from very general to highly specialized. For example, Magnoliidae with simple floral structure, such as Winteraceae, may be pollinated by beetles (*Exospermum*), flies (*Pseudowintera*), Thysanoptera (*Pseudowintera*), and even by pollen-chewing micropterigid moths (*Sabatinca* pollinating *Zygogynum*; Thien *et al.*, 1985). Certain other Magnoliidae including Annonaceae, *Calycanthus*, Magnoliaceae and Nymphaeaceae have generally large and elaborate flowers. Floral syndromes in flowers of these taxa include floral movements, large size, flattening and large numbers of floral parts that make them very well adapted to pollination by various, often specific, beetles (Grant, 1950; Thien, 1974; Schatz & Young, 1985; Schneider, 1979; Meeuse & Schneider, 1980). The Magnoliaceae have often been considered as the most primitive extant angiosperms (Walker & Walker, 1984) and Magnoliidae with similar elaborate floral structure might be considered as having shared these primitive characters. However, pollen of the Magnoliaceae has not been discovered in Lower Cretaceous sediments, and the Aptian–Albian pollen of Winteraceae (Walker *et al.*, 1983) and that of the somewhat earlier Chloranthaceae (Walker & Walker, 1984) are the first bona fide evidence of any taxa of extant Magnoliidae. Recently, and consistent with the fossil record, there has been the suggestion that Magnoliidae with complex floral structure (possibly first appearing in the Late Albian; Friis & Crepet, this volume, Chapter 6) are derived, and simpler-flowered taxa, such as those of Winteraceae, are more primitive. This is based largely on the specialized adaptations of complex-flowered Magnoliidae to certain pollinators (Gottsberger, 1974, 1977; Carlquist, 1969; Schneider, 1979). Walker & Walker (1984), however, maintain a traditional position on the phylogenetic status of Magnoliaceae.

If Magnoliidae with large and complex flowers are derived rather than primitive within the subclass, then it is possible that self-compatibility arose in some of these taxa from self-incompatible ancestors in association with the evolution of highly specialized beetle pollination syndromes (e.g. Schatz & Young, 1985). Protogyny and protandry, then, might have been secondarily derived in Magnoliidae that had already become self-compatible. It is obvious that a better understanding of the relative phylogenetic positions of taxa of the Magnoliidae will require more objective bases for the determination of character-state polarities within the subclass than those presently being advocated.

It is interesting to consider what the reproductive biology of hypothetical early Magnoliidae might have been like. *Illicium*, while clearly a more derived taxon than Winteraceae, has simple and generalized floral structure, self-incompatibility and passive seed dispersal in its reproductive biology and may be similar to the primitive condition in the subclass. *Illicium* flowers synchronously and is pollinated by flies or occasionally by other insects, such as beetles (Thien *et al.*, 1983). Fruit set is low (2% to 3%) and there is a high concentration of pollen from other flowers of the same plant on the stigmas (Thien *et al.*, 1983). If early Magnoliidae did have reproductive characteristics in common with *Illicium*, as suggested by our interpretation of the fossil evidence, the implied premium on outcrossed offspring suggests that their environment was temporally or spatially varied. This has been suggested by Stebbins (1974), Hickey & Doyle (1977) and Doyle (1978).

The fossil record of insects is consistent with the early fossil record of the Magnoliidae (Figure 7.2). Beetles were already diverse and included a variety of flower-visiting taxa (Carpenter, 1976). Flies were also diverse by the Early Cretaceous, with the Empididae, the most important flower-visiting Brachycera, known from the Neocomian and Barremian (Hennig, 1970; Jarzembowski, 1984). Micropterigid moths similar to the pollinator of extant *Zygogynum* (Winteraceae) are also known from the Early Cretaceous (Whalley, 1977).

Hymenoptera are the most important insect pollinators of extant angiosperms. They (represented by the Symphyta or sawflies, Xyelidae) first appear in the Triassic (Burnham, 1978) and Xyelidae have also been reported from the Early Cretaceous (Krassilov & Rasnitzen, 1982). These fossils have numerous pollen grains preserved in their guts. Two specimens contain coniferous pollen and the third contains more generalized bisaccate pollen that may be pteridospermous. Sawflies are primitive hymenopterans (sister group to the Apocrita) that lack the constriction between the abdomen and thorax that defines the gaster of true wasps. They tend to pollinate small, unspecialized flowers today

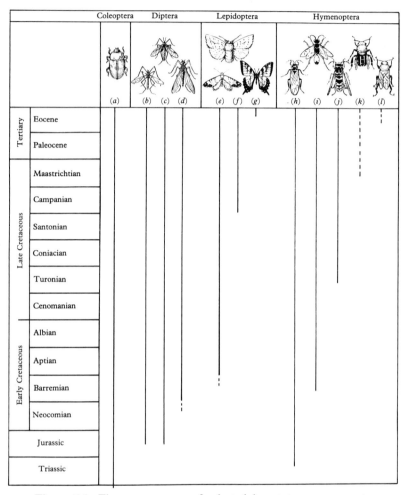

Figure 7.2. First occurrences of selected insect taxa germane to the evolution of insect pollination. Due to space limitations, there is no attempt to place all taxa of significance in the table. For example the bees become diverse in the fossil record around the Oligocene but all taxa are not shown (instead, refer to Burnham, 1978). The beeflies (Bombylidae) and hoverflies (Syrphidae), also important pollinators, are reported as diverse by the Tertiary (Rohdendorf, 1974), but fossil evidence for their earliest occurrence is not unequivocal. The times of first occurrences are based on the reference or discussions cited in the text. Dashed lines indicate probable range, but inferred rather than based on direct fossil evidence. (*a*) Coleoptera (Nitidulidae illustrated). (*b*) to (*d*) Diptera: (*b*), Tipulidae, (*c*) Mycetophilidae, (*d*), Empididae. (*e*) to (*g*) Lepidoptera: (*e*) Micropterigidae, (*f*) Noctuidae, (*g*) Papilionidae. (*h*) to (*l*) Hymenoptera: (*h*) Symphyta, (*i*) Sphecidae, (*j*) Vespoidea, (*k*) Meliponinae, (*l*) Anthophoridae.

(Proctor & Yeo, 1973). Within the Hymenoptera, the sphecid wasps are the sister group of the all-important bees (Apidae) and, according to Wilson (1971), 'Apoidea can be loosely characterized as sphecoid wasps that have specialized in collecting pollen instead of insect prey as larval food'. Morphologically, the sphecid wasps are only distinguished from the bees by plumose hairs and enlarged basitarsi that function as adaptations for collecting pollen (Burnham, 1978). These characters are not pronounced in the most primitive bees (Colletidae) and may be difficult to recognize in fossil material. Currently, there is no unequivocal evidence of Jurassic sphecoids (Burnham, 1978), but Early Cretaceous Sphecidae from southern England (Jarzembowski, 1984) and Canada (Evans, 1969, 1973) reveal an early level of advancement in the sister group of the Apoidea. This is of particular interest in light of the suggestion based on biogeography that the Apoidea and even the Apidae had a Late Cretaceous origin (Michener, 1979; Crepet, 1984; and see below). It is also significant that wasps, while relatively unimportant pollinators at higher latitudes (Proctor & Yeo, 1973) are more important in that role in the tropics (Kevan & Baker, 1983) and may have been more important pollinators in the milder climate of the Early Cretaceous.

Taken together, the available botanical and entomological evidence suggests that the earliest angiosperms had functionally more simple flowers than did modern Magnoliaceae, perhaps similar to those of Winteraceae or even simpler, like those of Chloranthaceae. Coleoptera were probably the major pollinators (V. Grant, unpublished), but it is also possible that early magnoliids were generalists and that flies, micropterigid moths, sawflies and even sphecid wasps participated in their pollination biology. Although one taxon of modern Chloranthaceae has been reported as being wind pollinated (Hammen & Gonzalez, 1960), other taxa seem to be well adapted for insect pollination and further studies are needed to evaluate fully their pollination biology.

Recently discovered Albian inflorescences provide new insights into angiosperm pollination in the Early Cretaceous. Androecial remains of Chloranthaceae (Friis, Crane & Pedersen, 1986) confirm the presence of the family in the Early Cretaceous and inflorescences of the Platanaceae (Crane, Friis & Pedersen, 1986) and Rosidae (Friis & Crepet, this volume, Chapter 6) also confirm the antiquity of these taxa.

Reproductive biology of early Magnoliidae: implications

The early monosulcate pollen-bearing Magnoliidae were apparently never as successful or diverse as the tricolpate-derived pollen-bearing dicotyledons or derived monocotyledons (e.g. Niklas, Tiffney & Knoll, 1980; Tiffney, 1981; Muller, 1984) and the initial success of the angiosperms in displacing the gymnosperms comes well after the Aptian origin of tricolpate angiosperms (Doyle *et al.*, 1977; Crane, this volume, Chapter 5). Tricolpate-derived taxa have diversified in association with advanced often faithful, insect pollinators (Crepet, 1984) and have also adapted well to wind pollination. Why did monosulcate-bearing dicotyledons never succeed to the degree attained by taxa with tricolpate-derived pollen and why were they unable to develop plant–pollinator relationships with more advanced insects? One possible explanation is that flowers of certain taxa of the monosulcate Magnoliidae (e.g. the Magnoliaceae and Nymphaeaceae) were too specialized to shift to new adaptive modes that would have allowed floral structure to evolve in response to new types of pollinators. Of course, the Magnoliidae did succeed in adapting to new and advanced insect pollinators in the sense that they were ancestral to tricolpate and tricolpate-derived pollen-bearing taxa. However, it is easier to imagine magnoliid ancestors of tricolpate-bearing taxa that were less specialized in floral structure, like Winteraceae or Chloranthaceae. The latter family is a particularly attractive candidate because of its antiquity (Walker & Walker, 1984; Friis *et al.*, 1986), and because it includes species with inaperturate, monosulcate, and multiaperturate pollen. Ancestral taxa with relatively simple flowers have persisted through the Early Cretaceous and then radiated in response to new selective conditions including, perhaps, the advent of more advanced pollinators or seasonal dry periods. The latter might also favor more efficient harmomegathic and germination mechanisms such as the tricolpate condition (Doyle, 1978) and may have been important in the origin of wind pollination (Whitehead, 1969; Crepet, 1981).

The recent discovery of inflorescences of the Platanaceae in the Albian of North America (Crane *et al.*, 1986) illustrates early adaptation to wind dissemination of pollen. Flowers simple enough to have the proposed evolutionary flexibility to enter a co-evolutionary relationship with advanced insect pollinators would also be logical ancestors to wind-pollinated taxa such as those in the Hamamelidales. Evolutionary modification of initially simple flowers for anemophily, and advanced entomophily, is more easily imagined than drastic simplification from *Magnolia*-like ancestors.

Pollination in Cenomanian angiosperms

The radiation of insect-pollinated dicotyledons with tricolpate-derived pollen had progressed considerably by the Cenomanian. The tricolpate or tricolpate-derived pollen flora was already diverse (Brenner, 1976; Doyle, 1978) and flowers with significant evolutionary innovations appear in the Cenomanian of the United States (Basinger & Dilcher, 1984). These flowers have low numbers of cyclically arranged floral parts and share certain morphological features with flowers of the modern Rhamnaceae. Their general morphology and large disk, presumed to be nectariferous, suggest that they were well adapted to pollination by flies (Primack, 1979), but pollination by wasps or beetles cannot be ruled out. Other Cenomanian angiosperm remains include fructifications and floral parts very similar to those of the modern Magnoliales (Dilcher & Crane, 1984; Friis & Crepet, this volume, Chapter 6). These fossils suggest that highly specialized beetle pollination was also well developed by the Cenomanian.

Pollination in Turonian–Campanian angiosperms

The most significant evidence related to insect pollination in the angiosperms during this time interval is the appearance of aculeate vespoid wasps in the Turonian (Rasnitzen, 1977), and the first record of haustellate Lepidoptera (butterflies and moths) in the Campanian (Common, 1975; Gall & Tiffney, 1983). The Turonian record of aculeate vespoid wasps further documents an early diversification of the Hymenoptera. Aculeate hymenopterans have the advantage of defense, through a modified ovipositer, that is thought to have played an important role in the development of sociality that apparently evolved first in the vespoid wasps (Burnham, 1978). Extant vespoid wasps are significant pollinators of small, radially symmetrical flowers with various degrees of connation in perianth parts (Proctor & Yeo, 1973).

Megafossil evidence of angiosperm flowers is infrequent from the beginning of this time interval, but becomes progressively better through the remainder of the Cretaceous. Fossil evidence on the flower and fruit structure of Santonian–Campanian angiosperms reveals remarkable uniformity in some features (Friis, 1984, 1985*a*,*b*; Friis & Crepet, this volume, Chapter 6). These angiosperm flowers are all small in size, the insect-pollinated types are radially symmetrical, and they often have well-developed nectaries. Frequently, they have low numbers of cyclically arranged floral parts and the majority of the floral types are epigynous. Their small size and simple structure are typical of what might be expected in taxa radiating from ancestors with small, relatively simple flowers, but these features may also reflect canalization associated with the predominant insect pollinators of the time. These Santonian–Campanian flowers are typical of those now pollinated by flies and wasps (Proctor & Yeo, 1973), but are unspecialized in the sense that they do not have floral features that would exclude other types of pollinators. The high proportion of epigynous flowers and robust styles suggest that potentially destructive insects such as beetles may have remained significant pollinators (Friis, 1985b; Friis & Crepet, this volume, Chapter 6). Tricolpate/tricolporate pollen in these taxa is almost uniformly too small (less than 10 μm) for effective wind dispersal (Whitehead, 1969) and may reflect adaptation to the limitations of the mouthparts of pre-syrphid flies (Proctor & Yeo, 1973) and perhaps other pollinators. The frequent presence of pollenkitt-like substances is consistent with the small pollen size in suggesting insect pollination (Friis & Crepet, this volume, Chapter 6). Other flowers from the same sediments contain larger (14 to 20 μm) triporate Normapolles pollen types that are particularly well adapted for wind dispersal, according to the criteria suggested by Whitehead (1969, 1983).

Syncarpy is dominant in the Santonian–Campanian flowers and the common style appears for the first time, suggesting intensified gametophytic competition and more efficient pollination (Mulcahy, 1979; Endress, 1982). Sympetally, a major character in flowers pollinated by nectar-feeding Lepidoptera and Hymenoptera, also makes its first appearance (Friis, 1985a; Friis & Crepet, this volume, Chapter 6). These features are important, since they are fundamental to further radiation with advanced insect pollinators. Angiosperm fossils other than flowers suggest that other derived floral features now associated with highly advanced pollinators were evolving during this time interval. Seeds of the Zingiberaceae have now been reported from the Campanian (Friis, 1988), wood of the Euphorbiaceae has been reported from the Coniacian (Mädel, 1962), and pollen of the Myrtaceae and Sapindaceae has been reported from the Santonian and early Senonian, respectively (Boltenhagen, 1976a,b; Belsky, Boltenhagen & Potonié, 1965).

The appearance of these characters might be evidence of increasing specialization in the flower-visiting insect taxa present since the Early Cretaceous, but might also reflect the emergence of major new groups of pollinators such as the haustellate Lepidoptera or even the bees.

Pollination in Maastrichtian angiosperms

Maastrichtian floral evidence is meager, but petrified myrtalean flowers (*Raoanthus*) from the Deccan Intertrappean Series are the first direct floral evidence of zygomorphy, although it is only slightly developed (Chitaley & Patel, 1975). Other palaeobotanical evidence including fruits, seeds, pollen and leaves reveals the presence of several families that have flowers specifically adapted for faithful pollinators or that have close relationships with such pollinators. These families include Caesalpinioideae (Van Campo, 1963; Monteillet & Lappartient, 1981), Proteaceae (Couper, 1960), Zingiberaceae (Hickey & Petersen, 1978) and Musaceae (Jain, 1963).

The Maastrichtian record of flower-visiting insects is more diverse, with a greater variety of lepidopterans and wasps (Common, 1975; Burnham, 1978), and with the implied presence of today's most important pollinating flies, the syrphids, before the end of the Cretaceous (Rohdendorf, 1974). Considering the early appearance of the sphecoid wasps (sister group to the bees) in the Early Cretaceous and the Turonian appearance of vespoid wasps, it is perplexing that bees are unknown in the Cretaceous fossil record.

The Maastrichtian appears to have been an important time in angiosperm evolution, with a major modernization beginning around that time and extending into the Early Tertiary (Muller, 1970, 1981, 1984; Niklas *et al.*, 1980; Tiffney, 1981), suggesting that we should evaluate carefully the possibility that bees evolved then or slightly earlier. There is now evidence to suggest such timing in the origin of bees (Michener, 1979; Crepet, 1984) and a Cretaceous origin of bees is widely hypothesized by hymenopterists on the basis of various kinds of evidence (e.g. Wilson, 1971; Burnham, 1978). A better perspective on the possibility that bees originated during the Campanian–Maastrichtian and on the probable status of other advanced pollinators, particularly the Lepidoptera, during the same interval can be gained by considering Paleocene–Eocene fossil floral structure.

Pollination in Paleocene angiosperms

There is considerable floral evidence to suggest that advanced lepidopterans and bees were important pollinators during the Early Tertiary (Crepet, 1984; Friis & Crepet, this volume, Chapter 6). Moderate-sized flowers with sympetalous funnelform corollas and narrow corolla tubes (possibly Rubiaceae) from the Middle Eocene suggest that lepidopterans with elongate mouthparts were well developed. This is confirmed by the presence of papilionid butterflies in deposits of similar age (Durden & Rose, 1978). Other sympetalous funnelform flowers of the same age have shorter and wider corolla tubes appropriate for pollination by long-tongued bees as well as by lepidopterans. Bilateral symmetry, a feature characteristically associated with bee pollination, occurs in small Middle Eocene flowers of unknown affinity and in flowers of probably zingiberaceous affinity of the same age (Crepet, 1984; Friis & Crepet, this volume, Chapter 6). Brush-type inflorescences that are typically pollinated by lepidopterans and by bees are present in Middle Eocene mimosoid legumes (Friis & Crepet, this volume, Chapter 6). Early Eocene open-funnelform sympetalous flowers of the Gentianaceae are typical of one type of bee pollination syndrome (Faegri & van der Pijl, 1971). The distinctive pollen from the anthers of these flowers, *Pistillipollenites*, is found dispersed from the Early Paleocene to the Eocene with no evidence of significant change in morphology, average size or ultrastructure (Crepet & Daghlian, 1981a), which may suggest that a similar floral structure extended back to the base of the Paleocene as well.

The legumes are one of the most diverse groups of flowering plants and their diversification into three subfamilies (or families) is thought to have involved increasing specialization within their bee pollinators (Raven & Polhill, 1981). Mimosoideae, and particularly Papilionoideae, have flowers that are very closely adapted to their bee pollinators. The presence of flowers of both these subfamilies in the Late Paleocene suggests that the diversification of legumes and their bee pollinators began even earlier.

Malpighiaceous flowers of Middle Eocene age provide more specific insights into bee pollination during the Early Tertiary (Taylor & Crepet, 1987). They are distinctive in having sepals with large, paired, abaxial oil glands and clawed petals (Figure 7.3(a)). The flowers have ten stamens in two whorls and pollen is tricolporate with reticulate supratectal ornamentation and unusual granular-anastomosing exine infrastructure.

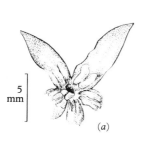

(a)

(b)

Figure 7.3. (a) Illustration of a fossil calyx of an Eocene mal-
pighiacean flower with paired elaiophores. (b) *Centris*, an anthophorid bee shown pollinating a modern flower of the Malpighiaceae
((a) After Taylor & Crepet, 1987; (b) after Vogel, 1974.)

Flowers with these features are confined to the modern Malpighiaceae,
a predominantly South American family with relatively few species in
the Old World tropics. The extemely close relationship that apparently
exists between the Malpighiaceae and three tribes of anthophorid bees
may be a major cause of the uniform floral structure evident in this
family. The bees pollinate the flowers by reaching around the clawed
bases of the petals to collect oil from the sepal glands (elaiophores) (Figure
7.3(b)). These pollinators are confined to the New World tropics where
all but a few derived taxa of Malpighiaceae have well-developed sepal
glands. In the Old World taxa, and in the absence of the anthophorid
oil-collecting pollinators, sepal glands are often reduced or entirely
absent. The presence of malpighiaceous flowers with well-developed
sepal glands in the Eocene of North America is not only evidence of bee
pollination, but also provides evidence of a particularly close association
between specific bee pollinators and certain flowers by the Eocene.

Another family of angiosperms with many bee-pollinated taxa is the
diverse and predominantly tropical Euphorbiaceae. Monoecious
staminate inflorescences of the tribe Hippomaneae are known from the
Late Paleocene and Middle Eocene (Crepet & Daghlian, 1981*b*) and
pseudanthia of the tribe Euphorbieae are also known from the Late
Paleocene (Friis & Crepet, this volume, Chapter 6). These two tribes are
the most highly advanced of the 40 extant tribes in Euphorbiaceae
(Hutchinson, 1969) and their presence as early as the Paleocene strongly
suggests that the family diversified, presumably in association with bee

pollinators, at least during the later part of the Cretaceous. This is consistent with the report of euphorbiaceous wood of Coniacian age (Mädel, 1982).

There is now considerable floral evidence that bees and advanced lepidopterans were important angiosperm pollinators during the Paleocene and Eocene. The diversity of apparently bee-pollinated angiosperms in the Early Tertiary along with the presence of highly derived specific floral morphologies closely associated with bee pollinators in extant plants strongly suggests that bees evolved no later than the Maastrichtian. This is also consistent with the fossil record of the Hymenoptera, which documents an Early Cretaceous origin of the sphecid wasps (sister group of the bees, see p. 187), and with interpretations of the biogeography of extant bees. Distribution patterns suggest an origin of bees, including the important stingless honeybees, the Meliponini, during the Late Cretaceous (Michener, 1979; Crepet, 1984). In this light, the appearance of certain derived floral features in a small proportion of the Campanian–Maastrichtian floral types (Friis & Crepet, this volume, Chapter 6) may well reflect the origin of the bees.

Even if the Paleocene–Eocene floral data are interpreted conservatively, they suggest that faithful pollinators, predominantly bees, were important during the major Maastrichtian–Paleocene radiation of the angiosperms that for the most part involved taxa with tricolpate or tricolpate-derived pollen. Our assessment of the available fossil evidence further suggests that many of the advantages accruing to the tricolpate pollen-bearing angiosperms, because of their association with bees and other faithful pollinators, were critical to their evolutionary success. The principal advantages were a possible relaxation of self-incompatibility due to the behavior of such pollinators, intensified gametophytic competition through higher percentages of appropriate exogenous pollen on the stigmas, energetic advantages, outcrossing in areas of low population density, and possible facilitation of the speciation process (Grant, 1949; Crepet, 1984; Friis & Crepet, this volume, Chapter 6).

Major evolutionary steps in insect pollination

(1) Circumstantial evidence suggests that insect pollination evolved in Upper Carboniferous seed ferns and was well developed in Jurassic Bennettitales. Coleoptera may have been the most significant early insect pollinators, but the size and structure of the cones of *Williamsoniella* suggest that they may have been pollinated by flies as well (Tipulidae and Mycetophilidae were contemporaneous).

(2) By the Early Cretaceous, specialization for coleopteran pollination had evolved in the bennettitalean genus *Cycadeoidea*.

(3) The earliest angiosperms were most likely insect pollinated and generalists with flowers simpler than those of the modern Magnoliaceae/Annonaceae and possibly self-incompatible. Possible pollinators include beetles, flies (Empididae), micropterigid moths, sawflies and sphecid wasps, all of which are now known from the Early Cretaceous.

(4) Specialization for beetle pollination evolved in the Magnoliidae by the Late Albian (Magnoliales) and was well-developed by the Cenomanian.

(5) Specialization for pollinators that fed on nectar as well as pollen had begun at least by the Cenomanian. There is evidence for the stabilization of number of floral parts, parts in whorls and nectariferous disks. Fossil evidence from later deposits suggests that the Cenomanian was the beginning of a radiation of primitive Rosidae that reached a maximum in the Santonian–Campanian. This radiation may have been in response to a greater variety of available nectar-feeding Hymenoptera (an additional group of wasps, the vespoids, appears in the Turonian) and Lepidoptera (haustellate forms appear in the Campanian). It may also have involved increasing specialization (including mouthparts adapted to allow nectar feeding) and diversity in beetle pollinators, because a high percentage of the flowers are epigynous with stout styles.

(6) Certain floral characteristics now associated with advanced pollinators began to appear in the Santonian–Campanian including fusion of floral parts and bisymmetrical flowers. The common style appears for the first time, suggesting more efficient pollination and increased gametophytic competition. The presence of zygomorphy is suggested by seeds of the Zingiberidae in the Campanian, but there is no confirming floral evidence.

(7) The Maastrichtian provides the first floral evidence of zygomorphy, albeit weakly developed, and palynological evidence suggests that several families that now have floral or inflorescence structure modified for advanced pollinators existed at that time. Biogeographic evidence suggests that bees originated before the end of the Cretaceous. This is consistent with the available palynological, floral and other megafossil evidence.

(8) The Maastrichtian was the beginning of a second major radiation or modernization of angiosperms. This radiation apparently involved advanced pollinators (bees and haustellate Lepidoptera)

and such advanced Rosidae as the Leguminosae and Euphorbiaceae among other major angiosperm taxa.

(9) Early Tertiary evidence is consistent with a Late Cretaceous origin of bees and a subsequent radiation of angiosperms. Advanced floral and inflorescence types appeared by the Late Paleocene, as did families whose diversification has been associated with bee pollinators.

(10) Refinements in plant–advanced-pollinator relationships characterize the Eocene–post-Eocene. The relationship between the Malpighiaceae and the anthophorid bees, for example, is well established by the Eocene.

Conclusions

The fossil record of insects and angiosperms provides a tentative chronology of important events in the history of insect pollination and raises certain questions, particularly about the relationship between specific angiosperm radiations and the advent of certain types of insect pollinators.

New fossil data will continue to improve our understanding of the chronology of events in the relationship between angiosperms and insect pollinators, and continuing studies of pollination ecology and population biology will enhance our appreciation of the significance of these interactions.

We thank W. G. Chaloner, P. R. Crane and C. Maier for helpful suggestions and constructive criticism. Thanks are also due to Mary Jane Spring for preparing the illustrations. The work has been supported by a Niels Bohr Fellowship (E. M. F.), the NATO Senior Fellowship Scheme (E. M. F.) and by the National Science Foundation Grant DEB810217 (W. L. C.).

References ·

Arber, E. A. N. & Parkin, J. (1907). On the origin of angiosperms. *Botanical Journal of the Linnean Society*, **38**, 29–80.

Baker, H. B. & Hurd, P. D. (1968). Intrafloral ecology. *Annual Review of Entomology*, **13**, 385–414.

Basinger, J. F. & Dilcher, D. L. (1984). Ancient bisexual flowers. *Science*, **224**, 511–13.

Belsky, D. Y., Boltenhagen, E. & Potonié, R. (1965). Sporae dispersae der Oberen Kreide von Gabun, Äqatoriales Afrika. *Paläontologische Zeitschrift*, **39**, 72–83.

Boltenhagen, E. (1976*a*). Pollens et spores Sénoniennes du Gabon. *Cahiers de Micropaléontologie*, **3**, 1–21.

Boltenhagen, E. (1976*b*). La microflore Sénonienne du Gabon. *Revue de Micropaléontologie*, **18**, 191–9.

Brenner, G. J. (1976). Middle Cretaceous floral provinces and early migrations of angiosperms. In *Origin and Early Evolution of Angiosperms* ed. C. B. Beck, pp. 23–47. New York: Columbia University Press.

Burnham, L. (1978). Survey of social insects in the fossil record. *Psyche*, **85**, 85–133.

Carlquist, S. (1969). Toward acceptable evolutionary interpretations of floral anatomy. *Phytomorphology*, **19**, 332–62.

Carpenter, F. H. (1976). Geological history and the evolution of the insects. *Proceedings of the 15th Congress of Entomology*, ed. D. White, pp. 63–70. Washington, DC: American Entomological Society.

Chitaley, S. D. & Patel, M. Z. (1975). *Raoanthus intertrappea*, a new petrified flower from India. *Palaeontographica B*, **153**, 141–9.

Common, I. F. B. (1975). Evolution and classification of the Lepidoptera. *Annual Review of Entomology*, **20**, 183–203.

Couper, R. A. (1960). New Zealand Mesozoic and Cainozoic plant microfossils. *New Zealand Geological Survey Paleontology Bulletin*, **32**, 1–87.

Crane, P. R. (1985). Phylogenetic analysis of seed plants and the origin of angiosperms. *Annals of the Missouri Botanical Garden*, **72**, 716–93.

Crane, P. R., Friis, E. M. & Pedersen, K. R. (1986). Angiosperm flowers from the Lower Cretaceous: fossil evidence on the early radiation of the dicotyledons. *Science*, **232**, 852–4.

Crepet, W. L. (1974). Investigations of North American cycadeoids: the reproductive biology of *Cycadeoidea*. *Palaeontographica B*, **148**, 144–59.

Crepet, W. L. (1979). Insect pollination: a paleontological perspective. *BioScience*, **29**, 102–8.

Crepet, W. L. (1981). The status of certain families of the Amentiferae during the Middle Eocene and some hypotheses regarding the evolution of wind pollination in dicotyledonous angiosperms. In *Paleobotany, Paleoecology and Evolution*, vol. 2, ed. K. J. Niklas, pp. 103–28. New York: Praeger.

Crepet, W. L. (1983). The role of insect pollination in the evolution of the angiosperms. In *Pollination Biology*, ed. L. Real, pp. 29–50. London: Academic Press.

Crepet, W. L. (1984). Advanced (constant) insect pollination mechanisms: pattern of evolution and implications *vis-à-vis* angiosperm diversity. *Annals of the Missouri Botanical Garden*, **71**, 607–30.

Crepet, W. L. & Daghlian, C. P. (1981*a*). Lower Eocene and Paleocene Gentianaceae: floral and palynological evidence. *Science*, **214**, 75–7.

Crepet, W. L. & Daghlian, C. P. (1981*b*). Euphorbioid inflorescences from the

Middle Eocene Clairborne Formation. *American Journal of Botany*, **69**, 258–66.

Dilcher, D. L. & Crane, P. R. (1984). *Archaeanthus*: an early angiosperm from the Cenomanian of the Western Interior of North America. *Annals of the Missouri Botanical Garden*, **71**, 351–83.

Doyle, J. A. (1977). Patterns of evolution in early angiosperms. In *Patterns of Evolution as Illustrated by the Fossil Record*, ed. A. Hallam, pp. 501–46. Amsterdam: Elsevier.

Doyle, J. A. (1978). Origin of angiosperms. *Annual Review of Ecology and Systematics*, **9**, 365–92.

Doyle, J. A., Biens, P., Doerenkamp, A. & Jardiné, S. (1977). Angiosperm pollen from the pre-Albian Lower Cretaceous of Equatorial Africa. *Bulletin de Centre de Recherche d'Exploration–Production Elf-Aquitaine*, **1**, 451–73.

Durden, C. J. & Rose, H. (1978). Butterflies from the Middle Eocene: earliest occurrence of fossil Papilionoidea (Lepidoptera). *The Texas Memorial Museum, Pearce-Sellards Series*, **29**, 1–25.

Endress, P. K. (1982). Syncarpy and alternative modes of escaping disadvantages of apocarpy in primitive angiosperms. *Taxon*, **31**, 48–52.

Evans, H. E. (1969). Three new Cretaceous aculeate wasps (Hymenoptera). *Psyche*, **76**, 251–61.

Evans, H. E. (1973). Cretaceous aculeate wasps from Taimyr, Siberia (Hymenoptera). *Psyche*, **80**, 166–78.

Faegri, K. & van der Pijl, L. (1971). *The Principles of Pollination Ecology*. Oxford: Pergamon Press.

Friis, E. M. (1984). Preliminary report of Upper Cretaceous angiosperm reproductive organs from Sweden and their level of organization. *Annals of the Missouri Botanical Garden*, **71**, 403–18.

Friis, E. M. (1985a). *Actinocalyx* gen nov., sympetalous angiosperm flowers from the Upper Cretaceous of Southern Sweden. *Review of Palaeobatany and Palynology*, **45**, 171–83.

Friis, E. M. (1985b). Structure and function in Late Cretaceous angiosperm flowers. *Biologiske Skifter Danske Videnskabernes Selskab*, **25**, 1–37.

Friis, E. M. (1988). *Spirematospermum chandlerae* sp. nov., an extinct species of Zingiberaceae from the North American Cretaceous. *Tertiary Research*, **9**, 7–12.

Friis, E. M., Crane, P. R., & Pedersen, K. R. (1986). Floral evidence for Cretaceous chloranthoid angiosperms. *Nature*, **320**, 163–4.

Gall, L. F. & Tiffney, B. H. (1983). A fossil noctuid moth egg from the Late Cretaceous of Eastern North America. *Science*, **219**, 507–9.

Givnish, T. J. (1982). Outcrossing versus ecological constraints in the evolution of dioecy. *American Naturalist*, **119**, 849–65.

Gottsberger, G. (1974). The structure and function of the primitive angiosperm flower – a discussion. *Acta Botanica Neerlandica*, **23**, 461–71.

Gottsberger, G. (1977). Some aspects of beetle pollination in the evolution of flowering plants. *Plant Systematics and Evolution, Supplement,* 1, 211–26.

Grant, V. (1949). Pollinating systems as isolating mechanisms. *Evolution,* 3, 82–97.

Grant, V. (1950). The pollination of *Calycanthus occidentalis. American Journal of Botany,* 37, 294–97.

Hammen, T. van der & Gonzalez, E. (1960). Upper Pleistocene and Holocene climate and vegetation of the Sabana de Bogota (Columbia, South America). *Leidse Geologische Mededelingen,* 25, 261–315.

Handlirsch, A. (1906–8). *Die fossilen Insekten und die Phylogenie der Rezenten Formen.* Leipzig: Engelmann.

Hennig, W. (1970). Insektenfossilien aus der unteren Kreide. II. Empididae (Diptera, Brachycera). *Stuttgarter Beiträge zur Naturkunde,* 214, 1–12.

Hickey, L. J. & Doyle, J. A. (1977). Early Cretaceous fossil evidence for angiosperm evolution. *The Botanical Review,* 43, 3–104.

Hickey, L. J. & Peterson, R. K. (1978). *Zingiberopsis,* a fossil genus of the ginger family from Late Cretaceous to Early Eocene sediments of Western Interior North America. *Canadian Journal of Botany,* 56, 1136–52.

Hutchinson, J. (1969). Tribalism in the Euphorbiaceae. *American Journal of Botany,* 56, 738–58.

Jain, R. K. (1963). Studies in Musaceae. 1. *Musa cardiosperma* sp. nov., a fossil banana fruit from the Deccan Intertrappean Series. *The Palaeobotanist,* 12, 45–8.

Jarzembowski, E. A. (1984). Early Cretaceous insects from Southern England. *Modern Geology,* 9, 71–93.

Kevan, P. C. & Baker, H. G. (1983). Insects as flower visitors and pollinators. *Annual Review of Entomology,* 28, 407–53.

Krassilov, V. A. & Rasnitzen, A. P. (1982). A unique find: pollen in the intestine of Early Cretaceous saw flies. *Paleontological Journal,* 16, 80–97.

Mädel, E. (1962). Die fossilen Euphorbiaceen-Hölzer mit besonderer Berücksichtigung neuer Funde aus der Oberkreide Süd-Afrikas. *Senckenbergiana lethaea,* 43, 283–321.

Meeuse, B. J. D. & Schneider, E. L. (1980). *Nymphaea* revisited: a preliminary communication. *Israel Journal of Botany,* 28, 65–79.

Michener, C. D. (1979). Biogeography of the bees. *Annals of the Missouri Botanical Garden,* 66, 277–347.

Monteillet, J. & Lappartient, J. R. (1981). Fruits et graines du Crétacé Supérieur des carrières de Paki (Sénégal). *Review of Palaeobotany and Palynology,* 34, 331–44.

Mulcahy, D. L. (1979). The rise of the angiosperms: a genecological factor. *Science,* 206, 20–3.

Muller, J. (1970). Palynological evidence on early differentiation of angiosperms. *Biological Reviews,* 45, 417–50.

Muller, J. (1981). Fossil pollen records of extant angiosperms. *The Botanical Review*, **47**, 1–142.

Muller, J. (1984). Significance of fossil pollen for angiosperm history. *Annals of the Missouri Botanical Garden*, **71**, 419–43.

Niklas, K. J., Tiffney, B. H. & Knoll, A. (1980). Apparent changes in the diversity of fossil plants. *Evolutionary Biology*, **12**, 1–89.

Primack, R. B. (1979). Reproductive biology of *Discaria toumatou* (Rhamnaceae). *New Zealand Journal of Botany*, **17**, 9–13.

Proctor, M. & Yeo, P. (1973). *The Pollination of Flowers*. London: Collins.

Rasnitzen, A. P. (1977). New Jurassic and Cretaceous hymenopterans of Asia. *Paleontological Journal*, **11**, 349–57.

Raven, P. H. & Polhill, R. M. (1981). Biogeography of the Leguminosae. In *Advances in Legume Systematics*, vol. 1 *Proceedings of the International Legume Conference, Kew, 1978*, ed. R. M. Polhill & P. H. Raven, pp. 27–34. Kew: Royal Botanical Gardens.

Rohdendorf. B. (1974). *The Historical Development of Diptera*. Canada: University of Alberta Press.

Schatz, G. E. & Young, H. J. (1985). Patterns of visitation by dynastine scarab pollinators in a Costa Rican tropical wet forest. In *Third International Congress of Systematic and Evolutionary Botany*, Abstract p. 167.

Schneider, E. L. (1979). Pollination biology of the Nymphaceae. In Proceedings of the IVth International Symposium on Pollination. *Maryland Agricultural Experimental Station Special Miscellaneous Publications*, **1**, 419–29.

Scott, A. C. & Taylor, T. N. (1983). Plant/animal interactions during the Upper Carboniferous. *The Botanical Review*, **49**, 259–307.

Stebbins, G. L. (1974). *Flowering Plants: Evolution above the Species Level*. Cambridge, Massachusetts: Belknap.

Taylor, D. W. & Crepet, W. L. (1987). Fossil floral evidence of Malpighiaceae and an early plant–pollinator relationship. *American Journal of Botany*, **74**, 274–86.

Thien, L. B. (1974). Floral biology of *Magnolia*. *American Journal of Botany*, **61**, 1037–45.

Thien, L. B., Bernhardt, P., Gibbs, G. W., Pellmyr, O., Bergström, G., Groth, I. & McPherson, G. (1985). The pollination of *Zygogynum* (Winteraceae) by a moth, *Sabatinca* (Micropterigidae): an ancient association? *Science*, **227**, 540–2.

Thien, L. B., White, D. A. & Yatsu, L. Y. (1983). The reproductive biology of a relict – *Illicium floridanum* Ellis. *American Journal of Botany*, **70**, 719–27.

Tiffney, B. H. (1981). Diversity and major events in the evolution of land plants. In *Paleobotany, Paleoecology and Evolution*, vol. 2, ed. K. J. Niklas, pp. 193–230. New York: Praeger.

Tiffney, B. H. (1984). Seed size, dispersal syndromes, and the rise of the angiosperms: evidence and hypothesis. *Annals of the Missouri Botanical Garden*, 71, 551–76.

Upchurch, G. R. (1984*a*). Cuticle evolution in Early Cretaceous angiosperms from the Potomac Group of Virginia and Maryland. *Annals of the Missouri Botanical Garden*, 71, 522–50.

Upchurch, G. R. (1984*b*). Cuticular anatomy of angiosperm leaves from the Lower Cretaceous Potomac Group. I. Zone I leaves. *American Journal of Botany*, 71, 192–202.

Van Campo, M. (1963). Quelques réflexions sur les pollens de *Sindroa*. *Grana Palynologica*, 4, 361–6.

Vogel, S. (1974). Ölblumen und Ölsammelnde Bienen. In *Tropische Subtropische Pflanzenwelt*, vol. 7, ed. W. Rauh, pp. 265–80. Wiesbaden: Steiner.

Walker, J. W., Brenner, G. J. & Walker, A. G. (1983). Winteraceous pollen in the Lower Cretaceous of Israel: early evidence of a magnolialean family. *Science*, 220, 1273–5.

Walker, J. W. & Skvarla, J. (1975). Primitively columellaless pollen: a new concept in the evolutionary morphology of angiosperms. *Science*, 187, 445–7.

Walker, J. W. & Walker, A. G. (1984). Ultrastructure of Lower Cretaceous angiosperm pollen and the origin and early evolution of flowering plants. *Annals of the Missouri Botanical Garden*, 71, 464–521.

Whalley, P. (1977). Lower Cretaceous Lepidoptera. *Nature*, 266, 526.

Whitehead, D. R. (1969). Wind pollination in the angiosperms: evolutionary and environmental considerations. *Evolution*, 23, 28–35.

Whitehead, D. R. (1983). Wind pollination: some ecological and evolutionary perspectives. In *Pollination Biology*, ed. L. Real, pp. 97–108. London: Academic Press.

Wilson, E. O. (1971). *The Insect Societies*. Cambridge, Massachusetts: Belknap.

Zavada, M. S. (1984). The relation between pollen exine sculpturing and self-incompatibility mechanisms. *Plant Systematics and Evolution*, 147, 63–78.

Zavada, M. S. & Taylor, T. N. (1985). The role of self-incompatibility and sexual selection in the Gymnosperm–Angiosperm transition: a hypothesis. *Abstracts of the Botanical Society of America*, p. 904.

8

Interactions of angiosperms and herbivorous tetrapods through time

S. L. WING AND B. H. TIFFNEY

Herbivory, along with climate and interplant competition, is one of the central factors influencing the survival of plant species. Herbivores span a great range of sizes and dietary types, from tiny monophagous insects to giant polyphagous tetrapods; vertebrate herbivory, however, is especially significant in affecting the structure of vegetation and the dispersal of angiosperm seeds (Crawley, 1983; Janzen, 1983). The importance of this interaction is reflected in the great diversity of herbivore-related structural adaptations in plants (toxins, spines, attractants, basal meristems etc.), and in the wide range of feeding morphologies and behaviors seen in vertebrates. Although co-evolution of plants with pollinators and insect herbivores may involve a close and reciprocal relationship between two species or lineages, co-evolution of angiosperms with vertebrate herbivores and dispersers appears to be a less precise phenomenon, with distantly related groups of plant and animal species having similar evolutionary effects on one another. This latter concept has been termed 'diffuse co-evolution' (Herrera, 1985). Still, morphological structures in both plants and animals bespeak the importance to, and effect of, each on the other, reflecting a dynamic interaction of long standing and great evolutionary importance.

In looking at the evolution of plants, palaeobotanists have tended to assume that the primary selective forces were those of climate and competition between different lineages of plants. Herbivory, if considered at all, appears to have been treated as a stable pressure, the same in the past as today. This approach overlooks the possibility that (1) past herbivores may have exerted different kinds of selective pressures on plants, and (2) the relative proportions of different kinds of herbivory may have changed through time, creating different selective regimes. In short, the selective pressures exerted by extinct herbivore faunas may not have been equivalent to those exerted by living faunas: such differences could have molded past plant lineages and communities in ways very different from those observed in present communities.

In this paper we examine the Cretaceous and Tertiary fossil record of angiosperms and vertebrate herbivores to ascertain the history of interaction between the groups, and the effects of this interaction on the evolution of both. We conclude that changing patterns of herbivory have had a strong influence on the structure and composition of terrestrial communities, as well as on the evolutionary history of the component lineages.

Methods and data

Herbivory is a process involving behavior and population dynamics of both plants and animals. Palaeontology, by contrast, is largely a morphological undertaking, restricted to those parts of plants and animals that can be preserved as fossils. Fortunately, these fossilizable aspects of shape also convey information about size and diversity; all three are significant data in present-day ecological studies. Hence, it is possible to infer past ecological interactions through morphological analogies with extant organisms and systems. Such inferences can be made independent of systematic affinity, and permit us to escape the potential trap of assuming ecological features and community structure on the basis of the taxonomic affinities of the organisms in question.

Five morphological aspects of past plants and vertebrate herbivores were chosen as being indicative of plant–herbivore interactions in the Cretaceous and Tertiary. Each is recorded in fossil remains and is important in present-day communities.

Several qualifications apply to our fossil data. The data are widely spaced stratigraphically and thus display only long-term trends. Further, they are almost exclusively Northern Hemispheric, largely North American, in origin. The climatic bias inherent in this latitudinal restriction is not as severe as might appear at first, because Cretaceous and Paleogene climates were more equable than at present, and climatic gradients were greatly reduced (Wolfe, 1985; Upchurch & Wolfe, this volume, Chapter 4). When present, data from the Southern Hemisphere do not contradict our conclusions.

Herbivore body size

Body size is ecologically revealing because it is highly correlated with metabolic rate, which in turn influences diet and many aspects of herbivore distribution and life history (Damuth, 1981; Western, 1979). Large herbivores have greater total energy requirements and lower metabolic rates than do small herbivores (Jarman & Sinclair, 1979; Peters, 1983). In order to meet their large overall requirements, large herbivores tend to feed more or less continuously on a wide range of plant parts; much of the material consumed is high in fibre and low in energy. Low metabolic rate permits large herbivores to derive energy from cellulose by retaining it in the gut for long periods of microbial fermentation (Janis, 1976; Demment & Van Soest, 1985).

Because of their high metabolic rates small herbivores effectively have less gut capacity with which to meet their energetic needs. Consequently they must consume higher energy foodstuff (Demment & Van Soest, 1985), and may focus on fruits, seeds or supplemental animal protein. This requirement forces them to feed selectively, rather than to consume large quantities of low energy plant material.

These relationships are general across all living herbivores, and should apply as well to tetrapod herbivores of the past (Damuth, 1982). As a result, we will assume that herbivore body size provides an approximate measure of the nature and quality of herbivore diet.

The distribution of past herbivore body sizes is displayed in Figure 8.1. The weights of extinct herbivores were estimated only to order of magnitude, because greater precision was difficult to achieve for many species, and because small differences in body size have little effect on diet.

Late Jurassic and Early Cretaceous herbivore faunas (Dodson *et al.*, 1980; Ostrom, 1970; D. B. Weishampel, personal communication) were dominated by species of dinosaurs in the largest-size categories (> 1000 kg); diversity of intermediate-sized animals (1 to 1000 kg) was low, and only a few species of small (< 1 kg) herbivores were present. (The relatively high diversity of species under 100 g in the Morrison fauna probably represents oversplitting of the mammal genus *Docodon*, according to Jenkins (1969).) Because of their large body sizes, these herbivores would presumably have had low metabolic rates (Ostrom, 1980; Spotila, 1980; Hotton, 1980; Weaver, 1983) and therefore a large consumption of low-quality fodder and slow passage of material through

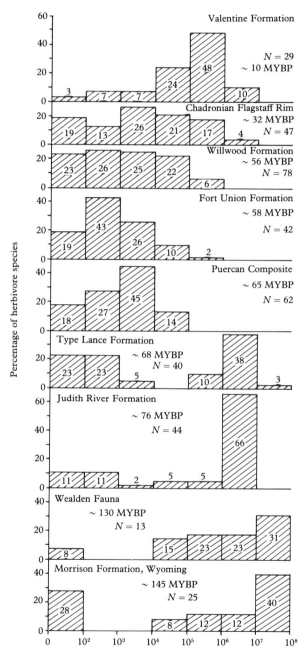

Figure 8.1. Distribution of body sizes (in grams) of herbivorous tetrapod faunas during the last 160 million years. N, the number of herbivore species in a given fauna. Note the sharp contrast between the body size distribution of the Mesozoic and early Cenozoic faunas. MYBP, million years BP.

the gut. The low diversity of all but the largest-size classes clearly distinguishes these from subsequent faunas.

Campanian and Maastrichtian herbivore faunas maintained a high diversity of species, mostly ornithopods, above 1000 kg, but some smaller herbivorous dinosaurs were also present. Ankylosaurs and pachycephalosaurs probably fell within the 100 to 1000 kg range, while hypsilophodonts and fabrosaurs were small enough (10 to 100 kg) that some species may have had a high metabolic rate and a specialized diet. Diversity of small (< 1 kg) mammalian herbivores, mostly marsupials and multituberculates, also increased during this period. The Late Maastrichtian Lance fauna (Estes & Berberian, 1970; Clemens, 1973) had almost equal diversity of small mammalian and large dinosaurian herbivores, and one species of multituberculate that exceeded 1 kg in weight. The small size of these mammals suggests that, if they were herbivorous, they probably consumed high-energy food such as fruits and seeds.

The demise of the dinosaurs at the Cretaceous–Tertiary (K/T) boundary drastically changed this pattern; all herbivores greater than 10 kg were eliminated. During the Paleocene and Eocene, mammals radiated to fill a variety of niches, but, in spite of this diversification, it was not until the latest Eocene or Early Oligocene that a large portion of the herbivore fauna exceeded 100 kg in size. Although conveyed by only three faunas in Figure 8.1, this appears to be a world-wide phenomenon. This change in body size suggests a sudden rise in the importance and consumption of high-energy food sources, particularly fruits and seeds.

Oligocene and Miocene (Emry, 1973; Webb, 1969) faunas show a trend to larger herbivores (the near-absence of small herbivores in the Burge fauna is probably a taphonomic bias). The high diversity of animals between 10 and 1000 kg is typical of modern faunas of open vegetation, e.g. the Serengeti, but not of forested areas (Fleming, 1973; Figure 8.2). While small herbivores dependent on high-energy sources remained diverse in these faunas, increasing diversity in larger size categories suggests the resurgence of generalist herbivores with lower metabolic rates.

Herbivore diet and locomotion

Tooth morphology distinguishes herbivores from carnivores, and, within herbivores, may reflect general diet types such as frugivore/omnivore, folivore/browser, and grazer. With these three broadly defined

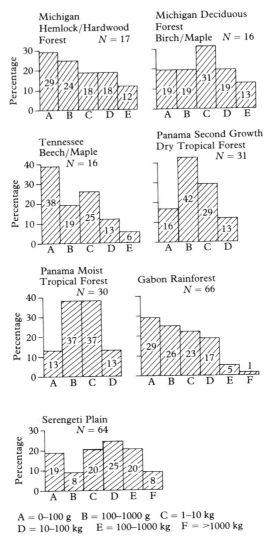

A = 0–100 g   B = 100–1000 g   C = 1–10 kg
D = 10–100 kg   E = 100–1000 kg   F = >1000 kg

Figure 8.2. Distribution of body sizes of terrestrial herbivorous/
omnivorous mammals in a variety of present-day vegetational types.
N, the number of species in a given fauna. Note the greater diversity
of large herbivorous mammals in the open vegetational setting of the
Serengeti. (North and Central American data of Fleming (1973);
Gabon data of Emmons, Gautier-Hion & Dubost (1983); Serengeti
data from field guides.)

herbivorous dietary types in mind, we have assembled information on diet from the palaeozoological literature. Herbivore locomotor ability also affects the kind of plant food that can be reached. Volant and arboreal forms may feed on plant parts in the upper canopy that are generally less accessible to terrestrial herbivores. As with dentally defined diet type, we have assembled information from the literature on probable locomotor style in our sequence of fossil faunas.

The dominant herbivores of the Late Jurassic were giant sauropods. Their dentition indicates that they fed on foliage, garnered by drawing branches through their rake-like teeth (Coombs, 1975). The absence of wear from tooth-to-tooth occlusion suggests that food was not chewed in the mouth, but processed in a gastric mill. All these observations are in keeping with the large-volume, low-quality diet presumed for these animals from their size. The large bodies of Jurassic and Early Cretaceous sauropods suggest a mobile life in an open habitat. The long necks of some members would have permitted browsing 10 to 12 m above the ground (Bakker, 1978). Stegosaurs were terrestrial forms of much lower stature that also lacked chewing teeth; together, sauropods and stego-saurs made up 95% of the biomass of tetrapod communities of the Middle and Late Jurassic (Bakker, 1978; see also Coe *et al.*, this volume, Chapter 9).

Late Jurassic mammals included possible herbivores (multituber-culates and docodonts), but their dentition is too generalized to identify specific diet (Kron, 1979; Clemens & Kielan-Jaworowska, 1979). The small size of these organisms suggests dependence on energy-rich food sources.

Early Cretaceous faunas demonstrate the beginnings of a shift from the dominance of sauropods to that of ornithopods. The ornithopods were smaller and shorter than sauropods, and possessed occluding dentitions suggestive of an ability to comminute food before swallowing (Weishampel, 1984). These differences in size and dentition suggest the possibility that ornithopods may have had higher metabolic rates than sauropods, but their large size indicates that they were still generalist feeders.

Ornithopods, especially hadrosaurs and ceratopsians, dominated Late Cretaceous herbivore communities. Both possessed complex batteries of posterior occluding teeth coupled with a toothless, anterior beak (Ostrom, 1964; Weishampel, 1984; Norman & Weishampel, 1985). Bakker (1978) has suggested that their lower stature was in response to feeding on angiosperms, which are presumed to have been small plants, at least in their early history.

Multituberculates were the most diverse and abundant group of possibly herbivorous latest Cretaceous mammals. Some Paleocene members of this group had dentitions suited for eating fruits and seeds (Krause, 1982) and postcranial skeletons suggesting arboreality (Krause & Jenkins, 1983), but Cretaceous forms are poorly known. A possibly frugivorous marsupial, *Glasbius*, is reported from the Maastrichtian Lance fauna (Clemens, 1966, 1973).

The dentitions of the radiating mammals of the Paleocene and Eocene provides evidence of rapid ecological specialization. Multituberculates, primates and rodents all evolved gliriform dentitions suggestive of fruit/seed specialization. Some condylarths, dermopterans and pantodonts were probably folivorous (Rose, 1981; Rose & Krause, 1982). Some larger taeniodonts may have specialized in rooting (Coombs, 1983), but other larger animals (some condylarths and pantodonts) retained generalized dentitions.

It is of note that both bats (Rose, 1984) and birds (Tiffney, 1984; Olson, 1985) also radiated in the early Tertiary. While evidence of frugivory is sparse among early Tertiary birds and absent in bats until the Pleistocene, the diversification of these two groups of small, specialist, herbivores at the same time as small mammals is consistent with the hypothesis that they were responding to a common stimulus.

The post-cranial remains of Tertiary mammals parallel the story told by dentition. Paleocene and Eocene faunas include many clearly arboreal members, although the representation of arboreal forms dropped off in later Eocene faunas of western North America (Stucky & Krishtalka, 1985). During the Paleocene and most of the Eocene, the larger elements of the fauna were not diverse, and many were low-built animals with short legs. Larger, longer-limbed herbivores, presumably living in open habitats, diversified in the Oligocene and Miocene (Webb, 1977).

Diaspore size

In living plants, diaspore size is correlated with the habitat of growth of the parent plant and the mode of dispersal of the diaspore. Endosperm represents stored energy. Seedlings arising from small seeds must commence photosynthesis immediately in order to survive. Hence, taxa with small seeds tend to live in open, light-rich areas. Seedlings arising from large seeds may grow to a substantial size, supported jointly by endosperm and photosynthesis. As a result, larger-seeded taxa may exist

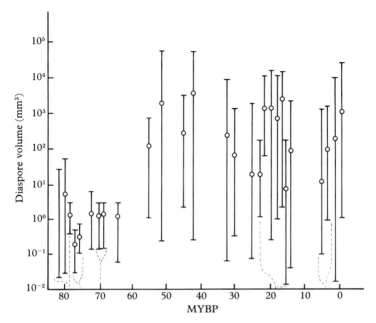

Figure 8.3. Distribution of angiosperm diaspore sizes in Cretaceous and Tertiary floras. Note the dominance of very small forms in the Cretaceous and the great increase in range of size beginning in the Paleocene. (Data from Tiffney, 1984, figure 2.) Empty circles are means for assemblages; horizontal bars are maxima and minima. MYBP, million years BP.

in closed, light-poor communities, where seedlings of small-seeded forms would perish (Harper, Lovell & Moore, 1970; Ng, 1978). The relationship between seed size and dispersal is more complicated, but, in general, large seeds tend to be transported by vertebrates, while smaller seeds may be transported by animals or abiotically (by wind or water; cf. Ridley, 1930; van der Pijl, 1982). Data from seed size in the Cretaceous and Tertiary (Figure 8.3) come from Tiffney (1984).

The oldest fossil angiosperm diaspores are individual, small (1 to 3 mm) fruits and seeds reported from scattered localities. Fruit and seed floras were first reported from the Campanian of Sweden (Friis, 1984) and Massachusetts, USA (Tiffney, 1977, and unpublished material). Both floras contain about 50 entities; no fruit or seed exceeds 3 mm in diameter. In both cases, the sedimentary context and other fossil remains preclude simple taphonomic bias. Small fruits and seeds dominate the record until the early Tertiary, although larger diaspores (e.g. Monteillet

& Lappartient, 1981) began to appear in the Campanian–Maastrichtian interval. Many of these larger diaspores may not have been angiospermous or were large fruits containing seeds of unknown size. Only in the Late Paleocene did large seeds, and thus large endosperm reserves, become common constituents of floras (e.g. the London Clay Flora; Reid & Chandler, 1933; Chandler, 1961). Small seeds remained present; the change involved addition of larger seeds, not decrease in numbers of small ones. In the later Tertiary, large seeds became less numerous, but were still present in fruit and seed floras. This is probably a bias of cooling climates in the Northern Hemisphere. (For further data and references on the foregoing summary, see Tiffney, 1984.)

Change in dispersal syndromes is largely inferential. Cretaceous diaspores were small, and so could have been abiotically dispersed in the manner of many modern small diaspores. Small seeds may have been dispersed by animals if they were embedded in larger fruits, or if they were incidentally consumed during generalized browsing by large herbivores (cf. Janzen, 1984). However, not until the early Tertiary did fruits and seeds exhibit clear adaptations (thick walls, attractive flesh) to animal dispersal.

Angiosperm stature

Plant stature provides information about the structure of the vegetation and about the life history of the plant; larger trees are generally longer lived and reproduce more slowly than do small plants (Harper, 1977; Strauss & Ledig, 1985). Furthermore, the three-dimensional structure of vegetation influences the kind of fauna supported by an area (e.g. Dubost, 1979). Here we summarize scattered reports of fossil angiosperm wood in order to obtain a broad idea of the role of angiosperms in Cretaceous and Tertiary vegetation.

There are remarkably few reliable reports of Early Cretaceous angiosperm wood. Angiosperm wood has not been reported from angiosperm leaf localities in the Lower Cretaceous Potomac Group, although small and large fragments of conifer wood are present in the same deposits (Hickey & Doyle, 1977). Early Cretaceous wood from England (Stopes, 1912), Madagascar (Fliche, 1905) and Japan (Suzuki & Nishida, 1974) are of questionable age and provenance (Wolfe, Doyle & Page, 1975; Page, 1981). Wood reported by Nishida (1962) may not be angiospermous (Page, 1981), and wood reported by Spackman (1948) may be Late Cretaceous. The oldest well-dated angiosperm woods are small (in the range 5 to 10 cm width) fragments of *Paraphyllanthoxylon* and *Icacinoxylon* from the Late Albian Cedar Mountain Formation of Utah,

USA (Thayn, Tidwell & Stokes, 1983, 1985). The size of the logs that produced these fragments is not clear.

Cenomanian–Campanian angiosperm wood is also rare. Bailey (1924) reported *Paraphyllanthoxylon* logs over 1 ft (30 cm) in diameter from the Late Cretaceous of Arizona, USA. The largest angiosperm trunks from this interval are 1 m diameter logs of *Paraphyllanthoxylon* that occur abundantly in late Cenomanian, or younger, channel sands of the Gordo Formation in Alabama, USA (Cahoon, 1972).

A diverse angiosperm wood flora has been reported from Campanian–Maastrichtian marine deposits of California; 70% of the nearly 200 specimens are between 2.5 and 5 cm in diameter (Page, 1981, and references therein). The small size of these wood fragments may in part reflect water transport of the assemblage, because approximately coeval rocks in Illinois and New Mexico, USA contain large angiosperm trunks (E. Wheeler, personal communication; J. McClammer, personal communication).

Large angiosperm trunks are relatively abundant and diverse in many Tertiary deposits. Ten species of angiosperm woods were reported from probable Lower Paleocene marginal marine sediments in Niger (Koeniguer, 1971), but the oldest diverse flora of large angiosperm trunks is from the Early Eocene of Wyoming, USA (Wheeler, Scott & Barghoorn, 1977, 1978). Among many diverse assemblages of large angiosperm trunks in the later Tertiary are those from the Oligocene of Egypt (Kräusel, 1939; Bown *et al.*, 1982), and the Miocene of the northwestern USA (Prakash & Barghoorn, 1961).

Herbivores and angiosperms – interpretation

Most literature on dinosaurs and angiosperms has considered each group independently, generally from a systematic perspective. Here, we emphasize the dynamic and functional consequences of the interaction of herbivores and angiosperms. Few of the observations detailed above are original, but they have only rarely been considered in an ecological framework (e.g. Bakker, 1978).

We recognize four broad stages in the history of angiosperm–herbivore interaction (Table 8.1). These are not characterized just by changes in the diversity or taxonomic composition of higher taxa, but by what we infer to be changes in the ecological roles of angiosperms and herbivorous tetrapods.

Table 8.1. *Stages in the history of angiosperm–herbivore interaction*

	Herbivore fauna size distribution	Herbivore locomotion	Inferred feeding type	Angiosperm diaspore size (mm^3)	Angiosperm stature	Inferences
Late Eocene–Recent	Variable, 70–80% < 10 kg in forest; ~50% < 10 kg in open vegetation	Terrestrial, arboreal, and volant	Grazing, browsing, frugivory, granivory	Mean, 10–~2000. Max. ~30000	Broad range of sizes	Mixed. Relative importance of frugivory/dispersal varying by vegetation type and climate
Paleocene–mid-Eocene	Initially all < 10 kg, later 70–80% < 10 kg	Arboreal, volant, terrestrial	Frugivory, granivory, folivory	Mean, 100–4000 Max. ~80000	Broad range including many large species	Frugivory/dispersal dominant. Biotic disturbance levels low. Angiosperms *K*-selected, animal dispersed and forming closed forest
Campanian–Maastrichtian	Bimodal. 20–50% < 10 kg 50–80% > 1000 kg	Dominantly terrestrial	Generalized browsing, some frugivory	Mean ? Max. ~1000?	Few large, most small	Transitional. Possibly increased importance of tetrapod dispersal of angiosperms. Areas of closed angiosperm forest?
Barremian–Campanian	Almost all > 100 kg	Terrestrial	Generalized browsing	Mean < 10 Max. < 100	Most species small	Generalized herbivory dominant. Biotic disturbance levels high. Angiosperms *r*-selected and forming open vegetation

Barremian–Campanian

The vertebrate herbivore community was dominated by huge, generalist, dinosaurs. These tetrapodal vacuum cleaners consumed massive quantities of foliage, and may incidentally have eaten fruits and seeds. While their activities could have resulted in dispersal, it is unlikely that they would have selected for large seed size in angiosperms or for specialized dispersal attractant morphology (cf. Janzen, 1984). It is likely that such large herbivores would have caused major and consistent disturbance of the vegetation. Small herbivores were a minor part of the community; vectors for selective diaspore dispersal were thus rare or absent.

The effect of the lack of selective pressures and vectors for larger angiosperm diaspores is consistent with the dominance of angiosperms with small fruits and seeds during this time. Further, small seeds imply small amounts of endosperm, and consequently a light-rich habitat for growth. This feature fits with the proposed high level of vegetational disturbance caused by dinosaurian herbivory. While some early successional angiosperms do attain tree stature (e.g. *Platanus*, *Cecropia*), early successional angiosperms are generally shrubs or small trees, frequently with weak wood. This feature could explain the rarity of large angiosperm wood in the early and middle Cretaceous.

In sum, plant-herbivore interactions during this period were a self-sustaining system. Herbivores created disturbance while selecting against large angiosperm plants and diaspores. Angiosperms, having originated in physically disturbed habitats (Doyle & Hickey, 1976), spread into other environments, because they were preadapted to disturbance. The high productivity of angiosperms quickly made them an important dinosaur food source, perhaps fueling the success of low-browsing ornithopods (Bakker, 1978). However, herbivore pressure and small seed size restricted angiosperms to open or disturbed communities, and widespread closed forests of angiosperms, such as those seen in modern temperate and tropical areas, did not exist.

Campanian–Maastrichtian

The synergism of large herbivores and 'weedy' angiosperms established in the Early Cretaceous continued into the later Cretaceous, but with an undercurrent of change. Small herbivores became more diverse, though not dominant. The presumed diet of these small herbivores suggests the beginnings of a community of fruit and seed dispersers. This may be reflected in the increasing frequency of larger diaspores in the fossil record. While some of these may be large fruits containing small seeds, others might have harbored large endosperm reserves. The appearance of larger diaspores, and presumably larger endosperm reserves, together with the appearance of suitable animal vectors suggests the possibility that angiosperms could have colonized increasingly light-poor vegetation. If this inference is correct, then larger, more 'K-selected', angiosperm trees might have appeared. The more common occurrence of large angiosperm axes in the later Campanian and especially Maastrichtian may reflect this process.

In sum, the latest Cretaceous provides several lines of evidence that suggest a slow development of more specialized herbivory by small herbivores. The dispersal specificity, and capability to move large diaspores, provided by these herbivores was reflected in gentle changes in diaspore size and angiosperm stature. This suggests the early development, perhaps in a patchy manner, of the type of closed angiosperm vegetation that spread widely in the early Tertiary.

Paleocene–Eocene

The extinction of the dinosaurs, whether instantaneous or gradual, removed a major structuring force from terrestrial communities. With their demise, disturbance levels must have dropped, reducing the available resource represented by disturbed habitat. Similarly, the selective pressure they exerted against large angiosperm diaspores and trees disappeared. The herbivore community was then dominated by mammals and presumably birds, both groups that radiated in the early Tertiary. Both groups commenced from stocks of relatively small stature, and the physiological requirements of small size presumably dictated that both focused on fruits and seeds as a primary food source. These circumstances accelerated the process started in the latest Cretaceous; large seeds became increasingly common until they dominated Early Eocene floras. Angiosperms came to dominate late successional communities, and for the first time formed a closed forest vegetation of more modern aspect. Early successional angiosperms remained, but occupied sites created by physical, rather than biotic, disturbance. The closed structure of these communities restricted the evolution of large mammals. Nearly 20 million years elapsed before large (> 100 kg) herbivores made up as large a portion of the fauna as they had in the Cretaceous (see also Collinson & Hooker, this volume, Chapter 10).

Late Eocene–Recent

The closed communities, dominated by small herbivores and large angiosperms with large diaspores, began to break down in temperate latitudes in the Late Eocene and Early Oligocene as a result of world-wide cooling climate (Wolfe, 1978). This change introduced seasonality into mid-latitude environments, resulting in the spread of annual or biennial reproductive strategies and herbs with buried or near-ground perennating buds, including the dominant group, the grasses. In aggregate, these seasonally adapted herbs formed open communities with a high productivity of primary growth. The open space and great productivity of these communities favored the diversification of larger mammalian herbivores. The spread of open vegetation is reflected in the decrease in mean diaspore size through the mid-Tertiary, although large diaspores remained present in these floras. Presumably, in some areas the communities established in the early Tertiary were structurally little changed during the mid-Tertiary.

Discussion

In past, as in present, ecosystems angiosperm–tetrapod interactions presumably must have ranged from generalized browsing herbivory (bulk consumption) to frugivory/dispersal. However, our reconstructed history does show periods of many millions of years during which angiosperm–herbivore interactions were dominantly toward one or the other end of this spectrum.

Generalized herbivory was most dominant during the Cretaceous, when a high proportion of herbivorous tetrapod species were of large size, and most angiosperms were small (though perhaps woody) plants with small diaspores. Plant predation and vegetational disturbance by large herbivores simultaneously favored the individual success of weedy angiosperms and tended to create vegetation where weedy plants could prosper. The success and spread of weedy angiosperms in turn favored the continued success of large, browsing herbivores by creating a highly productive type of vegetation.

For a geologically significant period of time in the early Tertiary the pendulum swung in the opposite direction. Most vertebrate herbivores were small, and many of them were probably some form of frugivore/granivore/omnivore. Many angiosperms were large trees with large diaspores. The absence of larger herbivores in the earliest Tertiary resulted in decreased selection against K-selected angiosperms, and a decreased level of biotic disturbance in plant communities. Interplant competition and a great increase in the relative importance of fruit and seed eaters/dispersers was probably responsible for an early Tertiary diversification in angiosperm seeds and growth habits.

Both 'big herbivore/little angiosperm' and 'little herbivore/big angiosperm' systems were characterized by ecological relationships that created self-reinforcing selective pressures that in turn led to the long-term stability of those ecological relationships. As a result, change in these systems was slow. Small herbivores and large angiosperms with large diaspores may have diversified somewhat in the latest Cretaceous, but had large herbivores not been eradicated in the K/T extinction, it seems likely that change would have remained slow, perhaps never resulting in the transition to a Tertiary-style community.

A similar pattern is seen in the Late Eocene–Oligocene diversification of larger mammals. Here, an extrinsic forcing factor, climatic change, led to the evolution of more open angiosperm communities of high biomass productivity. Climatically forced changes in angiosperm growth habit and vegetation allowed a fairly rapid radiation of large, open-habitat herbivores. Had the climatic change not occurred, it seems unlikely that large herbivores would have diversified significantly in the context of closed angiosperm forests.

This perspective of a synergism in community dynamics raises another important observation. The evolutionary success of a clade is not only dependent on its inherent capabilities, but also on the dynamic of the ecosystem into which it evolves. Angiosperms are viewed as the most successful of vascular plants, yet they probably occupied a small part of their present range of growth habits until the early Tertiary, at least 60 million years into their history. While this pattern may reflect in part the accumulation of adaptations during the later Cretaceous, it also reflects their initial appearance in the large-herbivore-dominated, highly disturbed communities of the Cretaceous. The 'weedy' ecology of the earliest angiospems suited them to success in this community, but they were confined to a relatively narrow range of habitats and habits until the large-herbivore hegemony was broken. In this circumstance, herbivore pressure influenced angiosperm evolution.

The tables were turned in the early Tertiary, when closed angiosperm forests were a significant part of world vegetation (Wolfe, 1985). Vegetational structure and the resulting distribution of herbivore food resources militated against the diversification of large herbivores, and only when climatic change forced changes in angiosperm physiognomy and reproductive biology were large herbivores able to reappear in numbers. In this case, the potential for the evolution of large mammalian herbivores was restricted by angiosperms.

In summary, we suggest that there has been a striking reciprocal interaction of vertebrate herbivores and angiosperms in the Cretaceous and Tertiary. This interaction has dictated the general energy flow and physiognomic structure of terrestrial communities and influenced the evolution of both angiosperms and herbivorous tetrapods (and probably other groups as well).

We thank many vertebrate paleontologists who shared their expertise and unpublished data on body size and diet of extinct tetrapods: J. David Archibald, Thomas M. Bown, Richard Cifelli, Robert Emry, Christine Janis, David Krause, Storrs Olson, Kenneth D. Rose, Kathleen Scott, and David Weishampel. We also acknowledge commentary and advice from Jack A. Wolfe, Louise Emmons, David Norman, Nicholas Hotton III, and Mike Brett-Surman. S.L.W. acknowledges profitable conversations with J. Damuth and W. A. DiMichele. B.H.T. acknowledges support of research by NSF grant BSR 83 06002.

References

Bailey, I. W. (1924). The problem of identifying wood of Cretaceous and later dicotyledons: *Paraphyllanthoxylon arizonense*. *Annals of Botany*, **38**, 439–51.

Bakker, R. T. (1978). Dinosaur feeding behaviour and the origin of flowering plants. *Nature*, **274**, 661–3.

Bown, T. M., Kraus, M. J., Wing, S. L., Fleagle, J. G., Tiffney, B. H., Simons, E. L. & Vondra, C. F. (1982). The Fayum primate forest revisited. *Journal of Human Evolution*, **11**, 603–32.

Cahoon, E. J. (1972). *Paraphyllanthoxylon alabamense* – a new species of fossil dicotyledonous wood. *American Journal of Botany*, **59**, 5–11.

Chandler, M. E. J. (1961). *The Lower Tertiary floras of Southern England, vol. I, Palaeocene Floras. London Clay Flora (Supplement)*. London: British Museum (Natural History).

Clemens, W. A., Jr (1966). Fossil mammals of the type Lance Formation Wyoming. Part II. Marsupialia. *University of California Publications in Geological Sciences*, **62**, 1–122.

Clemens, W. A., Jr (1973). Fossil mammals of the type Lance Formation Wyoming. Part III. Eutheria and summary. *University of California Publications in Geological Sciences*, **94**, 1–102.

Clemens, W. A., Jr & Kielan-Jaworowska, Z. (1979). Multituberculata. In *Mesozoic Mammals: The First Two-Thirds of Mammalian History*, ed.

J. Lillegraven, Z. Kielan-Jaworowska & W. A. Clemens, Jr, pp. 99–149. Berkeley: University of California Press.

Coombs, W. P., Jr (1975). Sauropod habits and habitats. *Palaeogeography, Palaeoclimatology, Palaeoecology*, **17**, 1–33.

Coombs, M. C. (1983). Large mammalian clawed herbivores: a comparative study. *Transactions of the American Philosophical Society*, **73**, 1–96.

Crawley, M. J. (1983). *Herbivory, the Dynamics of Animal–Plant Interactions*. Berkeley & Los Angeles: University of California Press.

Damuth, J. (1981). Population density and body size in mammals. *Nature*, **290**, 699–700.

Damuth, J. (1982). Analysis of the preservation of community structure in assemblages of fossil mammals. *Paleobiology*, **8**, 434–46.

Demment, M. W. & Van Soest, P. J. (1985). A nutritional explanation for body-size patterns of ruminant and nonruminant herbivores. *American Naturalist*, **125**, 641–72.

Dodson, P., Behrensmeyer, A. K., Bakker, R. T. & McIntosh, J. S. (1980). Taphonomy and paleoecology of the dinosaur beds of the Jurassic Morrison Formation. *Paleobiology*, **6**, 208–32.

Doyle, J. A. & Hickey, L. J. (1976). Pollen and leaves from the Mid-Cretaceous Potomac Group and their bearing on early angiosperm evolution. In *Origin and Early Evolution of Angiosperms*, ed. C. B. Beck, pp. 139–206. New York: Columbia University Press.

Dubost, G. (1979). The size of African forest artiodactyls as determined by the vegetation structure. *African Journal of Ecology*, **17**, 1–17.

Emmons, L. H., Gautier-Hion, A. & Dubost, G. (1983). Community structure of the frugivorous–folivorous forest mammals of Gabon. *Journal of the Zoological Society of London*, **199**, 209–22.

Emry, R. J. (1973). Stratigraphy and preliminary biostratigraphy of the Flagstaff Rim area, Natrona County, Wyoming. *Smithsonian Contributions to Paleobiology*, **18**, 1–43.

Estes, R. & Berberian, P. (1970). Paleoecology of a Late Cretaceous vertebrate community from Montana. *Breviora*, **343**, 1–35.

Fleming, T. H. (1973). Numbers of mammal species in North and Central American forest communities. *Ecology*, **54**, 555–63.

Fliche, P. (1905). Note sur des bois fossiles de Madagascar. *Bulletin de Société Géologique de France*, **4**, 346–58.

Friis, E. M. (1984). Preliminary report of Upper Cretaceous angiosperm reproductive organs from Sweden and their level of organization. *Annals of the Missouri Botanical Garden*, **71**, 403–18.

Harper, J. L. (1977). *Population Biology of Plants*. London, New York & San Francisco: Academic Press.

Harper, J. L., Lovell, P. H. & Moore, K. G. (1970). The shapes and sizes of seeds. *Annual Review of Ecology and Systematics*, **1**, 327–56.

Herrera, C. M. (1985). Determinants of plant–animal coevolution: the case of mutualistic dispersal of seeds by vertebrates. *Oikos*, **44**, 132–41.

Hickey, L. J. & Doyle, J. A. (1977). Early Cretaceous fossil evidence for angiosperm evolution. *The Botanical Review*, **43**, 3–104.

Hotton, N. III. (1980). An alternative to dinosaur endothermy: the happy wanderers. In *A Cold Look at the Warm-Blooded Dinosaurs*, American Association for the Advancement of Science Symposium, no. 28, ed. R. D. K. Thomas & E. C. Olson, pp. 311–50. Washington, DC: American Association for the Advancement of Science.

Janis, C. (1976). The evolutionary strategy of the Equidae and the origins of rumen and caecal digestion. *Evolution*, **30**, 757–74.

Janzen, D. H. (1983). Dispersal of seeds by vertebrate guts. In *Coevolution*, ed. D. J. Futuyma & M. Slatkin, pp. 232–62. Sunderland, Massachusetts: Sinauer Associates.

Janzen, D. H. (1984). Dispersal of small seeds by big herbivores: foliage is the fruit. *American Naturalist*, **123**, 338–53.

Jarman, P. J. & Sinclair, A. R. E. (1979). Feeding strategy and the pattern of resource-partitioning in ungulates. In *Serengeti, Dynamics of an Ecosystem*, ed. A. R. E. Sinclair & M. Norton-Griffiths, pp. 130–63. Chicago: University of Chicago Press.

Jenkins, F. A., Jr (1969). Occlusion in *Docodon* (Mammalia, Docodonta). *Postilla*, **139**, 1–24.

Koeniguer, J. C. (1971). Sur les bois fossiles du Paléocène de Sessao (Niger). *Review of Palaeobotany and Palynology*, **12**, 303–23.

Krause, D. W. (1982). Jaw movement, dental function, and diet in the Paleocene multituberculate *Ptilodus*. *Paleobiology*, **8**, 265–81.

Krause, D. W. & Jenkins, F. A., Jr (1983). The postcranial skeleton of North American multituberculates. *Bulletin of the Museum of Comparative Zoology*, **150**, 199–246.

Kräusel, R. (1939). Ergebnisse der Forschungreisen Prof. E. Stromers in den Wüsten Ägyptens IV. Die fossilen Floren Ägyptens. *Abhandlungen der Bayerischen Akademie der Wissenschaften, Mathematisch-Naturwissenschaftliche Abteilung, Neue Folge*, **47**, 1–140.

Kron, D. G. (1979). Docodonta. In *Mesozoic Mammals: The First Two-Thirds of Mammalian History*, ed. J. Lillegraven, Z. Kielan-Jaworowska & W. A. Clemens, Jr, pp. 91–98. Berkeley: University of California Press.

Monteillet, J. & Lappartient, J.-R. (1981). Fruits et graines du Crétacé supérieur des Carrières de Paki (Sénégal). *Review of Palaeobotany and Palynology*, **34**, 331–44.

Ng, F. S. P. (1978). Strategies of establishment in Malayan forest trees. In *Tropical Trees as Living Systems*, ed. P. B. Tomlinson & M. H. Zimmerman, pp. 129–62. Cambridge: Cambridge University Press.

Nishida, M. (1962). On some petrified plants from the Cretaceous of

Choshi, Chiba Prefecture, Japan. *Japanese Journal of Botany*, 18, 87–104.

Norman, D. B. & Weishampel, D. B. (1985). Ornithopod feeding mechanisms: their bearing on the evolution of herbivory. *American Naturalist*, 126, 151–64.

Olson, S. L. (1985). The fossil record of birds. In *Avian Biology*, *VIII*, pp. 79–237. New York: Academic Press.

Ostrom, J. H. (1964). A reconsideration of the paleoecology of hadrosaurian dinosaurs. *American Journal of Science*, 262, 975–97.

Ostrom, J. H. (1970). Stratigraphy and paleontology of the Cloverly Formation (Lower Cretaceous) of the Bighorn Basin Area, Wyoming and Montana. *Peabody Museum of Natural History Bulletin*, 35, 1–234.

Ostrom, J. H. (1980). The evidence for endothermy in dinosaurs. In *A Cold Look at the Warm-Blooded Dinosaurs*, American Association for the Advancement of Science Selected Symposium, no. 28, ed. R. D. K. Thomas & E. C. Olson, pp. 15–54. Washington, DC: American Association for the Advancement of Science.

Page, V. M. (1981). Dicotyledonous woods from the Upper Cretaceous of central California. III. Conclusions. *Journal of the Arnold Arboretum*, 62, 437–55.

Peters, R. H. (1983). *The Ecological Implications of Body Size*. Cambridge: Cambridge University Press.

Prakash, U. & Barghoorn, E. S. (1961). Miocene fossil woods from the Columbia basalts of central Washington. *Journal of the Arnold Arboretum*, 17, 165–95.

Reid, E. M. & Chandler, M. E. J. (1933). *The London Clay Flora*. London: British Museum (Natural History).

Ridley, H. N. (1930). *The Dispersal of Plants throughout the World*. Ashford, Kent: L. Reeve & Co., Ltd.

Rose, K. D. (1981). The Clarkforkian land-mammal age and mammalian faunal composition across the Paleocene–Eocene boundary. *University of Michigan Papers on Paleontology*, 26, 1–197.

Rose, K. D. (1984). Evolution and radiation of mammals in the Eocene, and the diversification of modern orders. In *Mammals, Notes for a Short Course*, University of Tennessee Department of Geological Sciences Studies in Geology, no. 8, ed. P. D. Gingerich & C. E, Badgley, pp. 167–81. University of Tennessee.

Rose, K.D. & Krause, D. W. (1982). Cyriacotheriidae, a new family of early Tertiary pantodonts (Mammalia) from western North America. *Proceedings of the American Philosophical Society*, 126, 26–50.

Spackman, W., Jr (1948). A dicotyledonous wood found associated with the Idaho Tempskyas. *Annals of the Missouri Botanical Garden*, 35, 107–15.

Spotila, J. R. (1980). Constraints of body size and environment on the temperature regulation of dinosaurs. In *A Cold Look at the Warm-Blooded Dinosaurs*, American Association for the Advancement of Science Selected Symposium, no. 28, ed. R. D. K. Thomas & E. C. Olson, pp. 233–52. Washington, DC: American Association for the Advancement of Science.

Stopes, M. C. (1912). Petrifactions of the earliest European angiosperms. *Philosophical Transactions Series B*, **203**, 75–100.

Strauss, S. H. & Ledig, F. T. (1985). Seedling architecture and life history evolution in pines. *American Naturalist*, **125**, 702–15.

Stucky, R. K. & Krishtalka, L. (1985). Evolution of Eocene mammals: faunal level patterns. In *Abstracts of Papers, 4th International Theriological Congress*, Abstract 0605. Edmonton, Canada.

Suzuki, M. & Nishida, M. (1974). *Chionanthus mesozoica*, sp. nov., a dicotyledonous wood from the Lower Cretaceous of Choshi, Chiba Prefecture with reference to comparison with recent *Chionanthus*. *Journal of Japanese Botany*, **49**, 47–54.

Tiffney, B. H. (1977). Dicotyledonous angiosperm flower from the Upper Cretaceous of Martha's Vineyard, Massachusetts. *Nature*, **265**, 136–7.

Tiffney, B. H. (1984). Seed size, dispersal syndromes, and the rise of the angiosperms: evidence and hypothesis. *Annals of the Missouri Botanical Garden*, **71**, 551–76.

Thayn, G. F., Tidwell, W. D. & Stokes, W. L. (1983). Flora of the Lower Cretaceous Cedar Mountain Formation of Utah and Colorado. Part I. *Paraphyllanthoxylon utahense*. *Great Basin Naturalist*, **43**, 394–402.

Thayn, G. F., Tidwell, W. D. & Stokes, W. L. (1985). Flora of the Lower Cretaceous Cedar Mountain Formation of Utah and Colorado. Part III. *Icacinoxylon pittiense*, n. sp. *American Journal of Botany*, **72**, 175–81.

van der Pijl, L. (1982). *Principles of Dispersal in Higher Plants*, 3rd edn. Berlin: Springer-Verlag.

Weaver, J. C. (1983). The improbable endotherm: the energetics of the sauropod dinosaur *Brachiosaurus*. *Paleobiology*, **9**, 173–82.

Webb, S. D. (1969). The Burge and Minnechaduza Clarendonian mammalian faunas of north-central Nebraska. *University of California Publications in Geological Sciences*, **78**, 1–191.

Webb, S. D. (1977). A history of savanna vertebrates in the New World. Part 1. North America. *Annual Reviews of Ecology and Systematics*, **9**, 393–426.

Weishampel, D. B. (1984). The evolution of jaw mechanics in ornithopod dinosaurs (Reptilia: Ornithischia). *Advances in Anatomy, Embryology and Cell Biology*, **87**, 1–110.

Western, D. (1979). Size, life history and ecology in mammals. *African Journal of Ecology*, **17**, 185–204.

Wheeler, E. A., Scott, R. A. & Barghoorn, E. S. (1977). Fossil dicotyledonous woods from Yellowstone National Park. *Journal of the Arnold Arboretum*, **58**, 280–302.

Wheeler, E. A., Scott, R. A. & Barghoorn, E. S. (1978). Fossil dicotyledonous woods from Yellowstone National Park. II. *Journal of the Arnold Arboretum*, **59**, 1–26.

Wolfe, J. A. (1978). A paleobotanical interpretation of Tertiary climates in the Northern Hemisphere. *American Scientist*, **66**, 694–703.

Wolfe, J. A. (1985). Distribution of major vegetational types during the Tertiary. *Geophysical Monographs*, **32**, 357–75.

Wolfe, J. A., Doyle, J. A. & Page, V. M. (1975). The bases of angiosperm phylogeny: paleobotany. *Annals of the Missouri Botanical Garden*, **62**, 801–24.

9

Dinosaurs and land plants

M. J. COE, D. L. DILCHER, J. O. FARLOW, D. M. JARZEN AND
D. A. RUSSELL*

Land plants surely formed the trophic foundation for the vast populations of dinosaurs that inhabited the surface of the Earth during
Mesozoic time. In the past, much attention has been given to the
functional anatomy of the skull in plant-eating dinosaurs, and less
emphasis was placed on dinosaur–plant interactions of a more generalized
nature and of longer duration. This chapter will explore how some of
these interactions, which were sustained through many tens of millions
of years, may have affected the structure of the dinosaurs and their plant
food sources. We will attempt to illustrate this by suggesting
convergences in the body form of herbivorous dinosaurs and the large
herbivorous mammals of the present day and by proposing that the
stature of the plants that sustained vertebrate browse has generally
decreased through time. Shorter-term events, such as the origin and
radiation of the angiosperms and the extinction of the dinosaurs, are
examined in order to speculate on the dynamics of plant–dinosaur
co-evolution.

Dinosaur versus plant in the fossil record

There is no unequivocal evidence that plants and dinosaurs interacted
with each other. No frond has ever been found with clear impressions
of dinosaur teeth preserved on it, nor has any thorn been discovered
embedded in what was once the flesh of a dinosaur. The evidence is

_____

* The authors of this chapter are listed in alphabetical order. The consensus herewith
presented was derived from contributions on parallel evolution (M.J.C), plant
macrofossils (D.L.D.), dinosaurian digestion (J.O.F.), plant microfossils (D.M.J.) and
dinosaurian trophic relationships (D.A.R.).

entirely circumstantial, in that some dinosaurs did have blunt teeth suited for cropping plant materials and some plants did have large spines that must have restrained the appetites of herbivorous dinosaurs (for a Late Jurassic thorn from the Morrison Formation of Utah, see Chandler, 1966). Footprints of bipedal herbivorous dinosaurs preserved in the roofs of coal mines in the Blackhawk Formation (Upper Cretaceous) of Utah are superimposed on the roots of tree stumps and fallen logs, or side-by-side, facing where the trunk of a tree once stood (Parker, 1980). Were these animals browsing?

It is not easy to identify a *corpus delicti* with which a charge of herbivory could be sustained against any dinosaur (for illustrations of representative dinosaur skulls, see Figure 9.1). Stomach contents are rarely found. Krassilov (1981) observed that the contents of the rib cage of an *Edmontosaurus* from the Lance Formation (Maastrichtian, Wyoming) consisted of fruits and seeds mixed with shoots of conifers and angiosperms. He speculated that the hadrosaur may have been feeding on peat prior to its own death. The abdominal region of a sauropod skeleton collected from the Morrison Formation (Upper Jurassic, Wyoming) contained stems, bits of leaves and other plant fragments which may have been part of the dinosaur's last meal (Bird, 1985). A calcareous mass associated with a sauropod skeleton from the same geologic unit in Utah (Stokes, 1964) contained abundant fragments of plant tissues including stems about 1 cm in diameter. There was also evidence of clay and bone material that had been ingested by the animal, perhaps in order to satisfy its needs for inorganic minerals.

All of the major groups of herbivorous dinosaurs possessed teeth, and stomach stones have been reported within or near the stomach region of a wide variety of skeletons (Brown, 1907, 1941; Janensch, 1929; Bond, 1955; Raath, 1974). The residence time of plant materials in the gut of modern herbivorous reptiles is longer than in modern herbivorous mammals of similar body size, although the efficiency of digestion is approximately the same (Parra, 1978; Hamilton & Coe, 1982; Troyer, 1984). Thus, how long food remained in the body is related to the question of whether or not dinosaurs had metabolic rates comparable to those of reptiles or of mammals. The present consensus seems to be that, with the exception of some small carnivorous forms, dinosaurs had reptilian metabolic rates (Thomas & Olson, 1980).

Just as the record is sparse concerning the ingestion of plant material by dinosaurs, there is little direct evidence of the excretion of plant residues by dinosaurs. Small fossil faecal pellets (coprolites) containing

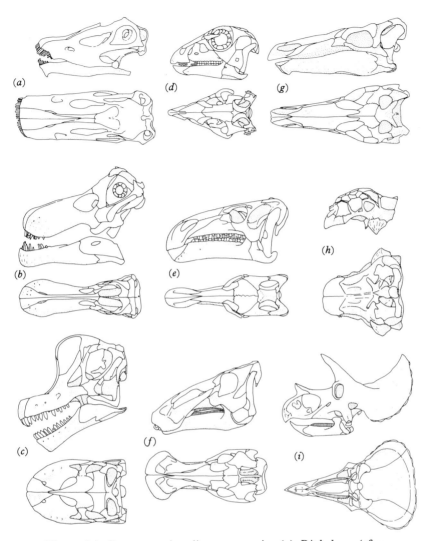

Figure 9.1. Representative dinosaur crania; (*a*) *Diplodocus* (after Holland), Sauropoda, Late Jurassic, USA; (*b*) *Nemegtosaurus* (after Nowiński), Sauropoda, Late Cretaceous, Mongolia; (*c*) *Brachiosaurus* (after Janensch), Sauropoda, Late Jurassic, Tanzania; (*d*) *Hypsilophodon* (after Galton), ornithopod, Early Cretaceous, UK; (*e*) *Iguanodon* (after Norman), ornithopod, Early Cretaceous, UK; (*f*) *Edmontosaurus* (after Lambe), hadrosaur, Late Cretaceous, Canada; (*g*) *Stegosaurus* (after Gilmore), stegosaur, Late Jurassic, USA; (*h*) *Euoplocephalus* (after Coombs), ankylosaur, Late Cretaceous, Canada; (*i*) *Triceratops* (after Hatcher), ceratopsian, Late Cretaceous, USA.

identifiable gymnosperm cuticles occur sporadically in the Yorkshire Deltaic beds of Middle Jurassic age. Bennettitalean tissues were abundantly preserved in one clump of 250 relatively large pellets (Hill, 1976). The original aggregate weight of the pellets was probably about 200 g, suggesting an animal the size of a large rabbit or small sheep. If the clump was indeed left behind by a small dinosaur, the reptile would have to have had a very complex intestine in order to separate faecal material into discrete spheroids (Karasov & Diamond, 1985). A terminal fragment of a coprolite was found preserved in association with tridactyl footprints in Wealden Strata in Spain (Viera & Torres, 1979), and a large coprolite, possibly of Late Cretaceous age, has been described from Hungary (Majer, 1923). In Dinosaur Provincial Park, Alberta, fluvial strata of Late Cretaceous (Campanian) age contain laminae composed of comminuted plant debris. Béland & Russell (1978) suggested these may represent reworked faeces of large ornithischian herbivores. Fibrous and poorly cohesive herbivore dung is seldom preserved intact in the fossil record.

It is probable, however, that plants did derive some benefits from dinosaur droppings. Like modern large herbivores, dinosaurs probably sought shade beneath the branches of trees during the heat of the day. Soil around the roots would thus have been enriched by nutrients derived from dung left behind on its surface. Dinosaurs probably disseminated viable seeds in the process of casually scattering their droppings over Mesozoic landscapes, as do modern large ground birds (Stocker & Irvine, 1983) and large mammals (Janzen & Martin, 1982; Collins & Uno, 1985). The flesh surrounding the seeds in *Ginkgo biloba*, which produces an allergic reaction in man, may have been evolved in association with a herbivorous dinosaur that did not chew its food well (Janzen & Martin, 1982). The seeds need not have been associated with fleshy structures, for small propagules may be ingested in the process of consuming bulky foliage to which they are attached and subsequently dispersed in dung (Janzen, 1984).

Plants must have figured significantly in dinosaurian locomotion. If a herbivore collided with a low branch in the process of trying to evade the attack of a carnivore, its escape would obviously have been impeded. Thus, the mean shoulder heights of mammalian forest herbivores are apparently selected to be somewhat less than the typical heights of the lower branches of trees (Dubost, 1979). In the same manner, the level of the lowest branches in understory plants of Mesozoic forests could be estimated by examining regularities in the height of the hips in various species of associated dinosaurs. Clearance is also a significant factor in locomotion across terrain upon which plants are growing or which are strewn with fallen woody stems. Thus, in the cases of two massive sauropods, the belly of *Apatosaurus* cleared the ground by slightly over a meter (Gilmore, 1936), while that of *Brachiosaurus* cleared the ground by nearly three meters (Janensch, 1961). Perhaps the habitat of the giraffoid *Brachiosaurus* was littered with fallen tree trunks over which the animal regularly had to step, which could have effectively blocked the movement of an *Apatosaurus*.

Convergent selection in ecosystems

Biologists have long suspected that similar environmental conditions promote the evolution of similar plants and animals (Louw & Seely, 1982). On a gross biogeographic level, terrestrial ecosystems exhibit marked similarities in structure and physiognomy on different continents (Whittaker, 1975; Pianka, 1978). If there is a limited number of efficient solutions to a set of functional demands, natural selection may force differing genotypes toward similar (convergent) phenotypes (Mayr, 1963). All viable phenotypes may not be realized within one ecosystem; for example, large pseudo-kangaroos have not evolved on northern continents. Nevertheless, many examples of convergence can be seen among organisms living under climatic conditions comparable to those that existed over broad areas of the planet during Mesozoic time.

Thus, clearings in humid West African forests are rapidly colonized by *Musanga cecropioides*, while similar habitats in South America are colonized by species of the genus *Cecropia*. These African and American species are remarkably convergent in their fast growth rates, soft stems, few terminal radiating branches and large palmate leaves. The ecological role of a pioneer apparently favors convergent solutions in rainforests of both regions. In the deserts and semi-arid grasslands of the Old and New World, demands of water conservation, high temperatures and solar radiation have produced convergences between the euphorbs of the former and the arborescent cacti of the latter.

The dominant mammalian land-dwelling terrestrial herbivores in West African forests are duikers and chevrotains, both of which possess the squat bodies, rounded backs and thick fur suited for life in dense secondary vegetation on or near the forest floor. In South America, capybaras and small deer are morphologically similar and fill similar ecological roles. Analagous convergences occur between Old World pangolins and New World armadillos (Bourlière, 1973). Both the golden moles of southern Africa and the marsupial mole of Australia, in response to the demands of a fossorial life in soft arid soil, have acquired fine silky fur, short and powerful digits in the forelimbs, and a horny prow on their snouts. In South America the fossorial fairy armadillo bears a striking resemblance to these species.

Competitive interactions among dinosaurian herbivores led to the evolution of a wide variety of body shapes, their statures (reaches) being to some extent constrained, as are those of mammals today, by the effect of rainfall on the stature of plants from arid grasslands, through low woodlands to the tall forests of the humid tropics (Coe, Cumming & Phillipson, 1976; Coe, 1980b). As with modern mammals, dinosaurs were probably also separated ecologically by the type of plant materials upon which they fed, their method of feeding, and seasonal patterns of dispersal and migration. Herbivorous dinosaurs may be separated, at least at the superficial physiognomic level, into quasi-familiar shapes (e.g. giraffe-like or elephant-like sauropods, rhino-like ceratopsians, camel-like hadrosaurs), which suggests that problems posed by Jurassic and Cretaceous vegetation to Mesozoic dinosaurs may have been rather similar to those posed by modern vegetation to modern mammals. The number of basic mechanical solutions adopted by large vertebrates in the gathering of plant food has probably not changed fundamentally since Jurassic and Cretaceous time, and these solutions have appeared at different times in totally different phylogenetic lineages.

Jarman (1974) noted that body size and muzzle shape in African antelopes are good predictors of feeding strategy. Small forest- and scrub-dwelling duikers seek energy-rich food items such as fruits, antelopes of medium size are mixed (graze and browse) feeders, and the largest antelopes generally graze on coarse vegetation. Concentrate feeders in intermediate size ranges, such as the giraffe, kudu and gerenuk, have narrow snouts and select nutrient-rich shoots, pods and fruits on bushes and trees (Hofmann, 1973). Bulk-roughage feeders, such as wildebeest, possess broad snouts. Concentrate feeding (shoot selector) strategies are suggested for many small ornithopods (e.g. *Hypsilophodon*) on the grounds of their narrow muzzles. Conversely, the broad rostrum of some large ornithopods (e.g. *Edmontosaurus*) resembles those of modern mammals that crop low swards of vegetation.

The foregoing examples illustrate the possible importance of con-vergent selective pressures in evolution. It is axiomatic that environ-mental constraints are critically important in determining which organisms survive in the struggle for existence. On a broader scale, more general convergences may become apparent. Thus, interactions with biological stresses may promote convergent increases in the size of the central nervous system in mammals and birds (Russell, 1983; Wyles, Kunkel & Wilson, 1983), and interactions with physical stresses may promote convergences in shell shape (Seilacher, 1984). With the current emphasis on random processes in evolution it is perhaps useful to consider that non-random evolutionary processes also occur; that is to say, some evolutionary pathways are more frequently followed than others.

The browse line

Speculations on adaptive interrelationships between plants and herbivorous dinosaurs are becoming more detailed and more firmly based on analogies with existing biologic systems. Some authors have argued that dinosaurian browse was indiscriminate and generally not focused upon plant reproductive structures. Only the smaller dinosaurs could have fed selectively on seeds or fruits (Hughes, 1976). Others have cited the existence of thorns, hairs and hard integuments as defensive adaptations of plant reproductive structures of Triassic and Jurassic age, before the rise of the angiosperms. The diversity of these defensive structures was taken as evidence of a degree of specificity in interactions between plants and animals, including dinosaurs (Weishampel, 1984). Yet even small dinosaurs must have been inferior generally to modern mammals and birds as dispersers of seed. Being neither as diversified nor as intelligent as modern homoiotherms, small dinosaurs would have been less well adapted to discrete niches and less clever in seeking out attractive propagules (Regal, 1977).

Hughes (1976) and Norman & Weishampel (1985) stress that specializations for herbivory in dinosaurs became increasingly refined through Mesozoic time, and imply that dinosaurs fed abundantly on angiosperm foliage. Furthermore, Bakker (1978) proposed that the feeding strategies of herbivorous dinosaurs changed from browsing preferentially on trees to browsing preferentially on shrubs and smaller plants between the Late Jurassic and Late Cretaceous. Bushy angiosperms were postulated to have been able to withstand browsing more successfully than gymnosperm saplings. Dinosaurian cropping thereby facilitated the expansion of primitive angiosperms at the expense of gymnosperm forests. This hypothesis could be extended and applied to a much longer interval of geologic time.

The Morrison Formation, of Late Jurassic age, is exposed in the eastern Cordillera and high plains of the United States and has produced one of the largest dinosaurian assemblages in the world (Dodson *et al.*, 1980). In most respects this assemblage is similar to others of Middle and Late Jurassic age around the globe, and it is taken here as typical of middle Mesozoic terrestrial vertebrates dwelling in semi-arid lowlands before the advent of the angiosperms.

Weight is a very important parameter in assessing the ecology of vertebrates. In the past it has been necessary to construct scale models of living dinosaurs in order to assess their body weights. Now, using a method devised by Anderson, Hall-Martin & Russell (1985), these weights can be estimated using measurements of limb bones. The weights obtained, in combination with information on the skeletal architecture of various Morrison dinosaurs together with estimates of the relative abundance of their remains in existing collections, can be used to calculate their relative abundance in their original populations, the preferred vertical range of feeding and the amount of plant material required to sustain them (see Appendix). The calculated total plant tissue consumed by all Morrison dinosaurian herbivores, apportioned in vertical 10-cm increments according to the feeding range of the individual herbivore taxa, is shown graphically in Figure 9.2 (*a*) (thus sauropods were the principal consumers of vegetation at ground level and beyond 3 m above ground level; stegosaurs were very important consumers near the 1-m level). The histogram may be considered as a browse profile. A comparable browse profile was constructed for mammals now inhabiting the Amboseli Basin in southern Kenya (Figure 9.2(*c*).

A comparison of the Morrison and Amboseli browse profiles suggests that one long-term trend in plant–vertebrate interactions is a reduction in the height of feeding levels. The effect would be independent of whether or not the vertebrates are dinosaurs or mammals, or whether the plants are gymnosperms or angiosperms. The dinosaurian assemblage from Dinosaur Provincial Park, Alberta, is intermediate in age (75 million years BP) between that of the Morrison Plain and the Amboseli Basin. Like other Late Cretaceous assemblages, it is dominated by ornithischian dinosaurs with body weights intermediate between those of most sauropods and most modern big-game mammals. Relevant data are listed in Table 9.8 (Appendix), derived according to the same procedure as for the Morrison dinosaurs (relative abundances have not been corrected for size effects because the herbivores are more nearly

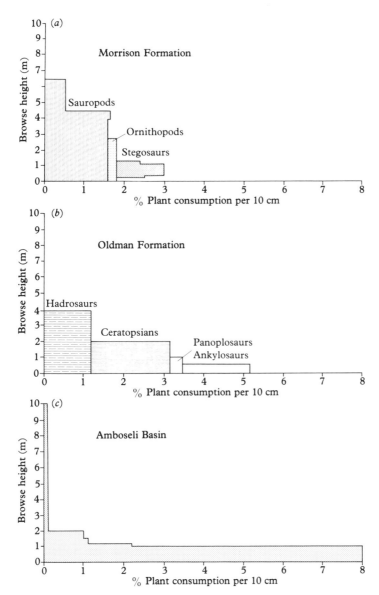

Figure 9.2. Browse profiles for (*a*) Morrison dinosaurs (150 MYBP), (*b*) Dinosaur Provincial Park (Oldman) dinosaurs (78 MYBP) and (*c*) large Amboseli mammals (Recent). For further explanation, see Appendix.

identical in weight; cf. data in Béland & Russell, 1978). The Dinosaur
Park browse profile (Figure 9.2 (*b*)), is intermediate in shape between
those calculated for the Morrison Plain and the Amboseli Basin. These
procedures illustrate a means of searching for changes in the co-evolution
of stature in plants and animals throughout a long interval of Earth
history. The postulated trend toward a reduction in the height of feeding
levels is secure only to the extent that the faunas of the three different
ages chosen are representative of their times.

Dinosaurian diversity and the origin of the angiosperms

Plants underwent fundamental evolutionary changes during the Mesozoic Era. Mesozoic vegetation was characterized by the abundance of
ferns, which were represented by eight modern families. The radiation
and final extinction of some seed ferns, bennettitaleans and czekanow-
skialeans took place, conifers differentiated into seven modern families,
cycads radiated and declined, and angiosperms made their appearance
and rapid initial radiation, all during Mesozoic time (see also Crane, this
volume, Chapter 5). It is widely held that megafloral remains (usually
leaf impressions) were derived from vegetation growing near the site of
preservation, while pollen and spores represent vegetation growing over
much larger areas such as an entire basin or watershed (Chaloner, 1968).
However, an element of transport is certainly reflected in plant materials
preserved in the fossil record (Potter, 1976; Roth & Dilcher, 1978). The
abundance with which an organ representing a plant was originally
produced is also relevant to the probability of its preservation. For
present purposes, the number of fossil species belonging to major groups
of plants (or their diversity) is taken as a crude indication of the group's
abundance in various ancient floras.

A Late Jurassic megaflora from southwestern Canada (see Figure 9.3;
Bell, 1956; Delevoryas, 1971) contains seven species of ferns, four
ginkgophytes, nine cycadophytes, five conifers and one pertaining to a
form of uncertain affinities. A palynomorph sample from sediments of
the same age (Pocock, 1962) contains 56 species of spore-bearing plants,
23 gymnosperms and 11 of uncertain affinities. From these data it can
be postulated that a few cycadophytes, ginkgophytes and conifers, as well
as a few species of ferns, may have been common in gallery forests along
water courses. On a regional scale, a large variety of herbaceous ferns
would be present in great abundance. Perhaps tree ferns and conifers
formed scattered clumps or forests within prairie-like expanses of ferns.

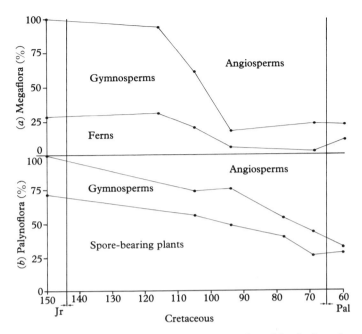

Figure 9.3. Palaeobotanical diversity in various North American sites according to (a) megafloral data, and (b) palynofloral data. Diversities are given according to the percentage of the total number of species that can be allocated to major plant groups. Jr, Jurassic; Pal, Paleocene.

An Aptian (late Early Cretaceous) megaflora, again from southwestern Canada (Bell, 1956), is essentially similar to floras of Late Jurassic age from this region. However, two species of angiosperms (representing 6% of the species identified) were living near the basin of deposition. An Albian (latest Early Cretaceous) palynoflora from Oklahoma (Hedlund & Norris, 1968) contains 42 species of spore-bearing plants, 14 gymnosperms, 19 angiosperms and five of uncertain affinity. Three western North American megafloral assemblages of the same (Albian) age contain an average of 5.7 species of ferns, no ginkgophytes, 3.3 cycadophytes, 7.3 conifers and 10.3 angiosperms (Bell, 1956; Berry, 1922; Delevoryas, 1971; Ward, 1900). Ferns and fern prairies may have been regionally important plant formations, as is further suggested by charred plant material dominated by ferns in the Wealden of England (Harris, 1981). Gymnosperms remained diverse, but the angiosperm radiation was under way. Megafloral species composition ranged from

0 to 57% angiosperms, and was evidently influenced by environmental conditions near the site of deposition.

In a palynofloral sample from the central United States (Linnenberger shale) of Cenomanian (early Late Cretaceous) age, 46% of the species present are ferns, while only 13% of the species belong to this group in a megafloral sample from the same unit. Fern spores evidently dominated the regional spore rain, but 63% of the more locally derived megafloral species are angiospermous. Between 63% and 85% of megafloral species from the same geographical region (Dakota Formation; Lesquereux, 1892), also of Cenomanian age, represent flowering plants. Angiosperm pollen increases in relative diversity from about 15% of species near the base of the unit to about 33% near the top (May & Traverse, 1973).

By the end of Cretaceous (Maastrichtian) time in the northern interior of the United States, the abundance of fern species had declined both in palynofloras (21%, Hell Creek Formation; Norton & Hall, 1969) and megafloras (3%, Fox Hills Formation; Brown, 1939). Angiosperms were dominant both regionally (55%) and locally (76%). Indeed some modern angiosperm genera can be identified from the palynofloral record in strata of slightly greater (Campanian) age (Jarzen, 1980, 1983). Except for a few bog-loving angiosperms and scouring rushes there is no record of any Late Cretaceous herbaceous plants other than ferns, and extensive fern savannas and prairies may have continued to exist well into Tertiary time (Elsik, 1968a, b; Hickey, 1977).

A shift in vegetation lasting through much of the Cretaceous period would probably have posed no serious threat to dinosaurian survival. Plants that were targets of heavy predation would have gradually developed either mechanical or chemical defenses analagous to the thorns of modern acacias, and anti-herbivory substances such as those of living poison ivy and Euphorbiaceae. There is no reason to suspect that any of the major groups of plants were entirely safe from browsing dinosaurs. Unfortunately, the record of herbaceous angiosperms is very poor in strata of Mesozoic age, and grasses do not appear until the Tertiary. Ferns constituted a very important element in Jurassic and Cretaceous vegetation. The vast grass prairies and savannas of geologically recent origin may unfortunately inhibit a proper appreciation of the previous ecological importance of fern prairies.

An analysis of changes in the diversity of dinosaurian species cannot at present be carried out because of the paucity of available data at a low level of taxonomic resolution. Data pertaining to the generic diversity of herbivorous dinosaurs in North America through Mesozoic time

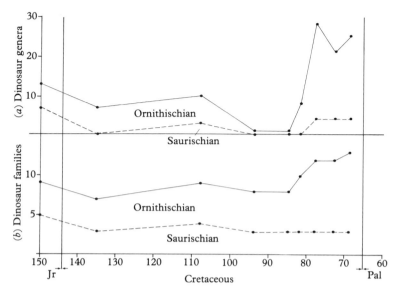

Figure 9.4 Diversity of North American dinosaurs, according to (*a*) genera, and (*b*) families listed in Russell (1984*b*). The total diversity in both cases is divided into the ornithischian and saurischian components. When a family has been recorded in younger and older strata, its presence has been listed in intermediate strata although it may not actually have been recorded there.

(Russell, 1984*a*, *b*) suggest a Jurassic diversification, followed by an Early Cretaceous decline and a final peak near the end of the Cretaceous period (Figure 9.4(*a*)). The Early Cretaceous decline is suspect, however, as it may only reflect smaller samples of dinosaurian specimens from strata of this age. At a family-group level, diversity is less sensitive to fluctuations in sample size. Herbivorous dinosaur families can be viewed as basic feeding types, with the component genera representing variations on a theme. Temporal trends in familial diversity would thus reflect changes in the number of basic strategies of herbivory. Data on familial diversity in herbivorous dinosaurs from North America (Russell, 1984*a*, *b*) suggest an early Mesozoic increase, a period of stasis during Early Cretaceous time, and a final rise, which was truncated by terminal Cretaceous extinctions (Figure 9.4(*b*)).

At both generic and familial levels, North American data suggest a gradual drop in the absolute and relative importance of saurischian herbivores and an increase in the absolute and relative importance of ornithischians. If different feeding strategies are at the root of this change they are not obvious, for the skeletal anatomy of the local sauropods is very poorly understood. The Late Cretaceous peak in the diversity of herbivorous dinosaurs is due to an increase in the number of ornithischian taxa, which appears to more than compensate for waning saurischian diversity. A global compilation of dinosaurian genera and families is not currently available, although some faunal lists of representative Early and Late Cretaceous faunas from various localities around the world have been published (Molnar, 1980). From these it would appear that there was no decline in the world-wide diversity of saurischians through Cretaceous time, and that the world-wide Late Cretaceous expansion of ornithischians was more modest than that suggested by the North American record.

There is no sudden increase in the taxonomic diversity of dinosaurs after the middle Cretaceous origin of the angiosperms that would indicate a radiation of herbivorous dinosaurs in response to a new source of food. The high diversity levels achieved by Late Cretaceous dinosaurian herbivores could be interpreted as a response to the increasing importance of angiosperms in plant communities, but world-wide data imply that the increase in diversity was part of a trend which began during Jurassic time. It might be countered, with equal validity, that herbivorous dinosaurs were undergoing a long-sustained radiation that would have occurred regardless of the taxonomic affinities of the plants upon which they fed.

The first angiosperms were early successional plants of moist habitats (Retallack & Dilcher, 1981). Modern early successional species tend to grow more rapidly and be less well defended chemically than late successional forms (Opler, 1978). If the same were true of ancestral angiosperms, these features would have rendered them attractive to large plant-eaters such as dinosaurs. Through feeding and other activities, dinosaurian herbivores may have opened up wooded areas to early stages of succession (Farlow, 1976), and brought angiosperm propagules to them in their guts. Long passage times of digesta imposed by low metabolic rates and large size (Parra, 1978; Demment & Van Soest, 1985; Karasov & Diamond, 1985) may have rendered this form of propagule dispersal hazardous. Nevertheless, it is not inconceivable, over the course of the ten million years required for angiosperms to spread around the globe from their area of origin (Retallack & Dilcher, 1981), that dinosaur-mediated dispersal was as important as that of abiotic agents such as wind and water.

There were changes in the gross skeletal morphology of dinosaurian

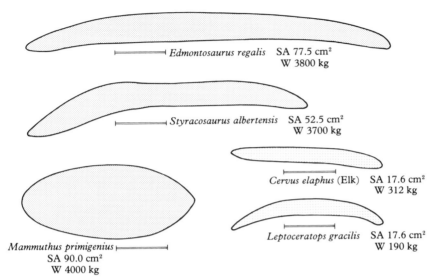

Figure 9.5. Surface areas (SA) and shapes of dental batteries of Cretaceous dinosaurs and subfossil or Recent mammals of comparable body weights (W). In each case the outline represents the occlusal surface of the dentition in one mandibular ramus. The comparisons suggest that, because the surface areas of the dental batteries and weights of *Edmontosaurus* (ornithopod), *Styracosaurus* (ceratopsian), and *Mammuthus* (mammoth), on one hand, and *Leptoceratops* (ceratopsian) and *Cervus* (elk), on the other, are similar, the metabolic rates within the two groups of herbivores may have also been similar. Weights were calculated according to equations 9.1 and 9.2 (see Appendix), and dental battery surface areas were measured directly from mandibles belonging to the same specimen as that on which the weight calculation was based.

herbivores between Late Jurassic and Late Cretaceous time. They too may well reflect general trends in the evolution of herbivorous vertebrates, rather than the special case of dinosaurian adaptation to angiosperm consumption. Large, Late Jurassic herbivores (sauropods, stegosaurs) tended to have enormous bodies, small heads, and cropping but not chewing teeth. Small Late Jurassic ornithopods (hypsilophodonts, camptosaurs) tended to have narrow beaks suggesting selectivity in cropping plant parts, and teeth that could crush plant tissues before they were swallowed.

By Late Cretaceous time the mean body size of herbivorous dinosaurs had declined substantially, and highly evolved grinding dentitions appeared within the larger, more muscular heads of advanced ornithopods (hadrosaurids) and horned dinosaurs (ceratopsids) (Figure 9.5). The greater taxonomic diversity of these two groups suggests a higher degree of specificity in animal–plant relationships than was the case with Jurassic dinosaurs. As in living herbivorous lizards, turtles, birds and mammals, most herbivorous dinosaurs probably employed a symbiotic gut microflora to ferment high-fiber plant materials, and perhaps even to detoxify poisonous plant secondary substances. By chewing their food, hadrosaurs and ceratopsids may have increased digestive rates and the rate of passage of food material through the gut. Higher activity levels, higher fermentation rates and smaller body masses are suggestive of a closer approach to endothermy than in the case of Late Jurassic dinosaurian herbivores (Ostrom, 1980). Late Cretaceous herbivores were intermediate in size between their small and large Late Jurassic counterparts, and were thus more uniform in weight. They differed from small Late Jurassic ornithischian herbivores in their greater size and broader beaks. Thus Late Cretaceous ornithopods evidently could not be as discriminating as their smaller forebears in the quality of the plant material they ingested. They may have been selective in the kind of plants they chose to feed on but not in what they cropped from them.

Sauropods may have adopted the strategy of stripping vegetation and retaining it in large quantities for slow fermentation. If so, this strategy would appear to have continued to be successful in some areas of the world during Late Cretaceous time, notably in South America and India (Molnar, 1980). Low metabolic rates (see Appendix) enabled sauropods to grow to enormous dimensions, and yet avoid serious surface–volume problems in heat dissipation that would have been encountered by endotherms of similar body weight. High digestive efficiencies could have been attained in the absence of mechanical food maceration simply by exposing the digesta to a more lengthy process of microbial fermentation. Without moving the center of mass, a long neck supported by a large skeletal frame could carry the head to crop plant material over a broad surface area of terrain or high into trees. Interestingly, the Late Cretaceous sauropod *Nemegtosaurus* possessed a long, distally expanded muzzle rather like those of hadrosaurs (Nowiński, 1971).

Many modern ferns are not eaten by animals because of poisonous secondary metabolites that accumulate in their fronds. These toxic compounds are probably the result of a long history of selective pressures, and may be more prevalent in ferns than in angiosperms simply because ferns are a much more ancient group. Such poisons may well have originated in response to attacks by phytophagous insects, but they are now quite effective in deterring excessive feeding by warm-blooded herbivores (Janzen, 1970). However, similar levels of chemical defense may not have protected plants from cold-blooded herbivorous dinosaurs. These animals were probably 'animated compost heaps', like the living giant tortoises, which ingest relatively small quantities of food and retain it in their bodies for up to 20 days. Tortoises will tolerate higher tannin concentrations in the plants they consume than will herbivorous homoiotherms (Swain, 1976). In accordance with the dictates of the Red Queen Hypothesis, whereby an organism must continue to improve in absolute adaptive fitness in an evolving world in order to maintain its relative evolutionary fitness (Van Valen, 1973), the biochemical sophistication of dinosaurian digestion must have kept pace with the biochemical defenses of angiosperms. In any case, Mesozoic ferns provided large quantities of vegetation that could easily have been cropped by grazing reptiles.

The geologic record thus shows that the dinosaurs continued a long-term increase in diversity coincident with the time of origin and initial diversification of angiosperms during middle and late Cretaceous time. Land plant diversity did increase during this interval, and gymnosperms steadily declined as the angiosperms diversified. Unlike the gymnosperms, however, the dinosaurian orders continued to diversify until the end of the Cretaceous Period.

Angiosperms and the extinction of the dinosaurs

The rise of the angiosperms has often been linked with environmental changes postulated to have driven the dinosaurs gradually to extinction. Swain (1976) proposed that angiosperms produce a variety of toxic chemicals that were stated to be generally absent in more primitive vascular plants, and that are less easily detected by reptiles than by mammals. The change in terrestrial floras was thus held to be at least partly responsible for determining which of the rival classes of terrestrial vertebrates would become dominant. In a similar manner, Krassilov (1981) linked the initial success and ultimate demise of the dinosaurs to the early Mesozoic expansion and late Mesozoic decline of sclerophyllous shrublands and fern marshes. Dinosaurian success was again considered to have been a function of plant evolution. The controversy surrounding the extinction of the dinosaurs has become particularly lively, with a division of opinion between supporters of more or less gradual stresses generated by terrestrially limited mechanisms, and supporters of catastrophic stresses generated by extraterrestrial mechanisms. It is not our intent here to take part in this debate, but simply to mention a few aspects of biotic changes that occurred during this interval as they could have related to dinosaur–plant interactions.

It has become apparent that a floral event was associated with the deposition of the Cretaceous–Tertiary boundary clay in some terrestrial environments (Hotton, 1984; Tschudy *et al.*, 1984; Nichols *et al.*, 1986). This event is marked by a great increase in the relative abundance of fern spores in sediments measuring only a few centimeters in thickness, and representing a very short interval of basal Tertiary time. This has been interpreted as evidence of the rapid recolonization of devastated mid-continental regions by wind-dispersed ferns following the extinction interval (Wolbach, Lewis & Anders, 1985). In itself, a vegetational simplification of this order would have produced a great reduction in the biomass of large vertebrate herbivores, which would probably have been accompanied by extinctions.

In the longer term, the identification of remains of thermophilic plants in Early Tertiary strata suggests that conditions in the physical environment were approximately the same during pre- and post-extinction time (Tschudy, 1970; Nichols *et al.*, 1986). Nevertheless, basal Tertiary floras in two regions of the Northern Hemisphere were characterized by an abundance of gymnosperms of modern aspect (Krassilov, 1978; Sweet & Hills, 1984). It has been suggested (Janzen & Martin, 1982) that, when the mammalian megafauna became extinct in Central America 10000 years ago, the ranges of several tree species which depended on large mammals for dispersing their fruits were reduced (but see Howe, 1985). By analogy, if some Late Cretaceous angiosperms also depended on dinosaurian herbivores for dispersing their propagules, they too would have been placed at a disadvantage in post-extinction floras. In this context, it is perhaps relevant to note a rapid increase in the mean size of plant propagules in earliest Tertiary time (Tiffney, 1984), which may have coincided with a shift from foliage-eating dinosaurs (where, in the model proposed by Janzen (1984) the foliage surrounding the seeds attracts large herbivores in the same manner as fruity tissues surrounding seeds attract small herbivores) to fruit-eating mammals and birds as dispersal agents. There may be evidence here of the ecological integration of dinosaurs with plant life beyond the stage of simply using plant vegetative organs for food. The abrupt and complete removal of large herbivores in itself would have resulted in altered forest compositions, as evidenced by the effects of the presence or absence of elephant in modern African woodlands (Eltringham, 1980).

It is doubtful that differences between Late Cretaceous and Early Tertiary vegetation were of sufficient magnitude to have been the primary cause of the extinction of the dinosaurs. Indeed, some of the vegetational changes that are observed may have largely resulted from the absence of dinosaurs in terrestrial ecosystems. The great herbivorous dinosaurs had evolved under rather constant and strong selective pressures that had prevailed during at least much of Mesozoic time. Could these selective pressures have continued broadly to direct the Tertiary radiation of large mammalian herbivores along pathways that resulted in large-scale structural convergences with dinosaurian predecessors?

Appendix: Derivation of browse profiles

According to Anderson *et al.* (1985) dinosaurian weights (W, g) can be estimated from the circumferences of the humerus (C_h, mm) and femur (C_f, mm) in quadrupeds, and the circumference of the femur in bipeds:

$$\log_{10} W = 2.73 \log_{10} (C_h + C_f) - 1.11 \tag{9.1}$$
$$\text{(for which } r^2 = 0.99\text{)}$$
$$\log_{10} W = 2.73 (\log_{10} C_f) - 0.76. \tag{9.2}$$

The correlation index, or coefficient of determination (r^2), is the proportion of the total variation accounted for by the fitted regression on the data given. The closer the individual data values lie to the linear regression, the nearer is the r^2 value to 1. Note that no r value is given in equation 9.2, because the equation was not derived through statistical procedures. Weights listed for Morrison dinosaurs (Table 9.1) were calculated for large specimens of each taxon largely on the basis of these relationships.

In order to obtain some appreciation for the relative abundance of different dinosaurs in the Morrison assemblage, minimum numbers of individuals (MNIs) were tallied from publications, from the field records of the American Museum of Natural History and from conversations with Dan Chure, Bruce R. Erickson, James A. Jensen, Robert A. Long, James H. Madsen Jr and John S. McIntosh, all of whom are authorities on Morrison dinosaurs. The percentage abundances of various genera are listed in Table 9.2, although it must be stressed that these numbers are working estimates that may be modified substantially in future revisions. Specimens of *Allosaurus* (Madsen, 1976) and *Diplodocus* (Brown, 1935) that occur in mass death associations have not been included, nor have those from the Dry Mesa Quarry in western Colorado that are currently being

Table 9.1. *Weights of Morrison dinosaurs*

Species	weight (kg)
Nannosaurus	0.75
Ornitholestes	7
Coelurus	13
Othnielia	19
Stokesosaurus	50
Elaphrosaurus	90
Marshosaurus	100
Dryosaurus	100
Ceratosaurus	675
Camptosaurus	1 000
Allosaurus	1 700
Nodosaur	2 000
Stegosaurus	5 000
Haplocanthosaurus	7 000
Diplodocus	16 000
Sauropod, undescribed	30 000
Camarasaurus	31 000
Barosaurus	40 000
Apatosaurus	42 500
Brachiosaurus	54 500
'*Ultrasaurus*'	90 000
'*Amphicoelias*'	100 000

studied by James A. Jensen. Otherwise, the Morrison dinosaurian assemblage is remarkably uniform over the entire area where skeletal remains have been collected.

The adult weights of different Morrison dinosaurs spanned six orders of magnitude. It is evident that skeletal remains of small animals are more often destroyed and are more frequently overlooked by collectors than are skeletal remains of large animals. Behrensmeyer, Western & Boaz (1979) provide data on the individual abundances of mammalian herbivores living on the semi-arid plains of the Amboseli Basin in southern Kenya and the number of skeletons they left behind. The ratio of animals counted (N) to skeletons observed (S) can be expressed as a dependent variable of body weight (W, kg; rhino data are excluded because of currently high death rates due to poaching):

$$\log_{10}[N/S] = 1.96 - 0.45 \log_{10} W, \tag{9.3}$$

$$\text{(for which } r^2 = 0.82).$$

The values of ratio N/S derived for animals of various weights listed in Table 9.1 are then multiplied by the MNI frequencies and normalized to 100% to obtain the 'corrected' MNI frequencies in Table 9.2. That the latter approximate the relative abundances of Morrison animals when living is suggested by the fact

Table 9.2. *Minimum number of individuals (MNI) of Morrison dinosaurs*

	Observed MNI (%)		Corrected MNI (%)
Camarasaurus	20.56	*Nannosaurus*	13.10
Apatosaurus	16.10	*Dryosaurus*	11.99
Diplodocus	14.09	*Othnielia*	11.71
Stegosaurus	13.03	*Allosaurus*	10.00
Allosaurus	11.61	*Coelurus*	9.34
Camptosaurus	8.06	*Camptosaurus*	8.81
Dryosaurus	3.89	*Stegosaurus*	6.91
Barosaurus	1.98	*Ornitholestes*	6.12
Ceratosaurus	1.90	*Camarasaurus*	4.80
Othnielia	1.80	*Diplodocus*	4.42
Haplocanthosaurus	1.73	*Apatosaurus*	3.33
Coelurus	1.21	*Stokesosaurus*	2.53
Brachiosaurus	0.99	*Ceratosaurus*	2.48
Ornitholestes	0.60	*Marshosaurus*	1.85
Marshosaurus	0.60	*Elaphrosaurus*	0.97
Stokesosaurus	0.60	*Haplocanthosaurus*	0.79
Nannosaurus	0.47	*Barosaurus*	0.41
Elaphrosaurus	0.30	Nodosaur	0.19
Nodosaur	0.24	*Brachiosaurus*	0.18
Sauropod, undescribed	0.24	Sauropod, undescribed	0.06

that the number of small juveniles becomes about equal to that of large adults, a circumstance typical of conditions of attritional mortality (Klein & Cruz-Uribe, 1984).

In Table 9.3 'corrected' MNIs and weights are multiplied and normalized to 100 % (or to 100 g) to calculate the relative biomass contributed by various taxa. The relationship between longevity and body weight has not been well studied in living reptiles, but a review of zoo data (for longevities see Bowler, 1977; weights were compiled from many sources) suggests that, with the exception of turtles, the relationship is approximately the same as for mammals (Sacher, 1975). If this is true, the values for the ratio productivity (P):biomass (B) may also be similar. However, there is a considerable amount of scatter in the relationship. For example, low values of $P:B$ in long-lived fishes (6 % to 7 %) contrast strongly with those in short-lived fishes, which may exceed 60 % (McNeill & Lawton, 1970). The annual turnover of turtle biomass on Aldabra is estimated to be just under 1 % for animals with a mean body weight of 22 kg (Hamilton & Coe, 1982). A population of herbivorous mammals with the same body weights would be expected to turnover as much as 80 % of their biomass each year (Coe, 1980a). The mammalian relationship (Farlow, 1976; p. 846; see also Banse & Mosher, 1980) between the $P:B$ ratio and body weight (W, g) is

Table 9.3 *Predator–prey relationships of Morrison dinosaurs*

Herbivores	% Biomass	Annual secondary productivity
Camarasaurus	32.4188	1.9069
Apatosaurus	30.8338	1.6677
Diplodocus	15.4076	1.0806
Stegosaurus	7.5273	0.7193
Camptosaurus	1.9194	0.2814
Barosaurus	3.5730	0.1964
Brachiosaurus	2.1373	0.1082
Haplocanthosaurus	1.2048	0.1053
Dryosaurus	0.2612	0.0706
Sauropod, undescribed	0.3922	0.0233
Othnielia	0.0485	0.0204
Nodosaur	0.0828	0.0101
Nannosaurus	0.0021	0.0021
Total	95.8088	6.1923

Carnivores	% Biomass	Annual Consumption
Allosaurus	3.7038	3.8310
Ceratosaurus	0.3647	0.4455
Marshosaurus	0.0403	0.0694
Coelurus	0.0265	0.0659
Stokesosaurus	0.0276	0.0539
Elaphrosaurus	0.0190	0.0334
Ornitholestes	0.0093	0.0259
Total	4.1912	4.5250

Total consumption for endothermic carnivores 23.7816

here applied to the dinosaurian biomasses and weights to calculate annual productivity:

$$\log_{10}[P/B] = 0.76 - 0.27 \log_{10} W \tag{9.4}$$

(for which $r^2 = 0.87$).

In Table 9.3 the normalized biomass percentages may be considered as gram weights of tissue that yield the listed gram weights of productivity on an annual basis. The annual consumptions of carnivores (C, g per g biomass listed in Table 9.3) are calculated according to a modification of the intake of reptilian carnivores by Farlow (1976), where kilocalorie intake (1 kcal = 4.184 kJ) per day is divided by 1.5 (or the kilocalorie content of an average gram of animal body mass (Farlow, 1976, p. 843)) and multiplied by 365 days to convert from kilocalories per day to grams per year:

$$C = \frac{[\text{antilog} \ (0.82 \log_{10} W - 1.25)][365/1.5]}{W}. \tag{9.5}$$

Table 9.4. *Daily food requirements of Morrison dinosaurs*

Fossil herbivores	(kg wet plants)	Present-day species (kg wet plants)
Nannosaurus	0.012	
Othnielia	0.165	
Dryosaurus	0.644	
Camptosaurus	4.25	
Stegosaurus	15.91	
Haplocanthosaurus	20.97	
Diplodocus	41.30	
Camarasaurus	71.04	
Barosaurus	87.57	
Apatosaurus	92.02	
Brachiosaurus	112.84	
'*Ultrasaurus*'	170.25	
'*Amphicoelias*'	185.61	*Loxodonta* 225

Fossil carnivores	(kg flesh)	Present-day species (kg flesh)
Ornitholestes	0.053	
Coelurus	0.089	
Stokesosaurus	0.267	
Elaphrosaurus	0.433	
Marshosaurus	0.472	
Ceratosaurus	2.26	
Allosaurus	4.82	*Panthera leo* 4.6

The biomass and productivity/consumption ratios between dinosaurian herbivores and carnivores are suggestive of reptilian metabolic rates in the case of the carnivores. This remains true even if the calculations are based on 'uncorrected' MNIs.

In Table 9.4 the intakes of herbivorous and carnivorous dinosaurs of different body weights (W, g) are calculated according to daily intake rates of carnivorous reptiles by Farlow (1976):

$$\log_{10}\text{Intake(kcal/day)} = 0.82 \log_{10} W - 1.25 \qquad (9.6)$$
$$\text{(for which } r^2 = 0.94).$$

and the results divided by the factors 1.5 and 1.1 to convert from kilocalories per day to grams of animal body mass per day and grams of wet plant material per day, respectively. Interestingly, maximum daily intakes are similar in dinosaurs and mammals, suggesting that maximum body size is limited ecologically by the availability of food rather than by the force of gravity acting upon the mechanical properties of flesh and bone.

Population densities of Morrison dinosaurs can be calculated using the mammalian populations of the Amboseli Basin as an ecological analog. The

248

Table 9.5. *Data for Amboseli mammals* (biomass = 4848 kg km^{-2})

	Number	Weight (kg)	Biomass (kg km^{-2})	Intake (kcal km^{-2})	Animals per km^2
Elephant	218	3000	751.4	20288.9	0.25
Hippo	75	1000	86.3	3132.5	0.09
Rhino	40	816	37.3	1433.5	0.05
Giraffe	115	772	101.8	3960.3	0.13
Buffalo	457	450	236.1	10624.4	0.52
Eland	37	363	15.5	740.0	0.04
Zebra	2200	200	505.6	28366.8	2.53
Oryx	88	167	16.0	942.3	0.10
Wildebeest	2473	165	468.8	27706.2	2.84
Kongoni	87	136	13.6	844.3	0.10
Grant's gazelle	1230	50	70.8	5768.6	1.42
Impala	700	45	36.4	3050.6	0.81
Thompson's gazelle	788	20	17.9	1872.7	0.90
Ostrich	142	114	18.4	1203.0	0.16
Cattle	9184	227	2394.9	129804.2	10.55
Shoat	2550	16	47.0	5215.1	2.94
Donkey	203	130	30.5	1924.2	0.23
Total			4848.3	246877.6	23.66

Jurassic and modern vertebrate faunas both lived in semi-arid environments. Weights and population numbers of Amboseli herbivores are by Western (1975) and Behrensmeyer *et al.* (1979), and estimates of hippo populations were provided by A. K. Behrensmeyer (personal communication). The relative intakes of different components of the Amboseli herbivore biomass were calculated according to Farlow's (1976) equation for endotherms:

$$\log_{10} \text{Intake (kcal/day)} = 0.23 + 0.72 \log_{10} W \qquad (9.7)$$
$$\text{(for which } r^2 = 0.95).$$

For a total Amboseli biomass of 4848 kg km^{-2} (Coe *et al.*, 1976), about 247 000 kcal of plant food are on the average removed from each square kilometer of the Basin per day (Table 9.5).

The dinosaurian herbivores of the Morrison Plain were predominantly sauropods. Sauropods may be considered as possessing reptilian metabolic rates on the basis of brain–body proportions (Hopson, 1980), surface–volume proportions and heat loss (Spotila, 1980), the amount of time necessary for feeding (Weaver, 1983), and the presence of thermally induced (seasonal) rings in their bones (Reid, 1981; Ricqlès, 1983). Modern analogs of mid-Mesozoic gymnosperms often have relatively slow rates of growth (Russell, Béland & McIntosh, 1980), but these may have been offset by higher concentrations of atmospheric carbon dioxide (Berner, Lasaga & Garrels, 1983; Rogers, Thomas & Bingham, 1983). Should differences in trophic relationships, such as a relative increase in primary productivity and a decrease in resource utilization by secondary consumers, cancel each other out between Amboseli and Morrison environments, the Morrison Plain would have sustained a biomass of large herbivores amounting to 93 000 kg km^{-2} (Table 9.6). The factor of nearly 20 separating the two biomass concentrations is due both to relatively lower reptilian metabolic rates and to the great difference in mean weights between the herbivores of the two ages (200 kg vs 7000 kg). In terms of numbers of individuals, Amboseli herbivore populations are twice as dense. However, in the Amboseli Basin, elephant population densities amount to only one animal per 4 km^2, while on the Morrison Plain over 50 animals, each weighing twice as much as a large elephant, would occur in the same area.

That such dense populations of giant herbivores could ever exist might hardly seem credible. The only potential modern analog for the huge, cold-blooded herbivorous dinosaurs are the giant tortoises that inhabit the Aldabra Atoll (240 km northwest of Madagascar), and the Galapagos Islands (800 km off the coast of Equador). Population studies (Bourn & Coe, 1978) have revealed that about 150 000 tortoises occur on Aldabra, 60% of which are concentrated in an area of 3360 ha. These figures translate into biomasses of 58 000 kg km^{-2} in the area of concentration, and 53 000 kg km^{-2} for the whole of their range. In African wildlife ecosystems dominated by large herbivorous mammals, the highest biomasses recorded are; 17 500 kg km^{-2} on the Rwindi Plain, Zaïre; 19 000 kg km^{-2} in the Manyara National Park, Tanzania; and 20 000 kg km^{-2} in

Table 9.6. *Morrison dinosaur–plant relationships*

	A	B	C
Dryosaurus	1778	251	2.51
Othnielia	444	47	2.47
Nannosaurus	25	1	1.93
Camptosaurus	8665	1852	1.85
Stegosaurus	25478	7279	1.46
Camarasaurus	79001	31350	1.01
Diplodocus	42315	14900	0.93
Apatosaurus	70953	29812	0.70
Haplocanthosaurus	3851	1167	0.17
Barosaurus	8320	3452	0.09
Brachiosaurus	4715	2068	0.04
Nodosaur	346	84	0.04
Sauropod, undescribed	963	379	0.01
Totals		92642	13.21

A. Percentage of Morrison herbivorous dinosaur intake multiplied by total Amboseli herbivore intake (246878 kcal day^{-1} km^{-2}).
B. Amount of biomass (kg km^{-2}) represented by intake listed in A.
C. Number of animals (n km^{-2}) represented by biomass listed in B.

the Rwenzori National Park, Uganda (Coe *et al.*, 1976). Thus, the biomass achieved by the giant tortoise, which is the only large herbivore on Aldabra (excluding feral goats), is three times higher than that attained by homoiothermous terrestrial herbivores in the African mainland, under the influence of a similar rainfall regime. The difference is due to the lesser (per unit weight) metabolic requirements of the tortoises. Large reptilian carnivores such as the Nile crocodile occur in relatively low biomasses typical of those of secondary consumers. Graham (1968) nevertheless reported that, under optimal habitat conditions in Lake Turkana in northwestern Kenya, crocodiles sustained biomasses of 3500–19000 kg km^{-2} in more or less linear ranges along the lake margin. Such observations clearly indicate that cold-blooded dinosaurs would also have been able to sustain very high biomasses.

In order to sketch the outline of a browse profile for Morrison dinosaurs and Amboseli mammals, the proportional kilocalorie intake of various animals in each fauna is normalized to 100 (%) and distributed evenly according to 10-cm increments of the feeding range of the various animals, approximated from skeletal proportions (although elephants are presumed to push over and consume tree foliage originally 10 m off the ground; some sauropods and ceratopsians also may have toppled trees but this possibility is not considered here). The approximate browsing ranges of the vertebrates are listed in Table 9.7. Relevant data for the Dinosaur Park vertebrates are listed in Table 9.8.

Table 9.7. *Vertical feeding range of herbivores*

	Range (m)
I. *Morrison dinosaurs*	
Nannosaurus	0– 0.4
Othnielia	0– 2.0
Dryosaurus	0– 2.0
Camptosaurus	0– 2.8
Stegosaurus stenops	0.3– 1.1
Stegosaurus ungulatus	0.4– 1.3
Haplocanthosaurus	0– 4.5
Diplodocus	0– 4.5
Apatosaurus	0– 4.5
Barosaurus	0– 4.5
Camarasaurus	0– 6.5
Brachiosaurus	0.4–12.0
II. *Amboseli mammals*	
Impala	0– 1.0
Thompson's gazelle	0– 1.0
Cattle	0– 1.0
Shoat	0– 1.0
Donkey	0– 1.0
Wildebeest	0– 1.2
Kongoni	0– 1.2
Grant's gazelle	0– 1.2
Hippo	0– 1.5
Oryx	0– 1.5
Rhino	0– 2.0
Buffalo	0– 2.0
Eland	0– 2.0
Zebra	0– 2.0
Ostrich	0– 2.0
Giraffe	0– 5.5
Elephant	0–10.0

Table 9.8. *Data for Dinosaur Park megafaunal herbivores*

	A	B	C	D	E
Euoplocephalus	14.69	2000	11.02	8.47	0– 0.50
Panoplosaurus	4.28	2250	3.61	3.06	0– 1.00
Ceratopsids	25.12	3500	32.97	40.01	0– 2.00
Hadrosaurids	55.91	2500	52.41	48.37	0– 4.00

A. Percentage of herbivore individuals.
B. Weight of individual herbivore (kg).
C. Percentage biomass of herbivore groups in assemblage.
D. Percentage of total kilocalories ingested by megafaunal herbivores per day (individual intake calculated from equation 9.6).
E. Vertical feeding range (m).

A brief digression is necessary to explain how the upper limit of sauropod browse was defined. Bakker (1978) proposed, as had several previous authors, that sauropods could thrust their necks high into trees by standing on their hind legs and bracing themselves in a tripodal posture with their tails on the ground. Alexander (1985) also suggested that sauropods could rotate their bodies about their hindlimbs, holding the forelimbs off the ground. His arguments are further supported by the extreme lightening of the sauropod neck (although taphonomic evidence suggests that if the neck was supported by a powerful dorsal tendon, the cervical vertebrae none the less became disassociated more rapidly than did other regions of the vertebral column). Elephants can also stand rather easily on their hind legs. However, they do not usually feed in this position, and sauropods probably did not often feed from a tripodal stance either. The upper limit of sauropod browse is set here by the maximum 'comfortable' level to which it is believed the head could be brought by the neck, without the animal rising on its back legs.

We are particularly grateful to William Chaloner and Peter Crane for their constructive counsel in the course of preparing this manuscript.

References

Alexander, R. McN. (1985). Mechanics of posture and gait of some large dinosaurs. *Zoological Journal of the Linnean Society*, **83**, 1–25.

Anderson, J. F., Hall-Martin, A. & Russell, D. A. (1985). Long-bone circumference and weight in mammals, birds and dinosaurs. *Journal of Zoology*, **207**, 53–61.

Bakker, R. T. (1978). Dinosaur feeding behaviour and the origin of flowering plants. *Nature*, **274**, 661–3.

Banse, K. & Mosher, S. (1980). Adult body mass and annual production/biomass relationships of field populations. *Ecological Monographs*, **50**, 355–79.

Behrensmeyer, A. K., Western, D. & Boaz, D. E. D. (1979). New perspectives in vertebrate paleoecology from a recent bone assemblage. *Paleobiology*, **5**, 12–21.

Béland, P. & Russell, D. A. (1978). Paleoecology of Dinosaur Provincial Park (Cretaceous), Alberta, interpreted from the distribution of articulated vertebrate remains. *Canadian Journal of Earth Sciences*, **15**, 1012–24.

Bell, W. A. (1956). Lower Cretaceous floras of western Canada. *Memoirs Geological Survey of Canada*, **285**, 1–331.

Berner, R. A., Lasaga, A. C. & Garrels, R. M. (1983). The carbonate–silicate geochemical cycle and its effect on atmospheric carbon dioxide over the past 100 million years. *American Journal of Science*, **283**, 641–83.

Berry, E. W. (1922). The flora of the Cheyenne Sandstone of Kansas. *U.S. Geological Survey Professional Papers*. **129I**, 199–225.

Bird, R. T. (1985). *Bones for Barnum Brown*, ed. V. T. Schreiber &
J. O. Farlow. Fort Worth: Texas Christian University Press.

Bond, G. (1955). A note on dinosaur remains from the Forest Sandsonte
(Upper Karoo). *Occasional Papers, National Museum of Rhodesia*, 2,
795–800.

Bourlière, F. (1973). The comparative ecology of rain forest mammals in
Africa and tropical America. In *Tropical Forest Ecosystems in Africa and
South America*, ed. B. J. Meggars, E. S. Ayensu & W. D. Duckworth, pp.
279–92. Washington, DC: The Smithsonian Institution.

Bourn, D. & Coe, M. J. (1978). The size, structure and distribution of the
giant tortoise population of Aldabra. *Philosophical Transactions of the Royal
Society, series B*, 282, 139–75.

Bowler, J. K. (1977). Longevity of reptiles and amphibians in North
American collections. *Society for the Study of Amphibians and Reptiles,
Miscellaneous Publications in Herpetology*, 6.

Brown, B. (1907). Gastroliths. *Science*, 25, 392.

Brown, B. (1935). Sinclair Dinosaur Expedition, 1934. *Natural History*, 42,
3–15.

Brown, B. (1941). The last dinosaurs. *Natural History*, 48, 290–5.

Brown, R. W. (1939). Fossil plants from the Colgate Member of the Fox
Hills Sandstone and adjacent strata. *U.S. Geological Survey Professional
Papers*, 189I, 239–75.

Chaloner, W. G. (1968). The paleoecology of fossil spores. In *Evolution and
Environment*. A Symposium presented on the Occasion of the 100th
Anniversary of the Foundation of the Peabody Museum of Natural History,
pp. 125–38. New Haven: Yale University Press.

Chandler, M. E. J. (1966). Fruiting organs from the Morrison Formation of
Utah, U.S.A. *Bulletin of the British Museum (Natural History) Geology*, 12,
137–71.

Coe, M. J. (1980a). African mammals in savanna habitats. In *International
Symposium on Habitats and their Influence on Wildlife*, pp. 83–109. Pretoria:
Endangered Wildlife Trust.

Coe, M. J. (1980b). The role of modern ecological studies in the
reconstruction of palaeoenvironments in sub-Saharan Africa. In *Fossils in
the Making*, ed. A. K. Behrensmayer & A. K. Hill, pp. 55–71. Chicago:
University of Chicago Press.

Coe, M. J., Cumming, D. H. & Phillipson, J. (1976). Biomass and production
of large African herbivores in relation to rainfall and primary production.
Oecologia, 22, 341–54.

Collins, S. L. & Uno, G. E. (1985). Seed predation, seed dispersal, and
disturbance in grasslands: a comment. *American Naturalist*, 125, 866–72.

Delevoryas, T. (1971). Biotic provinces and the Jurassic–Cretaceous floral
transition. *Proceedings of the North American Paleontological Convention*, ed.
E. L. Yochelson, vol. L, pp. 1660–74. Lawrence, Kansas: Allen Press.

Demment, M. W. & Van Soest, P. J. (1985). A nutritional explanation for body-size patterns of ruminant and nonruminant herbivores. *American Naturalist.* **125**, 641–72.

Dodson, P., Behrensmeyer, A. K., Bakker, R. T. & McIntosh, J. S. (1980). Taphonomy and paleoecology of the dinosaur beds of the Jurassic Morrison Formation. *Paleobiology*, **6**, 208–32.

Dubost, G. (1979). The size of African forest artiodactyls as determined by the vegetation structure. *African Journal of Ecology*, **17**, 1–17.

Elsik, W. C. (1968a). Palynology of a Paleocene Rockdale lignite, Milam County, Texas. I. Morphology and taxonomy. *Pollen et Spores*, **10**, 263–314.

Elsik, W. C. (1968b). Palynology of a Paleocene Rockdale lignite, Milam County, Texas. II. Morphology and taxonomy. *Pollen et Spores*, **10**, 599–664.

Eltringham, S. K. (1980). A quantitative assessment of range usage by large African mammals with particular reference to the effects of elephants on trees. *African Journal of Ecology*, **18**, 53–71.

Farlow, J. O. (1976). A consideration of the trophic dynamics of a late Cretaceous large-dinosaur community (Oldman Formation). *Ecology*, **57**, 841–57.

Gilmore, C. W. (1936). Osteology of *Apatosaurus*, with special reference to specimens in the Carnegie Museum. *Memoirs of the Carnegie Museum*, **11**, 171–298.

Graham, A. (1968). The Lake Rudolph crocodile (*Crocodylus niloticus* Laurenti) population. M.Sc. thesis, University of Nairobi.

Hamilton, J. & Coe, M. J. (1982). Feeding, digestion and assimilation of a population of giant tortoises (*Geochelone gigantea* Schweigger) on Aldabra Atoll. *Journal of Arid Environments*, **5**, 127–44.

Harris, T. M. (1981). Burnt ferns from the English Wealden. *Proceedings of the Geologists' Association*, **92**, 47–58.

Hedlund, R. W. & Norris, G. (1968). Spores and pollen grains from Fredericksburgian (Aptian) strata, Marshall County, Oklahoma. *Pollen et Spores*, **10**, 129–59.

Hickey, L. J. (1977). Stratigraphy and paleobotany of the Golden Valley Formation (Early Tertiary) of western North Dakota. *Memoirs of the Geological Society of America*, **150**, 1–183.

Hill, C. R. (1976). Coprolites of *Ptilophyllum* cuticles from the Middle Jurassic of north Yorkshire. *Bulletin of the British Museum (Natural History) Geology*, **27**, 289–94.

Hofmann, R. R. (1973). The ruminant stomach. *East African Monographs in Biology*, no. 2. Nairobi: East African Publishing House.

Hopson, J. A. (1980). Relative brain size in dinosaurs: implications for dinosaurian endothermy. *American Association for the Advancement of Science Selected Symposia*, **28**, 287–310.

Hotton, C. (1984). Palynofloral changes across the Cretaceous–Tertiary boundary in east central Montana, U.S.A. *Abstracts, Sixth International Palynological Conference, Calgary 1984*, p. 66.

Howe, H. F. (1985). Gomphothere fruits: a critique. *American Naturalist*, **125**, 853–65.

Hughes, N. F. (1976). *Palaeobiology of Angiosperm Origins*. Cambridge: Cambridge University Press.

Janensch, W. (1929). Magensteine bei Sauropoden der Tendaguru-Schichten. *Palaeontographica, Suppl.* **7**, *Erste Reihe*, Teil 2. 135–44.

Janensch, W. (1961). Die Skelettrekonstruktion von *Brachiosaurus brancai*. *Palaeontographica, Suppl.* **7**, *Erste Reihe*, Teil 3, 237–40.

Janzen, D. H. (1970). Herbivores and the number of tree species in tropical forests. *American Naturalist*, **104**, 501–28.

Janzen, D. H. (1984). Dispersal of small seeds by big herbivores: foliage is the fruit. *American Naturalist*, **123**, 338–53.

Janzen, D. H. & Martin, P. S. (1982). Neotropical anachronisms: the fruits the gomphotheres ate. *Science*, **215**, 19–27.

Jarman, P. J. (1974). The social organisation of antelope in relation to their ecology. *Behaviour*, **48**, 215–66.

Jarzen, D. M. (1980). The occurrence of *Gunnera* pollen in the fossil record. *Biotropica*, **12**, 117–23.

Jarzen, D. M. (1983). The fossil pollen record of the Pandanaceae. *Gardens' Bulletin of Singapore*, **36**, 163–75.

Karasov, W. H. & Diamond, J. M. (1985). Digestive adaptations for fueling the cost of endothermy. *Science*, **228**, 202–4.

Klein, R. G. & Cruz-Uribe, K. (1984). *The Analysis of Animal Bones from Archeological Sites*. Chicago: University of Chicago Press.

Krassilov, V. A. (1978). Late Cretaceous gymnosperms from Sakhalin, U.S.S.R., and the terminal Cretaceous event. *Palaeontology*, **21**, 893–905.

Krassilov, V. A. (1981). Changes of Mesozoic vegetation and the extinction of the dinosaurs. *Palaeogeography, Palaeoclimatology, Palaeoecology*, **34**, 207–24.

Lesquereux, L. (1892). The flora of the Dakota Group. *U.S. Geological Survey Monographs*, **17**.

Louw, G. N. & Seeley, M. K. (1982). *Ecology of Desert Organisms*. Harlow: Longman.

McNeill, S. & Lawton, J. W. (1970). Animal production and respiration in animal populations. *Nature*, **255**, 472–4.

Madsen, J. H. (1976). *Allosaurus fragilis*: a revised osteology. *Utah Geological and Mineral Survey, Bulletin*, **109**.

Majer, S. (1923). Spuren von dinosauriern der Oberkreide im liegenden des Kosder Eocänen Kohlenflötzes. *Földtani Közlony*, **51–2**, 66–75, 113–114.

May, F. E. & Traverse, A. (1973). Palynology of the Dakota Sandstone (middle Cretaceous) near Bryce Canyon National Park, southern Utah. *Geoscience and Man*, 7, 57–64.

Mayr, E. (1963). *Animal Species and their Evolution*. Cambridge, Massachusetts: Harvard University Press.

Molnar, R. (1980). Australian late Mesozoic terrestrial tetrapods: some implications. *Mémoires Société Géologique de France*, **139**, 131–43.

Nichols, D. J., Jarzen, D. M., Orth, C. J. & Oliver, P. Q. (1986). Palynological and iridium anomalies at the Cretaceous–Tertiary boundary, south-central Saskatchewan. *Science*, **231**, 714–17.

Norman, D. B. & Weishampel, D. B. (1985). Ornithopod feeding mechanisms: their bearing on the evolution of herbivory. *American Naturalist*, **126**, 151–64.

Norton, N. J. & Hall, J. W. (1969). Palynology of the Upper Cretaceous and Lower Tertiary in the type locality of the Hell Creek Formation, Montana, U.S.A. *Palaeontographica B*, **125**, 1–64.

Nowiński, A. (1971). *Nemegtosaurus mongoliensis* n. gen., n. sp. (Sauropoda) from the uppermost Cretaceous of Mongolia. *Palaeontologia Polonica*, **25**, 57–81.

Opler, A. (1978). Interaction of plant life history components as related to arboreal herbivory. In *The Ecology of Arboreal Folivores*, ed. G. G. Montgomery, pp. 23–31. Washington, DC: Smithsonian Institution Press.

Ostrom, J. H. (1980). The evidence for endothermy in dinosaurs. *American Association for the Advancement of Science, Selected Symposia*, **28**, 15–54.

Parker, L. R. (1980). Paleobotany (ed. L. J. Hickey). *Geotimes*, February 1980, p. 40.

Parra, R. (1978). Comparison of foregut and hindgut fermentation in herbivores. In *The Ecology of Arboreal Folivores*, ed. G. G. Montgomery, pp. 205–9. Washington, DC: Smithsonian Institution Press.

Pianka, E. R. (1978). *Evolutionary Ecology*. New York: Harper & Row.

Pocock, S. A. J. (1962). Microfloral analysis and age determination of strata at the Jurassic–Cretaceous boundary in the western Canada plains. *Palaeontographica B*, **111**, 1–95.

Potter, F. W. (1976). Investigations of angiosperms from the Eocene of southeastern North America: pollen assemblages from Miller Pit, Henry County, Tennessee. *Palaeontographica B*, **157**, 44–96.

Raath, M. (1974). Fossil vertebrate studies in Rhodesia: further evidence of gastroliths in prosauropod dinosaurs. *Arnoldia*, 7, 1–7.

Regal, P. J. (1977). Ecology and evolution of flowering plant dominance. *Science*, **196**, 622–9.

Reid, R. E. H. (1981). Lamellar-zonal bone with zones and annuli in the pelvis of a sauropod dinosaur. *Nature*, **292**, 49–51.

Retallack, G. J. & Dilcher, D. L. (1981). A coastal hypothesis for the dispersal and rise to dominance of the flowering plants. In *Paleobotany, Paleoecology, and Evolution*, vol. II, ed. K. J. Niklas, pp. 27–77. New York: Praeger Publishers.

Ricqlès, A. de (1983). Cyclical growth in the long limb bones of a sauropod dinosaur. *Acta Palaeontologica Polonica*, **28**, 225–32.

Rogers, H. H., Thomas, J. F. & Bingham, G. E. (1983). Response of agronomic and forest species to elevated atmospheric carbon dioxide. *Science*, **220**, 428–9.

Roth, J. L. & Dilcher, D. L. (1978). Some considerations in leaf size and leaf margin analysis of fossil leaves. *Courier Forschungs-Institut, Sekenberg*, **30**, 165–71.

Russell, D. A. (1983). Exponential evolution: implications for intelligent extraterrestrial life. *Advances in Space Research*, **3**, 95–103.

Russell, D. A. (1984a). The gradual decline of the dinosaurs – fact or fallacy? *Nature*, **307**, 360–1.

Russell, D. A. (1984b). A check list of the families and genera of North American dinosaurs. *Syllogeus*, **53**.

Russell, D. A., Béland, P. & McIntosh, J. S. (1980). Paleoecology of the dinosaurs of Tendaguru (Tanzania). *Mémoires de la Société Géologique de France*, **139**, 169–75.

Sacher, G. A. (1975). Maturation and longevity in relation to cranial capacity in hominid evolution. In *Primate Functional Morphology and Evolution*, ed. R. H. Tuttle, pp. 417–41. The Hague: Mouton Publishers.

Seilacher, A. (1984). Constructional morphology of bivalves: evolutionary pathways in primary versus secondary soft-bottom dwellers. *Palaeontology*, **27**, 207–37.

Spotila, J. R. (1980). Constraints of body size and environment on the temperature regulation of dinosaurs. *American Association for the Advancement of Science Selected Symposia*, **28**, 233–52.

Stocker, G. C. & Irvine, A. K. (1983). Seed dispersal by Cassowaries (*Casuarius casuarius*) in North Queensland's rainforests. *Biotropica*, **15**, 170–6.

Stokes, W. L. (1964). Fossilized stomach contents of a sauropod dinosaur. *Science*, **143**, 576–7.

Swain, T. (1976). Angiosperm–reptile co-evolution. *Linnean Society Symposia*, **3**, 107–22.

Sweet, A. R. & Hills, L. V. (1984). A palynological and sedimentological analysis of the Cretaceous–Tertiary boundary, Red Deer Valley, Alberta, Canada. *Abstracts, Sixth International Palynological Conference, Calgary* 1984, p. 160.

Thomas, R. D. K. & Olson, E. C. (eds.) (1980). A cold look at warm-blooded dinosaurs. *American Association for the Advancement of Science Selected Symposia*, **28**.

Tiffney, B. H. (1984). Seed size, dispersal syndromes, and the rise of the angiosperms: evidence and hypothesis. *Annals of the Missouri Botanical Garden*, 71, 551–76.

Troyer, K. (1984). Structure and function of the digestive tract of a herbivorous lizard *Iguana iguana*. *Physiological Zoology*, 57, 1–8.

Tschudy, R. H. (1970). Palynology of the Cretaceous–Tertiary boundary in the northern Rocky Mountain and Mississippi Embayment regions. *Geological Society of America Special Papers*, 127, 65–111.

Tschudy, R. H., Pillmore, C. L., Orth, C. J., Gilmore, J. S. & Knight, J. D. (1984). Disruption of the terrestrial plant ecosystem at the Cretaceous–Tertiary boundary, western interior. *Science*, 225, 1030–2.

Van Valen, L. (1973). A new evolutionary law. *Evolutionary Theory*, 1, 1–30.

Viera, L. I. & Torres, J. A. (1979). El Wealdico de la zona de Enciso (Sierra de los Cameros) y su fauna de grandes reptiles. *Munibe, Sociedad de Ciencias Aranzadi, San Sebastian*, 31, 141–57.

Ward, L. F. (1900). Status of Mesozoic floras of the United States. *U.S. Geological Survey, Twentieth Annual Report*, 2, 211–748.

Weaver, J. C. (1983). The improbable endotherm: the energetics of the sauropod dinosaur *Brachiosaurus*. *Paleobiology*, 9, 173–82.

Weishampel, D. B. (1984). Interactions between Mesozoic plants and vertebrates: fructifications and seed predation. *Neues Jahrbuch für Geologie und Paläontologie, Abhandlungen*, 167, 224–50.

Western, D. (1975). Water availability and its influence on the structure and dynamics of a savannah large mammal community. *East African Wildlife Journal*, 13, 265–86.

Whittaker, R. H. (1975). *Communities and Ecosystems*, 2nd edn. New York: MacMillan.

Wolbach, W. S., Lewis, R. S. & Anders, E. (1985). Cretaceous extinctions: evidence for wildfires and search for meteoritic material. *Science*, 230, 167–70.

Wyles, J. S., Kunkel, J. G. & Wilson, A. C. (1983). Birds, behavior and anatomical evolution. *Proceedings of the U.S. Academy of Sciences*, 80, 4394–7.

10
·

Vegetational and mammalian faunal changes in the Early Tertiary of southern England

M.E.COLLINSON AND J.J.HOOKER

Reconstructing the sequence of terrestrial ecosystems through time is frequently frustrated by the inadequacies of the fossil record. Occurrences of fossiliferous strata are often sporadic and there is qualitative and quantitative variation in fossil content. The influence of abiotic variables such as vulcanism, orogeny and difficulties in establishing synchrony of sedimentation may also complicate comparison between successive fossil biotas.

The Early Tertiary strata in southern England provide an example from the more adequate end of the spectrum of fossil occurrences. They range from Late Paleocene to Early Oligocene, spanning 25 million years, and were laid down in one depositional basin now separated by folding into the London and Hampshire tectonic basins (Figure 10.1). They accumulated in a low-lying coastal area (see Daley, 1972; King, 1981) and include all intermediate states from shallow marine through brackish and freshwater to subaerial environments (King, 1981; Plint, 1983, 1984; Buurman, 1980) (Figure 10.1). They outcrop at many coastal sections and are frequently exposed by quarrying or temporary building work. Several sections reveal long continuous sequences, e.g. at Alum and Whitecliff Bays on the Isle of Wight and the Dorset coast. Correlation between many isolated sections has been well established on lithostratigraphic and biostratigraphic evidence (e.g. Curry et al., 1978).

The strata contain abundant plant and vertebrate remains at numerous levels (Figure 10.1). However, preservation of plants in mainly calcareous beds and of mammals in non-shelly, organic-rich strata is unusual and generally remains of mammals are rarer than those of plants. Both are rare in the fully marine beds. In many sequences, however, lithology has favored the preservation of both, either in the same or in closely associated strata (Figure 10.1).

259

260

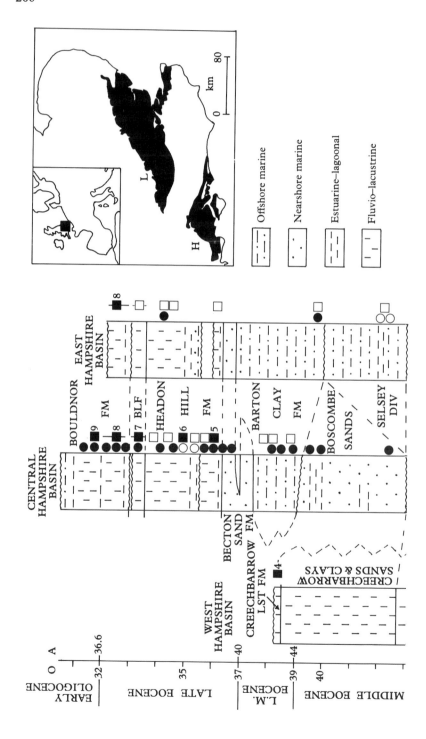

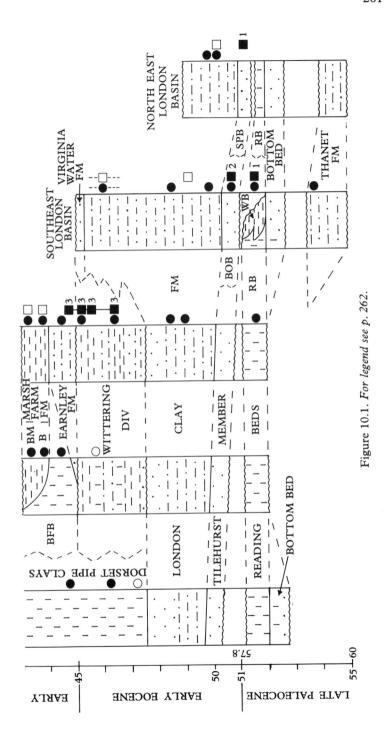

Figure 10.1. *For legend see p. 262.*

Fossil data base

The mammalian fossils consist mainly of teeth and jaw fragments but also include rarer, more complete, cranial and postcranial material. The plant fossils consist largely of fruits, seeds and megaspores here grouped under the term disseminules (Figure 10.2(*a*) to (*h*)).

The sediments are largely sands and clays and owe their unlithified nature to only shallow burial and the very limited effect of distant vulcanicity. They may therefore be easily disaggregated and wet-sieved, leading to total recovery of coherent fossil material (Collinson, 1983*a*; Hooker, 1986; Ward, 1984). Additional large material has been collected by prospecting (i.e. scrutinizing exposed surfaces for material *in situ*). This method normally biases against the smaller size categories (unless complete plants or skeletons are involved), whereas wet-sieving biases against the larger size categories because they have a lower relative abundance (Wolff, 1975). This bias can be reduced either by taking very large samples (e.g. at Creechbarrow; Hooker, 1986) or by much coarse as well as fine sieving (e.g. at Abbey Wood; Hooker, 1979). A combination of prospecting and sieving should produce a relatively well-balanced sample (see Table 10.1). Certain mammalian assemblages showing

---

Figure 10.1. Time-based stratigraphic logs in different major areas of the London and Hampshire Basins, showing depositional environ-ments and plant disseminule (circles) and mammal (squares) assemblages (closed symbols) and isolated occurrences (open sym-bols). Geochronological divisions essentially follow those of Aubry (1985). Two scales of absolute dates in millions of years BP follow the systems of Odin, Curry & Hunziker (1978) (O) and Aubry (1985) (A). Lithostratigraphy follows Ward (1977), King & King (1977), King (1981), Hooker (1986) and Insole & Daley (1985), for formal terms (i.e. Formations and Members), and Chandler (1962) and Hooker *et al.* (1980) for most informal terms. Dashed lines between columns indicate limits of units. Abbreviations: BFB, Bournemouth Freshwater Beds; BLF, Bembridge Limestone Formation; BMB, Bournemouth Marine Beds; BOB, Blackheath/Oldhaven Beds; DIV, division; FM, Formation; LST, Limestone; RB, Reading Beds; SPB, Suffolk Pebble Beds; WB, Woolwich Beds. Depositional environments mainly follow information in the lithostratigraphy references plus that given by Plint (1983, 1984). Numbers 1 to 9 refer to faunas in Table 10.1.

Inset: Outcrop map of Paleogene strata showing the London (L) and Hampshire (H) Basins (solid areas) in southern England.

Table 10.1. *Collecting methods and numbers of specimens and species for nine successive major southern English Paleogene mammalian faunas (excluding bats)*

	1	2	3	4	5	6	7	8	9
		Woolwich Shell Beds	Wittering/						
	Suffolk Pebble Beds	Abbey Wood	base Earnley	Creechbarrow	Hordle	Hatherwood Limestone	Bembridge Limestone	Bembridge Marls	Hamstead
No. of specimens	115	120	19	1500	1000	750	350	140	1500
No. of species	22	24	9	40	32	28	31	23	25
Screenwashing	+	+	+	+	+	+	+	+	+
Prospecting	−	−	+	−	+	+	+	+	+

Numbered faunas (1 to 9) are keyed to Figure 10.1. For details of faunas, localities and stratigraphic position, see Bosma, 1974; Hooker, 1986; Hooker *et al.*, 1980; Insole & Daley, 1985. Figures for Wittering/base Earnley are based on several small faunas from the Wittering division collected by wet-sieving (with additional localities given in King & Kemp, 1982; Kemp, 1984) and from Unit E4, Earnley Formation, Bracklesham Bay, collected by prospecting. Hordle comprises only lower sites i.e. in the Mammal and Crocodile beds (see Cray, 1973). Bembridge Limestone includes only material from Headon Hill. The approximately contemporaneous Suffolk Pebble Beds and Woolwich Beds sites are combined (Suffolk Pebble Beds) in the text and Figures 10.5 and 10.6 (SPB) to offset the collecting bias of each. For details see Appendix 2.

Figure 10.2. Fossil plant disseminules from the Early Tertiary of southern England: (*a*) *Acrostichum* fern sporangium, (*b*) Typhaceae seed, (*c*) *Ceriops* viviparous embryo (Rhizophoraceae), (*d*) Icacinaceae endocarp, (*e*) *Wetherellia* fruit (?Euphorbiaceae), (*f*) Potamogetonaceae endocarp revealing germinated seed, (*g*) Sapindaceae embryo cast, (*h*) lycopsid megaspore.
Scale bar on (*f*) represents (*a*) 63 μm; (*b*) 240 μm; (*c*) 15 mm; (*d*) 3.4 mm; (*e*) 4.3 mm; (*f*) 260 μm; (*g*) 1.9 mm; (*h*) 170 μm.
(*a*), (*b*) transmitted light micrographs; (*c*) to (*e*), (*g*) reflected light micrographs; (*f*), (*h*) scanning electron micrographs; (*c*), (*d*) and (*g*) pyrite permineralizations; others compressions.

evidence of predepositional size sorting have been excluded from this study.

Plant disseminules and mammalian teeth are advantageous for this study for several reasons. (*a*) They often permit unequivocal recognition of nearest relatives. (*b*) They are durable because they have survival and persistence functions during life and hence have high preservation potential. (*c*) Mature embryos inside seeds are known (Figure 10.2(*g*)) and evidence of germination is frequently found (Figure 10.2(*f*)), so clearly we are dealing with diminutive 'whole plants' that contributed to the ancient vegetation around the sites of deposition. (*d*) Some aspects of plant ecology may be deduced from disseminules (e.g. Harper, Lovell & Moore, 1970; Howe & Smallwood, 1982). (*e*) Adaptation for particular diets may be deduced from tooth morphology.

This chapter includes some previously published data (plants summarized by Chandler (1964)) but is largely based on new collecting from original and newly recognized sites. Although there are taphonomic biases (Scott & Collinson, 1983*a*, *b*; Behrensmeyer & Kidwell, 1985) influencing the interpretation of living communities from fossil assemblages, our assemblages are more or less comparable in presumably suffering the same kind of bias. Wherever possible, supporting evidence for ecological interpretations has been sought from lithofacies, other aspects of biofacies, oxygen isotope studies etc.

Interpretation of floral palaeoecology

Late Paleocene

Plant disseminule assemblages have been recovered from several sites (Appendix 1, p. 288), exposing shallow marine, lagoonal and fluvial flood-plain deposits, including many channel infills, rare lignites and occasional palaeosols (Buurman, 1980). Plant fossils include forms whose nearest living relatives are shrubs/trees (e.g. Burseraceae, Cornaceae, Myricaceae, Rutaceae, Theaceae), lianas (e.g. Icacinaceae, Menispermaceae, Vitaceae), herbaceous climbers (e.g. Cucurbitaceae) and herbs (e.g. the fern *Anemia poolensis* – Schizaeaceae). Aquatic plants are represented by Nymphaeaceae, *Stratiotes* (Hydrocharitaceae) and *Decodon* (Lythraceae), all of which are also typical of Late Eocene floras. Other aquatics, the heterosporous water ferns and lycopsids, represented by megaspores, are abundant and diverse. The water ferns recur in the Late Eocene but the lycopsid component is unique to the Paleocene flora.

Additional evidence of floras at this time is provided by palm stumps *in situ* in a lignite bed at Felpham, clearly part of a swamp community. Fern pinnules and rachides also occur at Felpham, the latter being similar to those of *Straelenipteris* from the Belgian early Middle Eocene and to unnamed forms from the London Clay (Collinson, 1983*b*).

Paleocene flood-plain environments included impersistent freshwater swamps, dominated by palms, and marshes across which rivers formed channels. Distant, more actively fluvial, areas of the flood-plain supported *r*-selective strategy plants, known from Newbury as associated organ compression floras on bedding planes, such as a *Cercidiphyllum*-like plant (Cercidiphyllaceae; Crane, 1984; Crane & Stockey, 1985) and a *Platanus*-like plant (Platanaceae). Other floral elements here included the *Palaeocarpinus* plant (Betulaceae; Crane, 1981), the *Casholdia* plant (Juglandaceae; Crane & Manchester, 1982), a *Sassafras*-like plant (Lauraceae) and a plant bearing leguminous leaves and pods (P. R. Crane, personal communication). The plant bearing the *Rhododendron* seeds (Ericaceae; Collinson & Crane, 1978) was probably also an associate of this community. The surrounding more stable areas included shrubs/ trees and lianas in patches of woodland or forest and also clearings in which herbaceous climbers, ferns and lycopsids thrived.

Early and early Middle Eocene

These plant disseminule assemblages are represented by samples from numerous sites in the London Clay (Collinson, 1983*b*) and from the complex of higher strata extending to the top of the early Middle Eocene (Figure 10.1, and see Appendix 1). All represent slight variations on a single basic theme and hence have been combined here.

At almost every site, except those of the London Clay lithofacies, fruits of '*Scirpus*' *lakensis* are found. They may sometimes be the only determinate plant fossils at a site and hence are strikingly characteristic of this flora. Their affinities with modern *Scirpus* (Cyperaceae) are doubtful, because very similar fruits occur within receptacular fruiting heads, totally unlike fruits of modern *Scirpus*, in the Eocene of Messel, West Germany (Collinson, 1982*a*). Their occurrences suggest that the parent plant grew along water courses and also around the Messel lake.

Limnocarpus and related extinct genera (Collinson, 1982*b*) are also typical components of this flora. They are absent from fully marine London Clay lithofacies but abundant in more saline depositional environments than '*S.*' *lakensis*. Nearest living relatives of *Limnocarpus* are aquatics of the families Ruppiaceae and Potamogetonaceae.

A third distinctive element in this flora consists of fruits of *Nipa* (Palmae) found in the London Clay Formation (Collinson, 1983*b*) and at one Dorset Pipe Clay locality, Swaythling, Gosport, Lee and two levels in Bracklesham Bay. In Bracklesham Bay, a bed of glauconitic sand, rarely exposed on the foreshore, reveals abundant fruits dispersed over a bedding plane, closely comparable in appearance with a modern strand area in *Nipa* vegetation. There is a similar occurrence in the Belgian early Middle Eocene (Stockmans, 1936). Evidence from the London Clay of Sheppey indicates that rhizophoraceous mangroves (e.g. *Ceriops*; Wilkinson, 1981) represented by viviparous embryos ('sea pencils') (Figure 10.2(*c*)) were associated with *Nipa*.

At several sites yielding *Nipa*, fossil fruits of *Wetherellia* have also been found, especially in the lagoonal/estuarine deposits of Bracklesham Bay. This suggests that they may have been an associate of the mangrove vegetation (see also Mazer & Tiffney, 1982). Coastal areas with extensive *Nipa*-dominated vegetation, occur today in the Philippines and Papua

New Guinea, where there is daily flooding in low-lying estuaries and where rivers carry high sediment loads into protected bays along gently shelving coastlines (Whitford, 1911; Paijmans, 1976; Womersley, 1978). Such vegetation extends inland up tidal rivers backed by riparian forest and habitats dominated by freshwater plants are generally absent. On slightly higher ground, rhizophoraceous mangroves including *Ceriops* are dotted amongst the *Nipa* stands (e.g. Paijmans, 1976, pl. 7, p. 36). Such a situation is consistent with the megafossil occurrences and with the restricted embayment proposed for at least part of this time on sedimentological grounds (Plint, 1983).

The remaining plant disseminules of this time include forms whose nearest living relatives are trees, shrubs or lianas, many of which are largely or exclusively tropical or subtropical in occurrence. The most common of these are members of the Anacardiaceae, Annonaceae, Burseraceae, Cornaceae, Hamamelidaceae, Icacinaceae, Lauraceae, Menispermaceae, Myricaceae, Nyssaceae, Palmae (= Arecaceae), Rutaceae, Staphyleaceae, Symplocaceae, Theaceae and Vitaceae. Magnoliaceae are also abundant in the London Clay. The *Cercidiphyllum*-like plant is represented by *Jenkinsella* fruits in the lower London Clay (Crane, 1984). *Anemia poolensis* is also represented by fertile pinnules.

Early and early Middle Eocene floras include some elements present in the Late Paleocene but are more diverse and more tropical in aspect. Furthermore, there are no elements whose nearest living relatives are exclusively freshwater aquatics.

Additional evidence of floras at this time may be drawn from beds at Alum Bay (Wittering division and Marsh Farm Formation) that yield foliage of the ferns *Osmunda*, *Acrostichum*, *Lygodium* and *Anemia poolensis* (*Ruffordia subcretacea*) and dicotyledons Lauraceae (Crane, 1977, 1978) and *Dryophyllum* (Fagaceae) (P. R. Crane, personal communication). Several of these also occur in leaf beds at Bournemouth. Twigs from the London Clay of Sheppey (Wilkinson, 1988) are a further, as yet scarcely exploited, source of data.

Collinson (1983*b*, p. 23) demonstrated that floras from the Early and early Middle Eocene in Britain included the largest proportion of potential tropical (up to 92%) and smallest proportion of potential temperate (up to 42%) nearest living relatives. Floras from the Late Paleocene and latest Eocene and Early Oligocene contain smaller proportions of potential tropical elements (e.g. 70% to 82%).

A great many of the near living relatives of the fossils co-exist today in paratropical rain forest in eastern Asia (Wolfe, 1979; Wang, 1961;

Keng, 1969; Whitmore, 1972; Womersley, 1978). Many are large forest trees, others lianas. The greatest diversity of these elements in the Early and early Middle Eocene floras suggests the presence of a dense forest vegetation. Paratropical rainforest does include elements often considered more temperate, such as occur with the tropical elements in the fossil assemblages. However, some, e.g. the *Cercidiphyllum*-like plant, have near living relatives more typical of mixed mesophytic forests.

In tropical rainforests there is a pronounced edge effect, where lianas, shrubs and herbaceous climbers abound, compared with the centre of the forest (Richards, 1952; Wang, 1961; Womersley, 1978). Such plants, along with those normally associated with more temperate vegetation, tend to occur along streamsides and in areas of secondary vegetation (Wolfe, 1979). Plant fossil assemblages might therefore be biased in favor of these elements. This suggests that the 'tropicality' of the Early and early Middle Eocene flora may have been under- rather than over-emphasized. It also implies that the less diverse, reduced tropical aspect, floras of Late Paleocene and early Late Eocene, with very evident lianas and climbers amongst their near living relatives, might be biased by this edge effect. However, in view of the overall floral changes, it seems likely that they do reflect an actual increase of 'edge area' within the local environment (e.g. a more open forest).

Independent evidence of environmental change is provided by oxygen isotope palaeotemperatures from the North Sea (Buchardt, 1978) and South Atlantic (Shackleton, 1984). According to these authors, maximum temperatures occurred in either Early or early Middle Eocene (see also Upchurch & Wolfe, this volume, Chapter 4). A cooling following the maximum is documented by both, but Shackleton (1984) shows some fluctuation in the Late Eocene. The Early Oligocene is marked by considerably cooler temperatures. These palaeotemperature curves show broad correspondence with our conclusions, indicating that, according to modern temperature parameters (see Wolfe, 1979), vegetation such as we propose could have existed during the Paleogene in southern England.

Beginning during deposition of the Earnley Formation, there is evidence of slight change in the flora. This does not affect ' *S.*' *lakensis*, *Limnocarpus* or *Nipa*, but is marked, for example, by the entry of *Potamogeton pygmaeus* and recurrence of *Stratiotes*, both characteristic elements in subsequent floras. Other changes also occur at this time (Bracklesham Group; Collinson, Fowler & Boulter, 1981). A very gradual decline in diversity occurs in those elements with tropical or subtropical nearest living relatives and the freshwater elements increase. Loss of *Nipa*, as documented by *Spinizonocolpites* pollen (Collinson *et al.*, 1981), may occur in low Earnley Formation or top Selsey division in different areas. Chandler (1960) referred to a single *Nipa* fruit thought to be from the basal beds at Hengistbury (top Boscombe Sands), last mentioned, but not figured, in 1894. However, by the close of the early Middle Eocene, the previously typical *Nipa* and ' *S.*' *lakensis* plant had disappeared from the floras.

Late Middle and early Late Eocene

Late Middle Eocene floras are rather poorly known, but do include some of the trees/shrubs and lianas of tropical aspect typical of the Early and early Middle Eocene floras (e.g. Annonaceae, Burseraceae, Cornaceae, Icacinaceae, Rutaceae, Symplocaceae, Theaceae and Vitaceae). Aquatic elements (e.g. *Stratiotes* and Nymphaeaceae) also occur.

Early Late Eocene floras (Appendix 1) mark the last occurrences, in much reduced diversity, of these previously dominant Early and early Middle Eocene tropical elements (Anacardiaceae, Burseraceae, Cornaceae, Hamamelidaceae, Icacinaceae, Lauraceae, Menispermaceae, Rutaceae, Sabiaceae, Symplocaceae, Theaceae and Vitaceae). These floras are more reminiscent of the Late Paleocene, having water ferns along with abundant freshwater aquatic and marsh elements typical of the latest Eocene and Early Oligocene floras. The *Cercidiphyllum*-like plant also reappears (Crane, 1984). *Limnocarpus* remained locally dominant but *Aldrovanda* (Droseraceae), Cyperaceae, *Decodon*, Nymphaeaceae and *Stratiotes*, along with numerous charophytes (freshwater algae; Feist-Castel, 1977), dominate early Late Eocene floras.

Additional evidence is provided by the occurrence of tree roots *in situ* from swamp trees similar to modern Taxodiaceae at two horizons (Fowler, Edwards & Brett, 1973). Taxodiaceous foliage and cones are associated with the roots at Hordle. Also, Crane & Plint (1979) document calcified angiosperm roots showing hydromorphic anatomy.

Latest Eocene and Early Oligocene

These floras (Appendix 1) are dominated by *Typha* (Typhaceae) and *Acrostichum* (the leather fern) forming a reed marsh with open-water bodies of broad shallow, fluvial, lagoonal or lacustrine nature. An extensive and diverse freshwater aquatic flora colonized these areas. Local areas of slight elevation probably supported small trees and shrubs, but these were rare and insignificant compared to the marsh and there is no evidence of swamp elements. Comparable environments occur today in southeastern North America where, *Typha* and *Acrostichum* form a minor association in areas dominated by *Mariscus* marsh (for discussion and references see Collinson, 1983*a*). The association is increasing (R. H. Hofstetter, personal communication), possibly as a result of drainage control, reducing water levels, discouraging or preventing peat formation.

In the Insect Limestone, near the base of the Bouldnor Formation (Reid & Chandler, 1926), the distinctive plant assemblage consists largely of wind-dispersed disseminules, including forms similar to modern *Clematis* and *Ranunculus* (Ranunculaceae), *Papaver* (Papaveraceae), *Abelia* (Caprifoliaceae) and *Engelhardia* (Juglandaceae). *Rubus* (Rosaceae), *Typha* and *Potamogeton* are more common, and the assemblage suggests the localized presence of small trees, shrubs and non-aquatic herbs, e.g. on 'tree islands'.

At some levels in the Early Oligocene Hamstead Member (Bouldnor Formation), coniferous leafy shoots and cones of *Sequoiadendron* and *Pinus* are quite abundant and some large logs are known. These coniferous elements, along with *Sequoia*-like forms, have a consistent but limited presence through the sequence in southern England although apparently more abundant at Hamstead. They do not appear to have been swamp elements in this flora and may perhaps have been distantly derived (Chandler, 1978).

Interpretation of mammalian palaeoecology

Data on body size, and locomotory and dietary adaptations have been used to reconstruct a series of ecological diversity spectra (EDS) (Figure 10.5) for the fossil assemblages (see Appendix 2, p. 289). These have been compared with those from modern communities, following the parameters of Andrews, Lord & Nesbit Evans (1979) and Nesbit Evans, Van Couvering & Andrews (1981), who modified and applied the methodology pioneered by Fleming (1973) to modern and late Cenozoic African faunas (Figure 10.5). Changes in dietary adaptation of two well-represented orders (Rodentia and Perissodactyla) have also been traced through time (Figure 10.6). Andrews *et al.* (1979) arranged the species composition of their sites into four categories: systematic position, body size, locomotory adaptation and dietary (= feeding) adaptation; and bats were excluded as they are usually rare as fossils. We have also excluded bats and, because of the vast difference in taxonomic composition between Early Tertiary and modern mammal faunas at least at generic and frequently at family and ordinal level, we have abandoned the 'systematic position' category.

The size categories (Appendix 2) follow those of Andrews *et al.* (1979). In the study by Andrews *et al.*, size was estimated using comparable body parts from related living animals. Here such a direct approach would have been misleading because of the large morphologic and taxonomic distance between the fossil and modern faunas. So we have used tooth size to extrapolate weight. For most of the primates, weights are provided by Gingerich, Smith & Rosenberg (1982). For other groups, we have extrapolated weight via head and body length (HBL) from the lower first molar area (length × width), following Creighton (1980), who calculated their static allometric relationship in various modern mammals. We have inferred weight from HBL, using data from Creighton (1980) and the ratio HBL^3/weight measurements obtained for various modern mammals by Corbet & Southern (1977) and Yablokov (1974). The accuracy of the fossil extrapolations has been confirmed using the ratio lower first molar area/HBL in various mammal skeletons from Messel.

We have used Andrews *et al.*'s (1979) six locomotory categories: large ground mammals (LGM), small ground mammals (SGM), arboreal, scansorial, aquatic and aerial; although no English Paleogene mammals apart from bats are judged to have been aerial. The difference between LGM and SGM is not purely one of size but relates to how terrestrial they are. LGMs may overlap in size with SGMs but they are restricted to the ground. SGMs 'also frequent lower branches of bushes and fallen trees' (Andrews *et al.*, 1979, p. 186). Because of the paucity of associated post-cranial and dental remains in the English Paleogene, we have had to extrapolate from the closest related fossil where locomotory adaptations are known (see Appendix 2). Where a relationship is as distant as family level (e.g. Cebochoeridae, Nyctitheriidae), we have deduced adaptation as being that dominant for the modern and fossil representatives of the relevant order. Where an extinct order has no known post-cranial remains (e.g. Apatotheria), we have extrapolated locomotory adaptations from the nearest modern dental analogs, in this case, the arboreal aye-aye (*Daubentonia*) and striped possum (*Dactylopsila*) (see Hooker, 1986).

We have followed all but one of Andrews *et al.*'s (1979) dietary (= feeding) classes: insectivore (I), frugivore (F), herbivore browser (HB), herbivore grazer (HG) and carnivore (C). In contrast to Fleming (1973), Andrews *et al.* (1979) and Nesbit Evans *et al.* (1981), we have

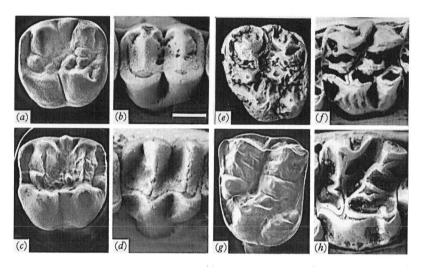

Figure 10.3. Fossils (*a*), (*c*), (*e*) and (*g*) from the Early Tertiary of southern England (BMNH Palaeontology Department) and modern (BMNH.ZD) mammal upper preultimate molars, shown as from left. (*a*) *Treposciurus helveticus*, pseudosciurid rodent (M35513); (*b*) *Hypsiprymnodon moschatus*, frugivorous marsupial (1939.2963); (*c*) *Suevosciurus authodon*, pseudosciurid rodent (M35281); (*d*) *Iomys horsfieldi*, frugivorous rodent (63.1593); (*e*) *Ailuravus stehlinschaubi*, paramyid rodent (M37193); (*f*) *Trogopterus xanthipes*, browsing rodent (22.9.1.46); (*g*) *Microchoerus wardi*, omomyid primate (M37162); (*h*) *Belomys pearsoni*, rodent (91.10.7.68).
   Scale bar in (*b*) represents (*a*) 0.8 mm; (*b*) 1.3 mm; (*c*) 0.6 mm; (*d*) 0.9 mm; (*e*) 1.9 mm; (*f*) 1.6 mm; (*g*) 1.0 mm; (*h*) 0.8 mm.
   (*a*), (*c*), (*g*) scanning electron micrographs; (*b*), (*d*) to (*f*), (*h*) light micrographs, ammonium chloride coated.

abandoned the omnivory class because, although indicating a mixed diet, the range is rarely categorized and can vary greatly. We have treated the individual subdivisions of omnivory separately, as did Andrews *et al.* (1979; p. 186) for dual combinations such as herbivore–frugivore.
   Literature is rather sparse on the dietary adaptations of mammalian taxa in the English Paleogene (except most of the primates and those from Messel where stomach contents are preserved; Appendix 2). Therefore the modern mammal collection in the Department of Zoology, British Museum (Natural History) (BMNH.ZD) and the literature have been searched for modern dental analogs. Despite certain problems in interpreting fossil diets (Stuart, 1970), the frequent appearance of the

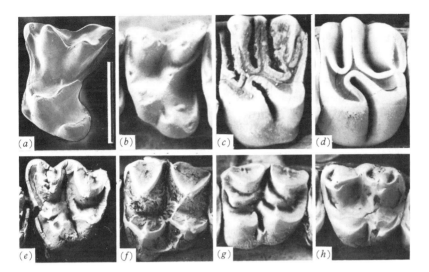

Figure 10.4. Fossils (*a*), (*c*), (*e*) from the Early Tertiary of southern England (BMNH.PD) and modern (BMNH.ZD) mammal upper preultimate molars shown as from left. (*a*) *Saturninia gracilis*, nyctitheriid lipotyphlan (M50600); (*b*) *Galago demidovii*, insectivorous primate (14.2.11.1.); (*c*) *Thalerimys headonensis*, theridomyid rodent (30159a); (*d*) *Thryonomys swinderianus*, frugivorous/hervivorous rodent (14.1.6.39); (*e*) to (*h*), selenodont teeth: (*e*) *Dacrytherium elegans*, anoplotheriid artiodactyl (M37717); (*f*) *Phascolarctos cinereus*, browsing marsupial (77.19); (*g*) *Pseudocheirus occidentalis*, browsing marsupial (57.3.6.3); (*h*) *Propithecus diadema*, frugivorous/browsing primate (76.1.31.10).

Scale bar on (*a*) represents (*a*) 1.0 mm; (*b*) 2.1 mm; (*c*) 1.7 mm; (*d*) 3.9 mm; (*e*) 9.0 mm; (*f*) 6.7 mm; (*g*) 3.8 mm; (*h*) 6.5 mm.

(*a*) scanning electron micrograph; others light micrographs, ammonium chloride coated.

same dental types with essentially the same diets in distantly related mammals is felt to be sufficient justification for this approach (see Figures 10.3, 10.4, 10.7). For instance, low-crowned bunodont cheek teeth associated dominantly with frugivory are found in monkeys (e.g. *Miopithecus talapoin*, *Cercocebus albigena*), squirrels (e.g. *Ratufa*, *Iomys*; Figure 10.3(*d*)), macropodid marsupials (e.g. *Hypsiprymnodon*; Figure 10.3(*b*); *Potorous*) and viverrid carnivores (e.g. *Arctictis*). Low-crowned selenodont cheek teeth associated dominantly with browsing occur in certain phalangerid marsupials (e.g. *Phascolarctos*, *Pseudocheirus*; Figures 10.4(*f*) to (*g*)) and indriid primates (e.g. *Indri*, *Propithecus*; Figure

274

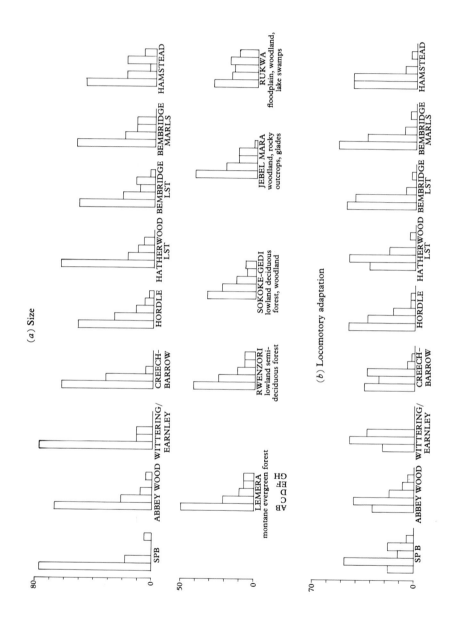

(a) Size

(b) Locomotory adaptation

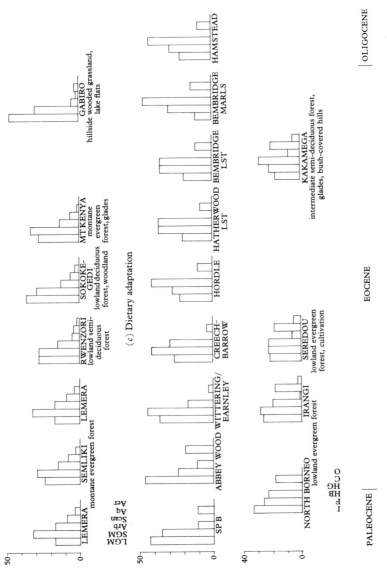

(c) Dietary adaptation

Fig. 10.5. *For legend see p. 276.*

10.4(*h*)). Heavily crested, fairly low-crowned teeth, with crenulate enamel, are found in the flying squirrels *Trogopterus* and *Belomys* and have dental patterns very similar to those of the fossils *Ailuravus stehlinschaubi* and *Microchoerus* spp., respectively, except that *Ailuravus* is lower crowned than *Trogopterus* (Figure 10.3(*e*) to (*h*)). Reports of the diets of these two modern genera are, however, few. According to Walker (1975), *Trogopterus* eats oak leaves, whereas the diet of *Belomys* is unknown (Jones, 1975). Apparent food material in tooth valleys of a specimen of *B. pearsoni* (BMNH.ZD.91.10.7.68) on superficial inspection appears to contain both fragments of leaf cuticle and pieces of seed coat.

The differences between insectivorous, frugivorous and herbivorous diets in primates can be characterized further by the degree of cresting on the cheek teeth (Kay, 1975). Similar criteria are applicable to other groups (Kay & Hylander, 1978). The accuracy of the modern analogy method is largely corroborated by published results obtained from the mammals with stomach contents from the Messel site (see Appendix 2).

Each EDS category is compared separately with spectra produced for individually named modern African sites by Andrews *et al.* (1979) and for Bornean lowland forest compiled from data in Davis (1962) (see Figure 10.5).

We do not intend that these direct comparisons be rigidly interpreted. The environments of the Paleogene in southern England are unlikely to have coincided with those of modern Africa and, furthermore, EDSs from many areas of the world are not yet available for comparison. Although these results must therefore be preliminary, we feel that the broad environmental patterns used here are reliable.

In the modern communities, a combination of the high ratio small/ large animals and SGMs/LGMs, high percentage of arboreal types, abundance of insectivores and frugivores, few herbivore browsers and

---

Figure 10.5. Ecological diversity spectra for nine successive southern English Paleogene mammal faunas (for explanation see Table 10.1) and for various modern localities in Africa (from Andrews *et al.*, 1979) and North Borneo (compiled from data in Davis, 1962). Figures indicate percentages. (*a*) Size: AB = < 1 kg; C = 1–10 kg; D = 10–45 kg; EF = 45–180 kg; GH = > 180 kg. (*b*) Locomotory adaptation: LGM, large ground mammal; SGM, small ground mammal; Arb, arboreal; Scan, scansorial; Aq, aquatic; Aer, aerial. (*c*) Dietary adaptation. I, insectivore; F, frugivore; HB, herbivore browser; HG, herbivore grazer; C, carnivore; O, omnivore.

virtually no grazers typifies closed forest/rainforest. A reversal of these ratios, few arboreal types, insectivores or frugivores, but many herbivores together typify an open environment. Intermediate assemblages indicate intermediate environments. In the next section we compare the results from the fossil mammals with the vegetational changes shown by the fossil floras and attempt to explain discrepancies.

It is worth noting that there are some sources of bias in the EDSs. All the faunas have more small and fewer large animals and fewer carnivores than the modern spectra. This would be expected in fossil assemblages because of the collecting bias mentioned earlier (Wolff, 1975). The size bias is probably most accentuated in the Suffolk Pebble Beds and Wittering/base Earnley plots. It may also cause the high SGM percentage in the Suffolk Pebble Beds. The small sample size of the Wittering/base Earnley fauna is probably the reason for absence of scansorial and aquatic and fewer than usual carnivorous mammals. Otherwise we judge that the EDSs convey faunal changes in a relatively reliable way.

Discussion
Environmental changes in southern England

Late Paleocene vegetation included reedswamp with palms, open freshwater areas with river channels, disturbed fluvial flood-plain environments with *r*-selective (colonizing) plants and more stable areas with patches of woodland/forest and clearings. Late Paleocene faunas have a high ratio of small to large mammals and of SGMs to LGMs, few arboreal forms, few herbivores and a high proportion of insectivores. The overall EDS patterns are closest to modern closed forest environments; the few arboreal forms are inconsistent with the high SGM/LGM ratio, but consistent with the palaeobotanical data.

Early and early Middle Eocene vegetation is characterized by widespread and dominant highly diverse forest of tropical aspect with a fringing coastal *Nipa*-dominated mangrove. Early Eocene mammalian faunas still show a high ratio of small to large and of SGMs to LGMs (although less marked), but also an increase in arboreal types indicative of the closed forest environment. By the end of the early Middle Eocene, floras show a reduced diversity with reduction in tropical aspect, return of a few freshwater elements, loss of the mangrove and loss of the '*S.*' *lakensis* plant.

By the late Middle Eocene, mammalian faunas are less skewed towards small-sized forms, show further increase in LGMs with reduction in

SGMs and insectivorous forms and an increase in herbivore browsers. In particular, the rodents show dominance of frugivory over insectivory/frugivory, whilst the perissodactyls consist of soft-browsing herbivores and mixed frugivore/herbivore browsers with an increased herbivorous component. The latter represent the culmination of trends begun during the Early Eocene. The overall patterns suggest a change to a more open environment, e.g. glades within forest.

Precise positioning of these floral and faunal changes is prevented by limited data during this period in southern England. However, this change may have begun early in the early Middle Eocene because the West German Messel fauna of this time produces EDS patterns like those of Abbey Wood (Early Eocene), whereas the slightly later East German Geiseltal fauna produces patterns more like those of Creechbarrow (late Middle Eocene). The floras of both these sites (Collinson, 1982a; Krumbiegel, Rüffle & Haubold, 1983) show some broad similarities to those from southern England (Collinson, 1983b). The environments of Geiseltal certainly include a mosaic vegetation compared to the more uniform circumstances at Messel. We accept, however, that changes may have been different in aspect or timing between continental Europe and southern England.

Early Late Eocene floras show a return to freshwater marshes and swamps but retaining patches of woodland/forest of partly tropical aspect. In the latest Eocene a *Typha–Acrostichum* dominated reedmarsh, with a few isolated 'tree islands', became established over a wide area and persisted into the Early Oligocene. Faunas from the early Late Eocene show a further reduction in small mammals and a further increase in large mammals, a dramatic increase in LGMs, reduction in arboreal and scansorial forms, increase in herbivore browsers and further slight reduction in insectivores. There is a drastic reduction in frugivorous rodents, especially pseudosciurids, and incoming of abundant frugivore/herbivore browsing theridomyids. Insectivorous/frugivorous dormice (Gliridae) enter and slightly later diversify and fluctuate in abundance. Earlier frugivore/herbivore browsing and soft-browsing herbivorous perissodactyls become extinct and are replaced by a radiation of the coarser-browsing herbivores *Palaeotherium* and *Plagiolophus*, some showing cursorial adaptations. These trends continued through the Late Eocene and into the Early Oligocene, with further increase in large mammals and reduction or loss of arboreal and scansorial forms, respectively. The rhinoceros *Ronzotherium* replaced *Palaeotherium* and *Plagiolophus* but had similar dietary adaptations. The overall patterns are consistent with the palaeobotanical data and suggest a more open environment with lakes and only scattered trees.

The Hatherwood Limestone (middle Late Eocene) EDS shows a slight indication of reversion to woodland within this otherwise smooth trend to a more open environment. However, this is derived mainly from a small lignite lenticle and may represent a local fauna that perhaps inhabited one of the woodland patches suggested by the palaeobotanical data.

The changes that we have documented could have been influenced by various factors, e.g. climate, palaeogeography, tectonics and floral and faunal migrations. Detailed consideration of these, however, is beyond the scope of this chapter.

Dietary modifications in southern England

The mammalian faunal changes reflect the increased coarse component, such as thick cell walls and fibrous tissues, in the reedmarsh vegetation of the less forested environments (e.g. monocotyledonous leaves compared with low-level tree browse). Dietary modifications in the rodents and perissodactyls provide appropriate documentation (Figure 10.6).

The high proportion of soft-fruit-eating frugivorous rodents (Figure 10.3(*a*),(*c*)) in the Early and Middle Eocene reflects the abundance and continuity of their food supply in the tropical aspect forest. Apart from nearest living relatives with likely fleshy fruits (represented in the fossil record by only fruit stones), evidence from Messel demonstrates directly the occurrence of soft fleshy fruits (Collinson, 1982*a*). In the earliest Oligocene, frugivorous rodents with adaptations for eating harder fruits and seeds (granivory) appeared in the form of eomyids and cricetids. The dramatic appearance of the semi-hypsodont Theridomyidae (Figure 10.4(*c*)), with teeth suggesting an important browsing component in the diet, at the beginning of the Late Eocene was foreshadowed by somewhat different low-crowned frugivore/herbivore browsing specializations in the Ailuravinae in the late Middle Eocene, just before their extinction (Figure 10.3(*e*)).

The perissodactyls show various shifts towards increased browsing and essential loss of the frugivorous component in the diet. A moderate browsing component was probably present along with frugivory in Early Eocene *Hyracotherium*. Subsequently, the development of selenolophodonty in its descendant *Propalaeotherium* produced longer crests for increased browsing (supported by Messel stomach contents, consisting of leaves and fruits: see Sturm, 1978; Koenigswald & Schaarschmidt,

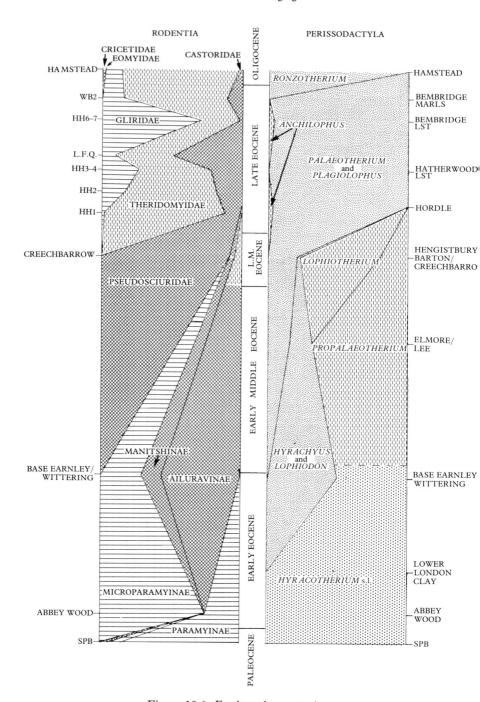

Figure 10.6. *For legend see opposite.*

1983). At the same time, larger perissodactyls with bilophodont teeth for specialized browsing appeared in the form of the helaletid *Hyrachyus* and members of the family Lophiodontidae. Later in the Eocene, trends in premolar molarization that provide increased surface area for mastication, affected both groups and also the newcomers *Lophiotherium*, *Palaeotherium* and *Plagiolophus*. The last two have a molar pattern similar to that of *Propalaeotherium* but with increased crown height and were probably specialized browsers. The end of the Middle Eocene saw extinction of most of these earlier types, but *Palaeotherium* and *Plagiolophus* radiated to become the dominant perissodactyls through the rest of the Eocene.

Apart from the overall trend from frugivory/herbivore browsing to strict herbivore browsing, there is also a major change in the dominant large browsing type at the end of the Middle Eocene from the bilophodont helaletids and lophiodontids to the higher-crowned selenolophodont *Palaeotherium* and *Plagiolophus*. To deduce the possible nature of the change, one can compare the fossil teeth with their nearest modern analogs: *Tapirus* for the bilophodont type and *Heterohyrax* for the semi-hypsodont selenolophodont type (Figure 10.7). The diet of *Tapirus* is almost totally restricted to leaves of forest trees, of which it may choose to eat just the young leaves and buds (Medway, 1974) or more mature leaves, sometimes in a dry state (Terwilliger, 1978). Cheek tooth function is virtually restricted to shearing and the consequent division of the leaves before reaching the gut is coarse, as Terwilliger (1978) noted that the faeces contained 'small squares of undigested leaf parts and pieces of twigs'.

Heterohyrax is also a browser but eats some grass in the wet season (Walker, Hoeck & Perez, 1978). Moreover, its necessary feeding on relatively coarse plant material in the dry season such as 'old leaves, twigs

---

Figure 10.6. Diagrams of percentage abundance (based on numbers of specimens) through time of two well-represented groups (rodents and perissodactyls), illustrating changing dietary adaptations through the southern English Paleogene. For explanation of faunas see Table 10.1. Horizontal lines, insectivore/frugivore; Crosshatching, frugivore; Coarse stipple, low herbivore component frugivore/herbivore browser; Dashed lines, high herbivore component frugivore/herbivore browser; Fine stipple, herbivore browser. Because of paucity of data, the exact timing of rodent changes between the end of the Early Eocene and the late Middle Eocene is somewhat speculative. HH, Headon Hill; LFQ, Laceys Farm Quarry; SPB, Suffolk Pebble Beds; WB, Whitecliff Bay.

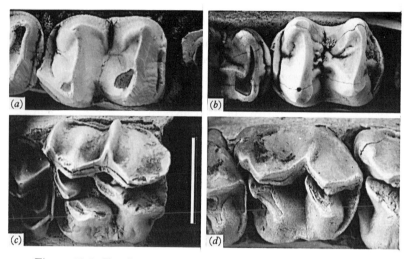

Figure 10.7. Fossils (a), (c) from the Early Tertiary of southern England (BMNH.PD) and modern (BMNH.ZD) mammal upper preultimate molars shown as from left: (a), (b) lowers, bilophodont; (c), (d) uppers, selenolophodont. (a) Hyrachyus stehlini, helaletid perissodactyl (M50194); (b) Tapirus indicus, soft-browsing herbivorous perissodactyl (1851.11.10.38); (c) Plagiolophus annectens, palaeotheriid perissodactyl (29715); (d) Heterohyrax brucei, coarse-browsing herbivorous hyracoid (88.12.6.1).

Scale bar on (c) represents (a) 8.3 mm; (b) 20.5 mm; (c) 10.0 mm; (d) 4.3 mm. Light micrographs, ammonium chloride coated.

or even bark' (Hoeck, 1975, p. 31) means that overall it is consuming more fibrous abrasive browse than did *Tapirus*, with generally lower nutritional value. To facilitate digestion, the plant material needs to be more finely divided by the cheek teeth (Rensberger, 1975). This is accomplished by a combination of shearing, crushing and grinding by selenolophodont teeth with longer enamel crests (Schmidt-Kittler, 1984) and more extensive lingual phase mastication (for its documentation in perissodactyls, see Butler, 1952). It can therefore be deduced that *Palaeotherium* and *Plagiolophus* were eating overall coarser, less nutritious browse than were *Hyrachyus* or *Lophiodon*. Incipient cursorial adaptations in *Plagiolophus* (Stehlin, 1939) would also mean increase in energy requirements leading to increased food intake, greater tooth abrasion and necessity for increased crown height and grinding efficiency (Rensberger, 1975).

Aspects of community evolution with global implications

We consider that some of the mammal changes documented in the English Paleogene represent more than just local events and demonstrate global community evolution *sensu* Olson (1966). Andrews *et al.* (1979) found little evidence for such evolution between the Miocene and Recent of East Africa. However, we are here dealing with much older biotas, nearer in time to the terminal Cretaceous events that included extinction of the dinosaurs (Coe *et al.*, this volume, Chapter 9).

Although the English Paleogene EDSs cannot be interpreted rigidly in terms of modern African habitats (Figure 10.5), the broad trends do corroborate the plant evidence. However, none is exactly like any modern community that has been studied.

The complete absence of herbivore grazers even in the more open habitats in the early Tertiary of southern England is consistent with a global absence of fossils referable to modern grassland grasses until the Late Oligocene (Thomasson, 1980). The larger number of frugivores in most faunas and of insectivores in the latest Paleocene and earliest Eocene of southern England, compared with the modern EDSs, could be enhanced by the paucity of herbivores. It could also be due to the worldwide absence of bats from Paleocene communities and to their low diversity in the Early to Middle Eocene (even when abundant as individuals as at Messel). The modern rapid increasing abundance, especially of frugivorous bats with decreasing latitude impressed Fleming (1973). The latest Paleocene and Early to early Middle Eocene faunas could therefore be biased somewhat towards insectivores and frugivores by omitting bats from the modern EDSs.

The latest Paleocene and Early Eocene EDSs from southern England show small numbers of large mammals and of herbivore browsers. This phenomenon is, however, typical of Paleocene mammal faunas from Europe, Asia and North and South America (see e.g. Russell, 1975; Rose, 1981; Savage & Russell, 1983) and therefore cannot simply be dismissed as being due to collecting bias, e.g. in the Suffolk Pebble Beds. The overall spectrum from the latest Paleocene of southern England with high numbers of small mammals, SGMs and insectivores and low numbers of large mammals, LGMs and herbivores is indicative of closed evergreen forest according to the modern distribution of these ecological types. However, the faunas contain few arboreal mammals. Rodent faunas show no component of herbivore browser until the Late Eocene and perissodactyl faunas show only a limited component of herbivory in the latest Paleocene compared to the late Middle Eocene (Figure 10.6). Late Paleocene vegetation in southern England does include elements typical of Early and early Middle Eocene tropical aspect forests but also indicates less forested, disturbed and open environments.

Are we to conclude, as is often done (e.g. Janis, 1982), that worldwide Late Paleocene and Early Eocene vegetation did comprise mainly dense forest? Wolfe (1980) envisaged broadleaved evergreen vegetation extending nearly to 60° N in the Paleocene, beyond 61° N in the Eocene, with broadleaved deciduous forests at higher latitudes and probably coniferous forests beyond these. Alternatively, did mammals take 10 to 15 million years to adapt to widespread invasion of the herbivore niche after the terminal Cretaceous archosaur extinction? It seems to have taken at least as long for complex herbivorous adaptations to develop in reptiles after earliest known records of the group (Milner, 1980). Selection pressure for a shift to a herbivorous diet might well have been low in the Paleocene in view of the comparatively low nutritional value of leaves and especially if distasteful or toxic plant secondary compounds had already evolved in the Mesozoic. Moreover, it is possible that, like the earliest herbivorous reptiles, the few Paleocene mammalian herbivores (mainly pantodonts, resembling oversized insectivores, at least in their teeth) had little herbivorous adaptation other than increased size, which presumably provided a larger gut for longer digestion. Our evidence of Late Paleocene vegetation with open and disturbed environments over the area of the London and Hampshire Basins, in conjunction with the near absence of browsing specializations in mammals here or elsewhere in the Late Paleocene, suggests that, by this time, few adaptations for mammalian herbivory had evolved (see also Wing & Tiffney, this volume, Chapter 8).

Our results document increasing diversification of mammalian herbivores through the Eocene. The gradual reduction of the forest through the course of the Middle and Late Eocene in southern England was probably related to cooling climate (e.g. Shackleton, 1984). It was followed by the rapid evolution and diversification of mammalian herbivores, a phenomenon also observable in other areas of the world. This adaptational change, once begun, is likely to have increased in intensity with progressive opening up of the environment. Higher-crowned teeth evolved in mammals in response to an increase in required energy levels (partly to escape predators by running) and therefore in dietary intake of a progressively less nutritious food source. These interactions culminated in the evolution of the highest-crowned teeth characteristic of grazers and of the plant type adapted to rapid and continued regrowth (by intercalary meristems) after being largely eaten, i.e. grassland grass.

An alternative life style in the trees and the air, avoiding predators largely by gliding and crypticity (Eisenberg, 1978), and thus requiring less energy, has been adopted by some mammals (e.g. flying squirrels, flying phalangers, dermopterans and various arboreal primates). These mammals provide most of the modern dental analogues (Figures 10.3(*b*), (*d*), (*f*), (*h*), 10.4(*f*) to (*h*)) for the Paleogene frugivores and herbivore browsers that lived on the ground.

Conclusions

The Paleogene sequence in southern England reveals floristically distinct periods (Late Paleocene, Early and early Middle Eocene, early Late Eocene and latest Eocene/Early Oligocene), based largely on fruit, seed and megaspore assemblages. Plant palaeoecology has been inferred from comparisons with nearest living relatives, supported by analyses of lithofacies and other biofacies. Vegetational history has been deduced from plant palaeoecology and fossil distribution and abundance in the various depositional environments. Broadly comparable vegetational changes have been deduced from statistical analyses of palynofloras by Boulter & Hubbard (1982) and Hubbard & Boulter (1983). Contemporaneous mammalian faunal change has been documented mainly from teeth and jaw fragments. As most of the taxa are extinct at family level, diet has been inferred from modern dental analogs supported by information from fossil stomach contents. Locomotion has been extrapolated mainly from closest related fossil taxa that have known postcranial remains supported by gait characters in modern representatives of the groups. These data are interpreted by comparison with ecological diversity spectra for various modern faunas.

In southern England dense forest of tropical aspect characterizes the Early to early Middle Eocene and mammals show a high ratio of small to large ground-dwelling forms, a relatively high percentage of arboreal species and adaptations to eating soft fruit and low fibre content leaves. By the early Late Eocene many tropical taxa have been lost and the vegetation is characterized by reedmarshes and swamps with patches of woodland or forest. In the latest Eocene, an extensive and persistent reedmarsh is developed. These changes are accompanied by extinctions of many earlier mammalian adaptive types, a reduction in arboreal

Table 10.2. *Summary of conclusions*

	MAMMAL FAUNAS	LITHOSTRATIGRAPHY	MAJOR FLORAS
Early Oligocene	Hamstead	Bouldnor Fm	Hamstead
	Bembridge Marls		Hamstead Ledge
Late Eocene	Bembridge Lst	Bembridge Lst Fm	Headon Hill
Late Eocene	Hatherwood Lst	Headon Hill Fm	
Late Eocene	Hordle	Headon Hill Fm	Headon Hill, Hordle
late Middle Eocene		Becton Sand Fm	
late Middle Eocene	Creechbarrow	Barton Clay Fm, Creechbarrow Lst	Barton & Hengistbury
early Middle Eocene	Elmore		Studley Wood
early Middle Eocene	Lee	Selsey div	Bracklesham
early Middle Eocene		Marsh Farm Fm	Alum Bay
early Middle Eocene		Earnley Fm	
	Base Earnley/Wittering	Wittering div	Swaythling
			Lake & Arne
			Sheppey
Early Eocene	Lower London Clay	London Clay Fm	Nursling
			Ockendon Walton
	Abbey Wood	Blackheath Beds	
Late Paleocene	Suffolk Pebble Beds	Woolwich and Reading Beds, Suffolk Pebble Beds	Felpham, St Pancras, Newbury
Late Paleocene		Thanet Fm	Herne Bay

VEGETATIONAL HISTORY	MAMMALIAN CHANGES
Freshwater reedmarsh dominated by *Typha* and *Acrostichum*; numerous freshwater aquatics in large shallow lakes/rivers; ?few shrubs on 'tree islands'	Further increase in large mammals and decrease in Arb types; finally, a few granivorous rodents appear
First abundant freshwater flora; Taxodiaceous swamp; last occurrence of previously dominant tropical elements	Drastic reduction of Arb and further reduction of small and I types; increase in large, especially LGM and HB types. Appearance of abundant F/HB theridomyids and reduction of F pseudosciurids; major diversification of coarser-browsing and extinction of soft-browsing perissodactyls.
	Reduction of small types; further reduction in I; further increase in LGM and HB types
Loss of mangrove and '*Scirpus*' *lakensis* plant.	Perissodactyls: a few coarser-browsing types appear; increase in herbivorous component of F/HB
Beginning of reduction in tropical elements and return of freshwater elements	
Forest with maximal tropical aspect; diversity of lianas, trees and shrubs; '*Scirpus*' *lakensis* plant by water margins; *Limnocarpus* in brackish settings; coastal area with *Nipa*-dominated mangrove and associated Rhizophoraceae	Arb types much increased; I reduction begins; slight increase in HB types. Rodents: F dominates over I/F. Perissodactyls: soft-browsing herbivores appear
	LGM and Arb types begin to increase
Woodland/forest + clearings; disturbed fluvial flood-plain environments. Freshwater swamp with palms and open water channels with distinctive floras; herbaceous lycopsids and water ferns common	Many small, SGM and I types. Few large, LGM, Arb or HB types

For abbreviations see Figures 10.1 and 10.5. Changes to be read from older to younger.

species and an increase in large ground dwelling forms, especially herbivores with specializations for a higher fiber diet. These general conclusions are summarized in Table 10.2.

Late Paleocene floras in southern England indicate open and disturbed environments with limited wooded/forested areas but mammalian communities (few large mammals or herbivore browsers) are typical of closed forest vegetation today. This incongruency suggests that these mammals had not then invaded the herbivorous niche. As Paleocene mammalian communities elsewhere are similar, this may have been a global phenomenon.

We thank the numerous enthusiastic field collectors who have generously assisted us in this study, especially D. A. Bone, P. Davey, R. L. E. Ford, R. Fowler, R. Gardner, W. George, D. J. Kemp, A. Lawson, W. J. Quayle, M. Salmon, S. Tracey, S. Vincent and A. and D. J. Ward. We especially thank D. A. Bone for drawing our attention to the plant fossils from Felpham. Q. Palmer contributed useful information on correlation in the West Hampshire Basin and D. L. Moore conducted a computer search for literature on modern mammalian diets and locomotory adaptations. We thank P. Andrews and B. V. Van Valkenburgh for helpful discussion and W. G. Chaloner, P. R. Crane, E. M. Friis, A. W. Gentry, B. H. Tiffney and S. Wing for valuable comments on earlier drafts of this paper. This work was conducted by M.E.C. under the tenure of a Royal Society 1983 University Research Fellowship, which is gratefully acknowledged. The photographs in Figure 10.3(*a*), (*c*) and (*g*) are reproduced by permission of the British Museum (Natural History), whose Photographic Studio also printed those in Figures 10.2(*h*), 10.3(*a*), (*c*), (*g*) and 10.4(*a*).

Appendix 1: Notes on sources for fossil floras

Late Paleocene

Plant disseminule assemblages have been recovered from the Thanet Formation at Herne Bay, Kent (Collinson *et al.*, 1985); from the Woolwich and Reading Beds at Felpham, West Sussex and St Pancras, London (Collinson in Bone, 1986); at Newbury, Berkshire (Collinson & Crane, 1978) and at Cobham, Kent (Martin, 1976).

Early and early Middle Eocene

Assemblages from early Middle Eocene strata include horizons at Bracklesham Bay (Chandler, 1961*b*); sites along the Dorset coast around Poole Harbour (e.g. Lake and Arne) and Bournemouth (Chandler, 1962, 1963*b*); two horizons at Alum Bay (Crane, 1977); one in Whitecliff Bay (bed V; Fisher, 1862); various temporary exposures and boreholes in Hampshire: Swaythling (Kemp, 1984),

Copythorne (King & King, 1977), Gosport (King & Kemp, 1982) and Swanwick (Curry & King 1965; King, 1981); and the foreshore at Lee, Hampshire (Kemp, 1985). All except the last four of these sites have been re-examined during this work.

Late Middle and early Late Eocene
These floras are described by Chandler (1960, 1961*a*). Much additional material has been gathered during this work especially from early Late Eocene strata on the Isle of Wight.

Latest Eocene and Early Oligocene
The main floras are from Hamstead Ledge on the Isle of Wight (Collinson, 1983*a*) but others occur at various levels (Figure 10.1) and localities all on the northern coast of the Isle of Wight between Alum and Whitecliff Bays. These include higher parts of the Headon Hill, the Bembridge Limestone and the Bouldnor Formations (Chandler, 1963*a*, 1978; Collinson, 1980; Reid & Chandler, 1926). Many of these have been re-investigated during the present work.

Appendix 2: Fossil mammal occurrences in the faunas listed in Table 10.1

The size, locomotory and dietary categories are deduced for each species level taxon (for abbreviations see Figure 10.5). Relevant references for the locomotory and dietary adaptations of these or related taxa are given. Subtle differences between the locomotory and dietary adaptations deduced here and those proposed in the references occur in some cases and result mainly from authors' differing definitions and terminologies. Asterisks indicate taxa newly recorded from England. Daggers indicate records at new levels. The list partly updates and corrects that of Hooker *et al.* (1980) and other sources include Bosma & de Bruijn (1979, 1982, 1989) and Hooker (1986). Cetacea (whales) are excluded. indet., indeterminate; undet., undetermined.

Appendix 2(1)

Taxon	Suffolk Pebble Beds	Woolwich Beds	Abbey Wood	Wittering/base Earnley	Creechbarrow	Hordle	Hatherwood Limestone	Bembridge Limestone	Bembridge Marls	Hamstead	Size	Locomotory adaptation	Dietary adaptation	References on locomotion	References on diet
Ectypodus childei			×								AB	Arb	I/F	Krause & Jenkins, 1983	Krause, 1982 (*Ptilodus*)
*Peratherium cf. matronense				×	×						AB	SGM/Arb	I-C/F		
Amphiperatherium cf. *maximum*				×	×						AB	SGM/Arb	I/F		
A. aff. *goethei*			×		×						AB	SGM/Arb	I-C/F	Crochet, 1980	Crochet, 1980
A. fontense					×	×					AB	SGM/Arb	I/F		
A. sp. 1						×	×	×	×	×	AB	SGM/Arb	I-C/F		
A. sp. 2						×	×	×	×	×	AB	SGM/Arb	I/F		
Didelphidae undet.	×										AB	SGM/Arb	I/C		
?*Palaeosinopa* sp.		×									AB	Aq	C	Koenigswald, 1980 (*Buxolestes*)	Koenigswald, 1980 (*Buxolestes*)
Cryptopithecus major						×					AB	Aq	C		
C. sp.									×		AB	Aq	C		
Dyspterna hopwoodi					×						C	Aq	C	Matthew, 1909 (*Pantolestes*)	
D. woodi										×	C	Aq	C		
Pantolestidae undet.					×						AB	SGM	I/C		
*Leptacodon sp.	×										AB	SGM	I		
*Pontifactor sp.	×			×							AB	SGM	I		
Scraeva woodi						×					AB	SGM	I		
S. hatherwoodensis							×				AB	SGM	I	Krumbiegel et al., 1983 (*Ceciliolemur*)	
Saturninia gracilis								×	×		AB	SGM	I		
'*S*'. sp. 1								×			AB	SGM	I		
'*S*'. sp. 2									×		AB	SGM	I		
Nyctitheriinae undet.					×						AB	SGM	I		
Amphidozotheriinae undet.				×	×						AB	SGM	I		
Butselia biveri						×					AB	SGM	I		
'*Adapisorex*' *anglicus*		×									AB	SGM	I		
†*Eotalpa anglica*						×					AB	SGM	I		

Taxon	AB	Substrate	I/F	Reference
Talpidae undet.	AB	SGM	I	
Paschatherium dolloi	AB	SGM	I/F	
Macrocranion sp.	AB	SGM	I	} Maier, 1979 (M. tupaiodon)
Neomatronella cf. luciannae	AB	SGM	I	
Gesneropithex figularis	AB	SGM	I/F	} Koenigswald & Storch, 1983 (Pholidocercus)
G. aff. grisollensis	AB	SGM	I/F	
Tetracus sp.	AB	SGM	I/F	
Arcius fuscus	AB	Scan	I	
Plesiadapis aff. tricuspidens	C	Scan	F/HB	} Szalay & Dagosto, 1980
P. aff. remensis	AB	Scan	F/HB	
*Teilhardina cf. belgica	AB	Arb	I	
Nannopithex sp. 1	AB	Arb	I	
Nannopithex sp. undet.	AB	Arb	I	
N. quaylei	AB	Arb	I	
Pseudoloris cf. crusafonti	AB	Arb	I	
P. aff. parvulus	AB	Arb	I	} Godinot & Dagosto, 1983 (Necrolemur)
Microchoerus wardi	AB	Arb	F/HB	Schmid, 1979 (Microchoerus)
M. creechbarrowensis	AB	Arb	F/HB	
M. erinaceus	AB	Arb	F/HB	
M. edwardsi	AB	Arb	F/HB	Rose & Walker, 1985
Cantius eppsi	C	Arb	I/F	(C. trigonodus)
Europolemur collinsonae	C	Arb	F/HB	Hooker, 1986
Leptadapis aff. magnus	C	Arb	HB	} Szalay & Dagosto, 1980 (L. magnus); Seligsohn & Szalay, 1978
L. stintoni	C	Arb	HB	Gingerich & Martin, 1981
Adapinae undet.	AB	Arb	F	
Paramys sp.	AB	SGM	I/F	
Microparamys cf. russelli	AB	SGM	I/F	} Wood, 1962
M. sp. undet.	AB	SGM	I/F	
*Pseudoparamys sp.	AB	SGM	F	} Hooker, 1986 (Plesiarctomys)
Plesiarctomys hurzeleri	C	SGM	F	

References (upper right): Godinot, 1984; Gingerich, 1976; Kay & Cartmill, 1977; Koenigswald & Storch, 1983 (Pholidocercus).

Appendix 2(2)

	Suffolk Pebble Beds	Woolwich Beds	Abbey Wood	Wittering/base Earnley	Creechbarrow	Hordle	Hatherwood Limestone	Bembridge Limestone	Bembridge Marls	Hamstead	Size	Locomotory adaptation	Dietary adaptation	References on locomotion	References on diet
P. curranti					×						AB	SGM	F		Hooker, 1986
Paramyinae undet.			×		×						AB	SGM	I/F		Hooker, 1986
Manitshinae undet.	×										AB	SGM	F		Hooker, 1986
Meldimys sp.				×							AB	Arb	I/F		
Ailuravus michauxi		×			×						AB	Arb	F	Wood, 1962	Hooker, 1986
A. stehlinschaubi					×						AB	SGM	F/HB	Weitzel, 1949 (A. macrurus)	
Sciuroides rissonei					×		×				AB	SGM	F		Hooker, 1986 (S. rissonei, T. helveticus, S. authodon)
Treposciurus helveticus						×	×				AB	SGM	F		
T. mutabilis							×	×			AB	SGM	F		
T. intermedius						×	×	×			AB	SGM	F		
Suevosciurus authodon					×						AB	SGM	F	Lavocat, 1955 (Suevosciurus ehingensis)	
S. sp. 1						×	×	×			AB	SGM	F		
S. sp. 2							×	×			AB	SGM	F		
Paradelomys quercyi					×		×				AB	SGM	F		
P. sp.						×					AB	SGM	F		
Pseudoltinomys gaillardi					×		×	×			AB	SGM	F/HB		
P. sp							×	×		×	AB	SGM	F/HB		
Thalerimys headonensis						×	×	×	×		AB	SGM	F/HB	Meulen & de Bruijn, 1982	Meulen & de Bruijn, 1982
T. fordi							×	×	×	×	AB	SGM	F/HB		
Isoptychus pseudosiderolithicus							×	×			AB	SGM	F/HB		
Ectropomys exiguus							×				AB	SGM	F/HB		
Gliravus priscus							×	×			AB	SGM	I/F		
G. devoogdi							×	×	×		AB	SGM	I/F		
G. jordi									×	×	AB	SGM	I/F		
G. minor					×	cf. ×					AB	SGM	I/F		
G. daamsi										×	AB	SGM	I/F		

Taxon			I/F	Reference
Bransatoglis bahloi			I/F	
Eomys sp.	cf. AB	SGM	I/F	Hooker, 1986
Eucricetodon atavus	AB	SGM	I/F	Hooker, 1986
Steneofiber sp.	AB	SGM	F/HB	Hooker, 1986
Didelphodus sp.	AB	SGM	I/C	Hooker, 1986 / Coombs, 1983
cf. *Apatemys* sp.	AB	Arb	I	
Heterohyus cf. *sudrei*	AB	Arb	I	
†*H. nanus*	AB	Arb	I	
H. morinionensis	AB	Arb	I	
H. sp. 1	C	LGM	HB	
Esthonyx sp.	GH	LGM	HB	
Coryphodon eocaenus	D	LGM	C	Coombs, 1983
Oxyaena sp.	D	LGM	C	
?*Palaeonictis* sp.	AB	SGM	C	
'*Prototomus*' sp.	C	LGM	C	
Hyaenodon minor	C	LGM	C	Mellett, 1977 (*Hyaenodon*)
H. brachyrhynchus	C	LGM	C	Mellett, 1977 (*Hyaenodon*)
H. sp.	C	LGM	C	
Pterodon dasyuroides	C	LGM	C	Springhorn, 1982 (*Proviverra*)
Creodonta undet.	C	Scan	C	
Quercygale angustidens	AB	Scan	I-C/F	Springhorn, 1980, 1982
Miacis sp.	AB	Scan	I-C/F	(*Paroodectes*, *Miacis kessleri*)
Miacidae undet.	AB	Scan	I-C/F	
Viverravidae undet.	C	LGM	C	
?*Cynodictis* sp.	AB	SGM	I-C/F	
Landenodon sp.	C	SGM	I-C/F	
Arctocyonidae undet.	AB	SGM	I/F	
Lessnessina packmani	AB	SGM	I/F	
Hyopsodus wardi	AB	SGM	I/F	Gazin, 1968
Microhyus musculus	AB	SGM	I/F	

Appendix 2(3)

	Suffolk Pebble Beds	Woolwich Beds	Abbey Wood	Wittering/base Earnley	Creechbarrow	Hordle	Hatherwood Limestone	Bembridge Limestone	Bembridge Marls	Hamstead	Size	Locomotory adaptation	Dietary adaptation	References on locomotion	References on diet
cf. *Phenacodus* sp.			x								C	LGM	F		
Vulpavoides cooperi					x						AB	SGM	I/C/F	Kitts, 1956	
Cymbalophus cuniculus	x										C	LGM	F/HB	⎫	
Hyracotherium aff. *vulpiceps*		x	x								C	LGM	F/HB	⎬ (N. American *Hyracotherium*)	
cf. *H.* sp.				x							C	LGM	F/HB	⎭	
Propalaeotherium aff. *parvulum*			x								C	LGM	F/HB		⎫ Sturm, 1978; Koenigswald & Schaarschmidt, 1983 (*P. messelense*)
Lophiotherium siderolithicum					x						C	LGM	F/HB		
Anchilophus radegondensis								x			C	LGM	F/HB		
A. dumasii						x					C	LGM	F/HB		
Plagiolophus curtisi					x						D	LGM	HB		
P. annectens						x	x	x			D	LGM	HB	⎫	
P. minor									x		D	LGM	HB		
P. fraasi								x			D	LGM	HB	⎬ Stehlin, 1939	
Palaeotherium magnum						x	x	x			GH	LGM	HB		
P. muehlbergi						x	x				EF	LGM	HB		
P. duvali						x	x				D	LGM	HB		
P. curtum								x	x		EF	LGM	HB		
P. medium								x	x		EF	LGM	HB	⎭	
Hyrachyus stehlini					x						D	LGM	HB	Franzen, 1981:a; Brunet, 1979	
Ronzotherium sp.										x	GH	LGM	HB		
Diacodexis 2 spp. undet.		x									AB	LGM	I/F	Franzen, 1981b (*Messelobunodon*)	
Dichobune leporinum			x					x			C	LGM	F		Richter, 1981 (*Messelobunodon*)
Hyperdichobune sp. 1				x	x						AB	LGM	F/HB		
H. sp. 2				x	x						C	LGM			

Taxon											Code	LGM	Type	Reference
Mixtotherium aff. gresslyi	×										C	LGM	HB	
M. sp.	×										AB	LGM	HB	
Cebochoerus robiacensis	×										C	LGM	F	
C. aff. fontensis		×									C	LGM	F	Hooker, 1986
C. sp. undet.			×								C	LGM	F	
Acotherulum campichii	×		×								C	LGM	F	
A. sp.	×										C	LGM	F	
Haplobunodon venatorum		×									C	LGM	F/HB	
H. lydekkeri		×									C	LGM	F/HB	
Amphirhagatherium sp.			×								C	LGM	F/HB	
*A. fronstettense				×							C	LGM	F/HB	
*Rhagatherium sp.		×		×							D	LGM	F/HB	Depéret, 1917
Choeropotamus depereti					×						D	LGM	F/HB	
C. aff. parisiensis	×										C	LGM	F/HB	
C. sp. indet.				×							EF	LGM	HB?	
Diplopus aymardi		×			×						EF	LGM	HB	Kowalevsky, 1874b
Elomeryx porcinus					×						EF	LGM	HB	
Bothriodon 3 spp.					×						EF	LGM	F/HB	Kowalevsky, 1873–
Anthracotherium alsaticum					×						D	LGM	F/HB	1874a (A. magnum)
Prominatherium dalmatinum			×								C	LGM	HB	
Catodontherium sp.		×									C	LGM	HB	
Dacrytherium elegans		×			×						C	LGM	HB	Depéret, 1917
D. ovinum						×					C	LGM	F/HB	
Tapirulus hyracinus				×	×						EF	LGM	HB	
Anoplotherium sp.			×								C	LGM	HB	Dor, 1938
Cainotheriidae undet.			×								C	LGM	HB	Hürzeler, 1936
Xiphodon gracile		×	×								AB	LGM	HB	Dechaseaux, 1967
Haplomeryx zitteli	cf. ×		×								C	LGM	HB	
Dichodon cervinus	cf. ×	×									D	LGM	HB	Depéret, 1917
D. cuspidatus	×	×									AB	LGM	F/HB	Sudre, 1978 (Amphimeryx)
Pseudamphimeryx hantonensis					cf.									Brunet, 1979
Entelodon magnus			×								GH	LGM	F/HB	

References

Andrews, P., Lord, J. M. & Nesbit Evans, E. M. (1979). Patterns of ecological diversity in fossil and modern mammalian faunas. *Biological Journal of the Linnean Society*, **11**, 177–205.

Aubry, M.-P. (1985). Northwestern European Palaeogene magnetostratigraphy, biostratigraphy, and paleogeography: calcareous nannofossil evidence. *Geology*, **13**, 198–202.

Behrensmeyer, A. K. & Kidwell, S. M. (1985). Taphonomy's contributions to paleobiology. *Paleobiology*, **11**, 105–19.

Bone, D. A. (1986). The stratigraphy of the Reading Beds (Palaeocene) at Felpham, West Sussex. *Tertiary Research*, **8**, 17–32.

Bosma, A. A. (1974). Rodent biostratigraphy of the Eocene–Oligocene transitional strata of the Isle of Wight. *Utrecht Micropaleontological Bulletins Special Publication*, **1**, 1–127.

Bosma, A. A. & de Bruijn, H. (1979). Eocene and Oligocene Gliridae (Rodentia, Mammalia) from the Isle of Wight, England. Part 1. The *Gliravus priscus–Gliravus fordi* lineage. *Proceedings of the Koninklijke Nederlandse Akademie van Wetenschappen, Series B*, **82**, 367–84.

Bosma, A. A. & de Bruijn, H. (1982). Eocene and Oligocene Gliridae (Rodentia, Mammalia) from the Isle of Wight, England. Part II. *Gliravus minor* n.sp., *Gliravus daamsi*, n.sp., and *Bransatoglis bahloi* n.sp. *Proceedings of the Koninklijke Nederlandse Akademie van Wetenschappen, Series B*, **85**, 365–80.

Bosma, A. A. & de Bruijn, H. (1989). Early and Middle Oligocene Eomyidae (Rodentia, Mammalia) from localities in England, Belgium and Spain. *Proceedings of the Koninklijke Nederlandse Akademie van Wetenschappen, Series B*, (in press).

Boulter, M. C. & Hubbard, R. N. L. B. (1982). Objective paleoecological and biostratigraphic interpretation of Tertiary palynological data by multivariate statistical analysis. *Palynology*, **6**, 55–68.

Brunet, M. (1979). *Les Grands Mammifères chefs de file de l'immigration oligocène et le problème de la limite Eocène–Oligocène en Europe*. Paris: Editions Singer-Polignac.

Buchardt, B. (1978). Oxygen isotope palaeotemperatures from the Tertiary period in the North Sea area. *Nature*, **275**, 121–23.

Butler, P. M. (1952). The milk-molars of Perissodactyla, with remarks on molar occlusion. *Proceedings of the Zoological Society of London*, **121**, 777–817.

Buurman, P. (1980). Palaeosols in the Reading Beds (Palaeocene) of Alum Bay, Isle of Wight, U.K. *Sedimentology*, **27**, 593–606.

Chandler, M. E. J. (1960). Plant remains of the Hengistbury and Barton Beds. *Bulletin of the British Museum (Natural History) Geology*, **4**, 193–238.

Chandler, M. E. J. (1961*a*). Flora of the Lower Headon Beds of Hampshire and the Isle of Wight. *Bulletin of the British Museum (Natural History) Geology*, 5, 93–157.

Chandler, M. E. J. (1961*b*). Post-Ypresian plant remains from the Isle of Wight and the Selsey Peninsula, Sussex. *Bulletin of the British Museum (Natural History) Geology*, 5, 15–41.

Chandler, M. E. J. (1962). *The Lower Tertiary Floras of Southern England*, vol. II *Flora of the Pipe Clay Series of Dorset (Lower Bagshot)*. London: British Museum (Natural History).

Chandler, M. E. J. (1963*a*). Revision of the Oligocene floras of the Isle of Wight. *Bulletin of the British Museum (Natural History) Geology*, 6, 323–83.

Chandler, M. E. J. (1963*b*). *The Lower Tertiary Floras of Southern England*, vol. III *Flora of the Bournemouth Beds, the Boscombe and the Highcliff Sands*. London: British Museum (Natural History).

Chandler, M. E. J. (1964). *The Lower Tertiary Floras of Southern England*, vol. IV *A Summary and Survey of Findings in the Light of Recent Botanical Observations*. London: British Museum (Natural History).

Chandler, M. E. J. (1978). Supplement to the Lower Tertiary Floras of Southern England. Part 5. *Tertiary Research Special Papers*, 4, 1–47.

Collinson, M. E. (1980). A new multiple-floated *Azolla* from the Eocene of Britain with a brief review of the genus. *Palaeontology*, 23, 213–29.

Collinson, M. E. (1982*a*). A preliminary report on the Senckenberg-Museum collection of fruits and seeds from Messel bei Darmstadt. *Courier Forschungsinstitut Senckenberg*, 56, 49–57.

Collinson, M. E. (1982*b*). A reassessment of fossil Potamogetoneae fruits with description of new material from Saudi Arabia. *Tertiary Research*, 4, 83–104.

Collinson, M. E. (1983*a*). Palaeofloristic assemblages and palaeoecology of the Lower Oligocene Bembridge Marls, Hamstead Ledge, Isle of Wight. *Botanical Journal of the Linnean Society*, 86, 177–225.

Collinson, M. E. (1983*b*). *Fossil Plants of the London Clay*. London: The Palaeontological Association.

Collinson, M. E., Batten, D. J., Scott, A. C. & Ayonghe, S. N. (1985). Palaeozoic, Mesozoic and contemporaneous megaspores from the Tertiary of southern England: indicators of sedimentary provenance and ancient vegetation. *Journal of the Geological Society of London*, 142, 375–95.

Collinson, M. E. & Crane, P. R. (1978). *Rhododendron* seeds from the Palaeocene of southern England. *Botanical Journal of the Linnean Society*, 76, 195–205.

Collinson, M. E., Fowler, K. & Boulter, M. C. (1981). Floristic changes indicate a cooling climate in the Eocene of southern England. *Nature*, 291, 315–17.

Coombs, M. C. (1983). Large mammalian clawed herbivores: a comparative study. *Transactions of the American Philosophical Society*, **73**, 1–96.

Corbet, G. B. & Southern, H. N. (eds.) (1977). *The Handbook of British Mammals*. 2nd edn. Oxford: Blackwell Scientific Publications.

Crane, P. R. (1977). The Alum Bay plant beds. *Tertiary Research*, **1**, 95–9.

Crane, P. R. (1978). Angiosperm leaves from the Lower Tertiary of southern England. *Courier Forschungsinstitut Senckenberg*, **30**, 126–32.

Crane, P. R. (1981). Betulaceous leaves and fruits from the British Upper Palaeocene. *Botanical Journal of the Linnean Society*, **83**, 103–36.

Crane, P. R. (1984). A re-evaluation of *Cercidiphyllum*-like plant fossils from the British early Tertiary. *Botanical Journal of the Linnean Society*, **89**, 199–230.

Crane, P. R. & Manchester, S. R. (1982). An extinct Juglandaceous fruit from the Upper Palaeocene of southern England. *Botanical Journal of the Linnean Society*, **85**, 89–101.

Crane, P. R. & Plint, G. (1979). Calcified angiosperm roots from the Upper Eocene of southern England. *Annals of Botany*, **44**, 107–12.

Crane, P. R. & Stockey, R. A. (1985). Growth and reproductive biology of *Joffrea speirsii* gen. et sp. nov., a *Cercidiphyllum*-like plant from the Late Paleocene of Alberta, Canada. *Canadian Journal of Botany*, **63**, 340–64.

Cray, P. E. (1973). Marsupialia, Insectivora, Primates, Creodonta and Carnivora from the Headon Beds (Upper Eocene) of southern England. *Bulletin of the British Museum (Natural History) Geology*, **23**, 1–102.

Creighton, G. K. (1980). Static allometry of mammalian teeth and the correlation of tooth size and body size in contemporary mammals. *Journal of Zoology London*, **191**, 435–43.

Crochet, J.-Y. (1980). *Les Marsupiaux du Tertiaire d'Europe*. Paris: Editions Singer-Polignac.

Curry, D., Adams, C. G., Boulter, M., Dilley, F. C., Eames, F. E., Funnell, B. M. & Wells, M. K. (1978). A correlation of Tertiary rocks in the British Isles. *Special Report of the Geological Society of London*, **12**, 1–72.

Curry, D. & King, C. (1965). The Eocene Succession at Swanwick, Hampshire. *Proceedings of the Geologists' Association*, **76**, 29–35.

Daley, B. (1972). Some problems concerning the early Tertiary climate of southern Britain. *Palaeogeography, Palaeoclimatology, Palaeoecology*, **11**, 177–90.

Davis, D. D. (1962). Mammals of the lowland rain-forest of North Borneo. *Bulletin of the National Museum, State of Singapore*, **31**, 1–129.

Dechaseaux, C. (1967). Artiodactyles des Phosphorites du Quercy. II Etude sur le genre *Xiphodon*. *Annales de Paléontologie* (Vértèbres), **53**, 25–47 (1–23).

Depéret, C. (1917). Monographie de la faune de mammifères fossiles du Ludien inférieur d'Euzet-les-Bains (Gard). *Annales de l'Université de Lyon*, Nouvelle Serie I. Sciences, Médecine, **40**, 1–290.

Dor, M. (1938). Sur la biologie de l'Anoplotherium. *Mammalia*, 2, 43–8.

Eisenberg, J. F. (1978). The evolution of arboreal herbivores in the class Mammalia. In *The Ecology of Arboreal Folivores*, ed. G. G. Montgomery, pp. 135–52. Washington DC: The Smithsonian Institution.

Feist-Castel, M. (1977). Evolution of the charophyte floras in the Upper Eocene and Lower Oligocene of the Isle of Wight. *Palaeontology*, 20, 143–57.

Fisher, O. (1862). On the Bracklesham Beds of the Isle of Wight Basin. *Quarterly Journal of the Geological Society of London*, 18, 65–94.

Fleming, T. H. (1973). Numbers of mammal species in north and central American forest communities. *Ecology*, 54, 555–63.

Fowler, K., Edwards, N. & Brett, D. W. (1973). *In situ* coniferous (Taxodiaceous) tree remains in the Upper Eocene of southern England. *Palaeontology*, 16, 205–17.

Franzen, J. L. (1981a). *Hyrachyus minimus* (Mammalia, Perissodactyla, Helaletidae) aus den mitteleozänen Ölschiefern der 'Grube Messel' bei Darmstadt (Deutschland, S-Hessen). *Senckenbergiana lethaea*, 61, 371–76.

Franzen, J. L. (1981b). Das erste Skelett eines Dichobuniden (Mammalia, Artiodactyla), geborgen aus mitteleozänen Ölschiefern der 'Grube Messel' bei Darmstadt (Deutschland, S-Hessen). *Senckenbergiana lethaea*, 61, 299–353.

Gazin, C. L. (1968). A study of the Eocene condylarthran mammal *Hyopsodus*. *Smithsonian Miscellaneous Collections*, 153, 1–90.

Gingerich, P. D. (1976). Cranial anatomy and evolution of Early Tertiary Plesiadapidae (Mammalia, Primates). *Papers on Paleontology, Museum of Paleontology, University of Michigan*, 15, 1–140.

Gingerich, P. D. & Martin, R. D. (1981). Cranial morphology and adaptations in Eocene Adapidae. II. The Cambridge skull of *Adapis parisiensis*. *American Journal of Physical Anthropology*, 56, 235–57.

Gingerich, P. D., Smith, B. H. & Rosenberg, K. (1982). Allometric scaling in the dentition of primates and prediction of body weight from tooth size in fossils. *American Journal of Physical Anthropology*, 58, 81–100.

Godinot, M. (1984). Un nouveau genre de Paromomyidae (Primates) de l'Eocène Inférieur d'Europe. *Folia Primatologica*, 43, 84–96.

Godinot, M. & Dagosto, M. (1983). The astragalus of *Necrolemur* (Primates, Microchoerinae). *Journal of Paleontology*, 57, 1321–4.

Harper, J. L., Lovell, P. H. & Moore, K. G. (1970). The shapes and sizes of seeds. *Annual Review of Ecology and Systematics*, 1, 327–56.

Hoeck, H. N. (1975). Differential feeding behaviour of the sympatric hyrax *Procavia johnstoni* and *Heterohyrax brucei*. *Oecologia*, 22, 15–47.

Hooker, J. J. (1979). Two new condylarths (Mammalia) from the early Eocene of southern England. *Bulletin of the British Museum (Natural History) Geology*, 32, 43–56.

Hooker, J. J. (1986). Mammals from the Bartonian (middle/late Eocene) of the Hampshire Basin, southern England. *Bulletin of the British Museum (Natural History) Geology*, **39**, 191–478.

Hooker, J. J., Insole, A. N., Moody, R. T. J., Walker, C. A. & Ward, D. J. (1980). The distribution of cartilaginous fish, turtles, birds and mammals in the British Palaeogene. *Tertiary Research*, **3**, 1–45.

Howe, H. F. & Smallwood, J. (1982). Ecology of seed dispersal. *Annual Review of Ecology and Systematics*, **13**, 201–28.

Hubbard, R. N. L. B. & Boulter, M. C. (1983). Reconstruction of Palaeogene climate from palynological evidence. *Nature*, **301**, 147–50.

Hürzeler, J. (1936). Osteologie und Odontologie der Caenotheriden, 1. *Abhandlungen Schweizerischen Paläontologischen Gesellschaft*, **58**, 1–89.

Insole, A. N. & Daley, B. (1985). A revision of the lithostratigraphical nomenclature of the late Eocene and early Oligocene strata of the Hampshire Basin, southern England. *Tertiary Research*, **7**, 67–100.

Janis, C. (1982). Evolution of horns in ungulates: ecology and paleoecology. *Biological Reviews*, **57**, 261–318.

Jones, G. S. (1975). Notes on the status of *Belomys pearsoni* and *Dremomys pernyi* (Mammalia, Rodentia, Sciuridae) on Taiwan. *Quarterly Journal of Taiwan Museum*, **28**, 403–5.

Kay, R. F. (1975). The functional adaptations of primate molar teeth. *American Journal of Physical Anthropology*, **43**, 195–216.

Kay, R. F. & Cartmill, M. (1977). Cranial morphology and adaptations of *Palaechthon nacimienti* and other Paromomyidae (Plesiadapoidea, ?Primates), with a description of a new genus and species. *Journal of Human Evolution*, **6**, 19–53.

Kay, R. F. & Hylander, W. L. (1978). The dental structure of mammalian folivores with special reference to Primates and Phalangeroidea (Marsupialia). In *The Ecology of Arboreal Folivores*, ed. G. G. Montgomery, pp. 173–91. Washington, DC: The Smithsonian Institution.

Kemp, D. J. (1984). M27 Motorway excavations near Westend Southampton (Hants). *Tertiary Research*, **6**, 157–63.

Kemp, D. J. (1985). The Selsey Division (Bracklesham Group) at Lee-on-the-Solent, Gosport (Hants). *Tertiary Research*, **7**, 35–43.

Keng, H. (1969). *Orders and Families of Malayan Seed Plants*. Kuala Lumpur: University of Malaya Press.

King, A. & King, C. (1977). The stratigraphy of the Earnley 'division' (Bracklesham Group) at Copythorne, Hampshire. *Tertiary Research*, **1**, 115–18.

King, C. (1981). The stratigraphy of the London Clay and associated deposits. *Tertiary Research Special Papers*, **6**, 1–158.

King, C. & Kemp, D. J. (1982). Stratigraphy of the Bracklesham Group in recent exposures near Gosport (Hants). *Tertiary Research*, **3**, 171–87.

Kitts, D. B. (1956). American Hyracotherium (Perissodactyla, Equidae). *Bulletin of the American Museum of Natural History*, **110**, 1–60.

Koenigswald, W. v. (1980). Das skelett eines Pantolestiden (Proteutheria, Mamm.) aus dem mittleren Eozän von Messel bei Darmstadt. *Paläontologische Zeitschrift*, 54, 267–87.

Koeningswald, W. v. & Schaarschmidt, F. (1983). Ein Urpferd aus Messel, das Weinbeeren frass. *Natur und Museum, Frankfurt*, 113, 79–84.

Koenigswald, W. v. & Storch, G. (1983). *Pholidocercus hassiacus* ein Amphilemuride aus dem Eozän der 'Grube Messel' bei Darmstadt (Mammalia, Lipotyphla). *Senckenbergiana lethaea*, 64, 447–95.

Kowalevsky, W. (1873/1874*a*). Monographie der Gattung Anthracotherium Cuv. *Palaeontographica* (NF) 2, 131–346.

Kowalevsky, W. (1874*b*). On the osteology of Hyopotamidae. *Philosophical Transactions of the Royal Society of London*, 163, 19–94.

Krause, D. W. (1982). Jaw movement, dental function and diet in the Paleocene multituberculate *Ptilodus*. *Paleobiology*, 8, 265–81.

Krause, D. W. & Jenkins, F. A. Jr (1983). The postcranial skeleton of North American multituberculates. *Bulletin of the Museum of Comparative Zoology, Harvard*, 150, 199–246.

Krumbiegel, G., Rüffle, L. & Haubold, H. (1983). *Das Eozäne Geiseltal, ein mitteleuropäisches Braunkohlenvorkommen und seine Pflanzen- und Tierwelt.* Wittenberg Lutherstadt: A. Ziemsen Verlag.

Lavocat, R. (1955). Sur un squelette de *Pseudosciurus* provenant du gisement d'Armissan (Aude). *Annales de Paléontologie* (Vertébrés), 41, 75–89.

Maier, W. (1979). *Macrocranion tupaiodon* an adapisoricid (?) insectivore from the Eocene of 'Grube Messel' (Western Germany). *Paläontologische Zeitschrift*, 53, 38–62.

Martin, A. R. H. (1976). Upper Palaeocene Salviniaceae from the Woolwich/Reading Beds near Cobham, Kent. *Palaeontology*, 19, 173–84.

Matthew, W. D. (1909). The Carnivora and Insectivora of the Bridger Basin Middle Eocene. *Memoirs from the American Museum of Natural History*, 9, 289–567.

Mazer, S. J. & Tiffney, B. H. (1982). Fruits of *Wetherellia* and *Palaeowetherellia* (?Euphorbiaceae) from Eocene sediments in Virginia and Maryland. *Brittonia*, 34, 300–33.

Medway, Lord (1974). Food of a tapir, *Tapirus indicus*. *Malayan Nature Journal*, 28, 90–3.

Mellett, J. S. (1977). Paleobiology of North American *Hyaenodon* (Mammalia, Creodonta). *Contributions to Vertebrate Evolution*, 1, 1–134.

Meulen, A. J. van de & de Bruijn, H. (1982). The mammals from the Lower Miocene of Aliveri (Island of Evia, Greece). Part 2. The Gliridae. *Proceedings of the Koninklijke Nederlandse Akademie van Wetenschappen, Series B*, 85, 485–524.

Milner, A. R. (1980). The tetrapod assemblage from Nýřany, Czechoslovakia. In *The Terrestrial Environment and the Origin of Land Vertebrates*,

Systematics Association Special Volume 15, ed. A. L. Panchen, pp. 439–96. London: Academic Press.

Nesbit Evans, E. M., Van Couvering, J. A. H. & Andrews P. (1981). Palaeoecology of Miocene sites in Western Kenya. *Journal of Human Evolution*, **10**, 99–116.

Odin, G. S., Curry, D. & Hunziker, J. C. (1978). Radiometric dates from NW European glauconites and the Palaeogene time-scale. *Journal of the Geological Society of London*, **135**, 481–97.

Olson, E. C. (1966) Community evolution and the origin of mammals. *Ecology*, **47**, 291–302.

Paijmans, K. (ed.) (1976). *New Guinea Vegetation*. Amsterdam: Elsevier.

Plint, A. G. (1983). Facies, environments and sedimentary cycles in the Middle Eocene, Bracklesham Formation of the Hampshire Basin: evidence for global sea-level changes? *Sedimentology*, **30**, 625–53.

Plint, A. G. (1984). A regressive coastal sequence from the Upper Eocene of Hampshire, southern England. *Sedimentology*, **31**, 213–25.

Reid, E. M. & Chandler, M. E. J. (1926). *Catalogue of Cainozoic plants in the Department of Geology*, vol. 1 *The Bembridge Flora*. London: British Museum (Natural History).

Rensberger, J. J. (1975). Function in the cheek tooth evolution of some hypsodont geomyoid rodents. *Journal of Paleontology*, **49**, 10–22.

Richards, P. W. (1952). *The Tropical Rain Forest: An Ecological Study*. Cambridge: Cambridge University Press.

Richter, G. (1981). Untersuchungen zur Ernährung von *Messelobunodon schaeferi* (Mammalia, Artiodactyla). *Senckenbergiana lethaea*, **61**, 355–70.

Rose, K. D. (1981). Composition and species diversity in Palaeocene and Eocene mammal assemblages: an empirical study. *Journal of Vertebrate Paleontology*, **1**, 367–88.

Rose, K. D. & Walker, A. (1985). The skeleton of early Eocene *Cantius*, oldest lemuriform primate. *American Journal of Physical Anthropology*, **66**, 73–89.

Russell, D. E. (1975). Paleoecology of the Paleocene-Eocene Transition in Europe. In *Approaches to Primate Paleobiology, Contributions to Primatology*, vol. 5, ed. F. S. Szalay, pp. 28–61. Basel: Karger.

Savage, D. E. & Russell, D. E. (1983). *Mammalian Paleofaunas of the World*. London: Addison-Wesley Publishing Company.

Schmid, P. (1979). Evidence of microchoerine evolution from Dielsdorf (Zürich region, Switzerland) – a preliminary report. *Folia Primatologica*, **31**, 301–11.

Schmidt-Kittler, N. (1984). Pattern analysis of occlusal surfaces in hypsodont herbivores and its bearing on morph-functional studies. *Proceedings of the Koninklijke Nederlandse Akademie van Wetenschappen, Series B*, **87**, 453–80.

Scott, A. C. & Collinson, M. E. (1983*a*). Investigating fossil plant beds.

Part 1. The origin of fossil plants and their sediments. *Geology Teaching*, 7, 114–22.

Scott, A. C. & Collinson, M. E. (1983*b*). Investigating fossil plant beds. Part 2. Methods of palaeoenvironmental analysis and modelling and suggestions for experimental work. *Geology Teaching*, 8, 12–26.

Seligsohn, D. & Szalay, F. S. (1978). Relationships between natural selection and dental morphology: tooth function and diet in *Lepilemur* and *Hapalemur*. In *Development, Function and Evolution of Teeth*, ed. P. M. Butler & K. A. Joysey, pp. 289–307. London: Academic Press.

Shackleton, N. J. (1984). Oxygen isotope evidence for Cenozoic climatic change. In *Fossils and Climate*, ed. P. Brenchley, pp. 27–34. New York: John Wiley & Sons.

Springhorn, R. (1980). *Paroodectes feisti* der erste Miacide (Carnivora, Mammalia) aus dem Mittel Eozän von Messel. *Paläontologische Zeitschrift*, 54, 171–98.

Springhorn, R. (1982). Neue Raubtiere (Mammalia: Creodonta et Carnivora) aus dem Lutetium der Grube Messel (Deutschland). *Palaeontographica A*, 179, 105–41.

Stehlin, H. G. (1939). Zur Charakteristik einiger Palaeotherium-arten des oberen Ludien. *Eclogae Geologicae Helvetiae*, 31, 263–92.

Stockmans, F. (1936). Végétaux Eocènes des environs de Bruxelles. *Mémoires du Musée Royal d'Histoire Naturelle de Belgique*, 76, 1–56.

Stuart, O. L., Jr (1970). The Rodentia as Omnivores. *Quarterly Review of Biology*, 45, 351–72.

Sturm, M. (1978). Maw contents of an Eocene horse (*Propalaeotherium*) out of the oil shale of Messel near Darmstadt. In *Advances in Angiosperm Palaeobotany, Courier Forschungsinstitut Senckenberg*, vol. 30, ed. Z. Kvaček & F. Schaarschmidt, pp. 120–2. Frankfurt: Senckenbergische Naturforschende Gesellschaft.

Sudre, J. (1978). Les artiodactyles de l'Eocène moyen et supérieur d'Europe occidentale (systématique et évolution). *Mémoires et Travaux de l'Institut de Montpellier*, 7, 1–229.

Szalay, F. S. & Dagosto, M. (1980). Locomotor adaptations as reflected on the humerus of Paleogene primates. *Folia Primatologica*, 34, 1–45.

Terwilliger, V. J. (1978). Natural History of Bairds Tapir on Barro Colorado Island, Panama Canal Zone. *Biotropica*, 10, 211–20.

Thomasson, J. (1980). Paleoagrostology: a historical review. *Iowa State Journal of Research*, 54, 301–17.

Walker, A., Hoeck, H. N. & Perez, L. (1978). Microwear of mammalian teeth as an indicator of diet. *Science*, 201, 908–10.

Walker, E. P. (1975). *Mammals of the World*, 3rd edn, 2 vols. Baltimore & London: Johns Hopkins University Press.

Wang, C. W. (1961). *The Forests of China*. Cambridge, Massachusetts: Harvard University Maria Moors Cabot Foundation, Publication no. 5.

Ward, D. J. (1977). The Thanet Beds exposure at Pegwell Bay, Kent. *Tertiary Research*, **1**, 69–76.

Ward, D. J. (1984). Collecting isolated microvertebrate fossils. *Zoological Journal of the Linnean Society*, **82**, 245–59.

Weitzel, K. (1949). Neue Wirbeltiere (Rodentia, Insectivora, Testudinata) aus dem mitteleozän von Messel-bei-Darmstadt. *Abhandlungen hrsg. von der Senckenbergischen Naturforschenden Gesellschaft*, **480**, 1–24.

Whitford, H. N. (1911). *The Forests of the Philippines*, part I *Forest Types and Products*. Manila: Department of the Interior, Bureau of Forestry Bulletin no. 10.

Whitmore, T. C. (1972). *The Flora of Malaya*, vol. 1. Kuala Lumpur: Longman.

Wilkinson, H. P. (1981). The anatomy of hypocotyls of *Ceriops* Arnott (Rhizophoraceae) Recent and Fossil. *Botanical Journal of the Linnean Society*, **82**, 139–64.

Wilkinson, H. P. (1988). Sapindaceous pyritised twigs from the Eocene of Sheppey, England. *Tertiary Research*, **9**, 81–6.

Wolfe, J. A. (1979). Temperature parameters of humid to mesic forests of Eastern Asia and relation to forests of other regions of the northern hemisphere and Australasia. *United States Geological Survey, Professional Paper*, **1106**, 1–37.

Wolfe, J. A. (1980). Tertiary climates and floristic relationships at high latitudes in the Northern Hemisphere. *Palaeogeography, Palaeoclimatology, Palaeoecology*, **30**, 313–23.

Wolff, R. G. (1975). Sampling and sample size in ecological analyses of fossil mammals. *Paleobiology*, **1**, 195–204.

Womersley, J. S. (ed.) (1978). *Handbooks of the Flora of Papua New Guinea*. Victoria: Melbourne University Press.

Wood, A. E. (1962). The Early Tertiary rodents of the family Paramyidae. *Transactions of the American Philosophical Society*, **52**, 1–261.

Yablokov, A. V. (1974). *Variability of mammals*. Washington, DC & New Delhi: Amerind Publishing Co. Pvt. Ltd.

Classification of plants and animals

---•---

CLASSIFICATION OF VASCULAR PLANTS

The classification of vascular plants includes taxa treated in the text (divisions, -PHYTA; classes, -opsida; subclasses, -idae; orders, -ales; families, -aceae). †indicates that a taxon is extinct, including all its subordinate taxa.

LYCOPHYTA
Lycopsida
  Lycopodiales
  Selaginellales: Selaginellaceae (*Selaginella*)
  Isoetales: Isoetaceae (*Isoetes, Nathorstiana*†)
SPHENOPHYTA
Sphenopsida
  Equisetales: Equisetaceae (*Equisetum*)
PTERIDOPHYTA
Filicopsida
  Cladoxylales
  Marattiales: Marattiaceae
  Osmundales: Osmundaceae (*Osmunda, Todites*†)
  Filicales: Schizaeaceae (*Anemia, Lygodium*), Gleicheniaceae, Matoniaceae (*Weichselia*†), Dipteridaceae, Dicksoniaceae, Dennstaedtiaceae (*Dennstaedtia, Onoclea*), Polypodiaceae (*Acrostichum*)
  Marsileales: (*Arcellites*†)
  Salviniales
  Fossil genera of uncertain relationship within the Filicopsida: *Tempskya*†, *Straelenipteris*†

PROGYMNOSPERMOPHYTA[†]

Progymnospermopsida

Aneurophytales: (*Aneurophyton, Eospermatites, Triloboxylon*)

Archaeopteridales: (*Archaeopteris, Svalbardia*)

PTERIDOSPERMOPHYTA[†]

Pteridospermopsida

Lyginopteridales: Lyginopteridaceae (*Heterangium, Lyginopteris*)

Medullosales: Medullosaceae (*Medullosa, Quaestora, Sutcliffia*)

Callistophytales: Callistophytaceae (*Callistophyton*)

Caytoniales: Caytoniaceae (*Caytonanthus, Caytonia, Sagenopteris*)

Corystospermales: Corystospermaceae (*Dicroidium, Pteruchus, Rhexoxylon, Umkomasia*)

Peltaspermales: Peltaspermaceae (*Antevsia, Lepidopteris, Peltaspermum*)

Glossopteridales: Glossopteridaceae (*Denkania, Ottokaria*)

CYCADOPHYTA

Cycadopsida

Bennettitales (Cycadeoidales)[†]: (*Ischnophyton, Vardekloeftia*), Williamsoniaceae (*Ptilophyllum, Zamites*), Wielandiellaceae (*Wielandiella, Williamsoniella*), Cycadeoideaceae (*Cycadeoidea, Monanthesia*)

Pentoxylales[†]: Pentoxylaceae (*Pentoxylon*)

Cycadales: Nilssoniaceae[†] (*Beania, Nilssonia*), Cycadaceae (*Cycas*)

Fossil genera of uncertain relationship within the Cycadophyta: *Ctenozamites*[†], *Exesipollenites*[†], *Taeniopteris*[†]

GINKGOPHYTA

Ginkgopsida

Ginkgoales: Ginkgoaceae (*Baiera*[†], *Ginkgo, Karkenia*[†], *Solenites*[†])

Czekanowskiales[†]: Czekanowskiaceae (*Czekanowskia*)

CONIFEROPHYTA

Cordaitopsida[†]

Cordaitales: Cordaitaceae (*Cordaianthus, Cordaites, Mesoxylon*)

Coniferopsida

Voltziales[†]: Lebachiaceae

Coniferales: Cheirolepidiaceae (Hirmerellaceae)[†] (*Classopollis*), Podozamitaceae[†], Pinaceae (*Pinus*), Taxodiaceae (*Glyptostrobus, Metasequoia, Parataxodium*[†], *Sequoia, Sequoiadendron*), Cupressaceae, Podocarpaceae, Cephalotaxaceae (*Cephalotaxus*), Araucariaceae (*Araucariacites*[†])

Taxales: Taxaceae (*Torreya*)

Fossil genera of uncertain relationship within the Coniferopsida:
Buriadia[†]

GNETOPHYTA

Gnetopsida

Gnetales: Welwitschiaceae (*Welwitschia*), Ephedraceae (*Ephedra*),
Gnetaceae (*Gnetum*)

Fossil genera of uncertain relationship within the Gnetopsida: *Elaterosporites*[†], *Galeacornea*[†]

ANGIOSPERMAE (ANTHOPHYTA)

Dicotyledones (Magnoliopsida)

Magnoliidae

Magnoliales: Winteraceae (*Exospermum, Pseudowintera, Zygogynum*), Magnoliaceae (*Magnolia*), Annonaceae

Illiciales: Illiciaceae (*Illicium*)

Laurales: Calycanthaceae (*Calycanthus*), Lactoridaceae, Lauraceae
(*Cinnamomum, Sassafras*)

Piperales: Chloranthaceae (*Ascarina*)

Nymphaeales: Nymphaeaceae

Ranunculidae

Ranunculales: Menispermaceae, Ranunculaceae (*Clematis, Ranunculus*)

Papaverales: Papaveraceae (*Papaver*)

Fossil genera and species of uncertain relationship within the
Magnoliidae/Ranunculidae: *Archaeanthus*[†], *Caspiocarpus*[†], *Clavatipollenites*[†], *Cocculophyllum*[†], *Hyrcantha*[†], *Lesqueria*[†], '*Liriodendron*', *Liriodendropsis*[†], *Liriophyllum*[†], *Magnoliaestrobus*[†],
Myrtophyllum[†], *Ranunculaecarpus*[†], *Triplicarpus*[†], '*Williamsonia*' *recentior*[†]

Hamamelididae

Trochodendrales: (*Nordenskioldia*[†]), Trochodendraceae (*Trochodendron*), Tetracentraceae (*Tetracentron*)

Cercidiphyllales: Cercidiphyllaceae (*Cercidiphyllum, Jenkinsella*[†],
Joffrea[†], *Nyssidium*[†])

Hamamelidales: ('*Aralia*', *Betulites*[†], *Debeya*[†], *Dewalquea*[†], '*Sassafras*', '*Viburnum*'), Hamamelidaceae, Platanaceae (*Credneria*[†],
Platanus, '*Sparganium*' *aspensis*[†], *Tricolpites*[†] *minutus*)

Urticales: Moraceae (*Cecropia, Musanga*), Urticaceae (*Forskohleanthium*[†])

Casuarinales

Fagales: (*Dryophyllum*[†]), Fagaceae (*Castanea, Fagus, Quercus*), Betulaceae (*Alnus, Betula, Palaeocarpinus*[†])

Myricales: Myricaceae

Juglandales: (*Antiquocarya*[†], *Caryanthus*[†], *Manningia*[†], *Plicapollis*[†], *Trudopollis*[†]), Juglandaceae (*Casholdia*[†], *Engelhardia, Platycarya*)

Caryophyllidae

Dilleniidae

  Paeoniales: Paeoniaceae

  Theales: Theaceae

  Violales: Passifloraceae, Cucurbitaceae

  Ericales: (*Actinocalyx*[†]), Ericaceae (*Rhododendron*)

  Ebenales: Symplocaceae

  Malvales: Tiliaceae, Sterculiaceae (*Sezannella*[†])

  Euphorbiales: Euphorbiaceae

  Primulales: (*Berendtia*[†], *Myrsinopsis*[†])

Rosidae

  Saxifragales: (*Scandianthus*[†])

  Rosales: Rosaceae (*Rubus*)

  Fabales: Fabaceae = Papilionaceae = Leguminosae (*Protomimosoidea*[†])

  Nepenthales: Droseraceae (*Aldrovanda*)

  Myrtales: (*Chitaleypushpam*[†], *Enigmocarpon*[†], *Raoanthus*[†], *Sahnianthus*[†], *Sahnipushpam*[†]), Lythraceae (*Decodon*), Myrtaceae, Rhizophoraceae (*Ceriops*)

  Rutales: Rutaceae, Meliaceae, Burseraceae, Anacardiaceae

  Sapindales: Staphyleaceae, Sapindaceae, Aceraceae, Sabiaceae

  Polygalales: Malpighiaceae

  Cornales: Nyssaceae (*Nyssa*), Cornaceae

  Araliales: Araliaceae (*Aralia*)

  Celastrales: Icacinaceae

  Santalales: (*Aquilapollenites*[†], *Triprojectus*[†])

  Rhamnales: Vitaceae

  Proteales: Proteaceae

  Fossil genera of uncertain relationship within the Rosidae: *Sapindopsis*[†], *Proteacidites*[†], *Wetherellia*[†], *Wodehouseia*[†]

Asteridae

  Gentianales: Rubiaceae, Gentianaceae (*Pistillipollenites*[†])

  Dipsacales: Caprifoliaceae (*Abelia, Sambucus, Viburnum*)

Monocotyledones (Liliopsida)
Alismatidae
Alismatales: Hydrocharitaceae (*Stratiotes*)
Najadales: Potamogetonaceae/Ruppiaceae (*Limnocarpus*[†], *Potamogeton*)
Liliidae
Cyperales: Cyperaceae (*Mariscus, Scirpus*)
Zingiberales: Musaceae (*Musa*), Zingiberaceae (*Spirematospermum*[†])
Arecidae
Arecales: Arecaceae = Palmae (*Manicaria, Nipa*)
Typhales: Typhaceae (*Typha*)
Fossil genera of uncertain relationship within the Arecidae: *Deccananthus*[†], *Palmocarpon*[†], *Spinizonocolpites*[†], *Tricoccites*[†], *Viracarpon*[†]

Fossil genera or species of uncertain relationship within the Angiospermae: *Acaciaephyllum*[†], *Afropollis*[†], *Atlantopollis*[†], *Asterocelastrus*[†], *Calycites*[†], *Cantia*[†], *Carpolithus*[†] *conjungatus*, *Celastrophyllum*[†], *Ceratocarpus*[†], *Citrophyllum*[†], *Complexiopollis*[†], *Conospermites*[†], *Cretovarium*[†], *Curvospermum*[†], *Dicotylophyllum*[†], *Ficophyllum*[†], *Halyserites*[†], *Hythia*[†], *Icacinoxylon*[†], *Kalinaia*[†], *Leptospermites*[†], '*Leptospermum*' *macrocarpum*[†], *Nelumbites*[†], *Paraphyllanthoxylon*[†], *Populophyllum*[†], '*Populus*', *Proteaephyllum*[†], *Protophyllum*[†], *Rogersia*[†], *Rutaecarpus*[†], *Sarysua*[†], '*Scirpus*' *lakensis*[†], *Taldysaja*[†], *Tricolpites*[†], *Vitiphyllum*[†], '*Zizyphus*'
Fossil genera or species of uncertain relationship ascribed to the Angiospermae: *Araliaecarpum*[†], *Caricopsis*[†], *Carpolithus*[†] sp., *Cyperacites*[†], *Erenia*[†], *Gurvanella*[†], *Kenella*[†], *Nyssidium*[†] *orientale*, *Nyssidium*[†] sp., *Onoana*[†], *Phyllites*[†], *Prisca*[†], *Problematospermum*[†], *Typhaera*[†]

CLASSIFICATION OF INSECTS, REPTILES AND MAMMALS

The classification of insects, reptiles and mammals includes taxa treated in the text (superfamilies, -oidea; families, -idae; subfamilies, -inae; tribe, -ini). [†] indicates that a taxon is extinct, including all its subordinate taxa.

CLASS INSECTA
Subclass Pterygota
order Thysanoptera
order Coleoptera
  suborder Polyphaga
    Elateroidea: Elateridae
    Cucujoidea: Nitidulidae
    Tenebrionoidea: Mycetophagidae
order Diptera
  suborder Nematocera
    Tipuloidea: Tipulidae
    Bibionoidea: Mycetophilidae
  suborder Brachycera
    Asiloidea: Bombyliidae
    Empidoidea: Empididae
    Syrphoidea: Syrphidae
order Lepidoptera
  suborder Zeugloptera
    Micropterigoidea: Micropterigidae (*Sabatinca*)
  suborder Ditrysia
    Papilionoidea: Papilionidae
    Noctuoidea: Noctuidae
order Hymenoptera
  suborder Symphyta
    Megalodontoidea: Xyelidae
  suborder Apocrita
    Vespoidea: Vespidae
    Sphecoidea: Sphecidae
    Apoidea: Colletidae, Anthophoridae (*Centris*), Apidae (Meliponinae, Meliponini)

CLASS REPTILIA
subclass Lepidosauria
order Eosuchia[†]
  suborder Younginiformes: Champsosauridae
subclass Archosauria
order Crocodilia
order Saurischia[†]
  suborder Theropoda: Coeluridae (*Coelurus, Ornitholestes*), Megalosauridae (*Marshosaurus, Stokesosaurus*), Ceratosauridae (*Cerato-*

saurus), Allosauridae (*Allosaurus*), Ornithomimidae (*Elaphrosaurus*)

suborder Sauropoda: Cetiosauridae (*Haplocanthosaurus*), Camarasauridae (*Camarasaurus*), Brachiosauridae (*Brachiosaurus*, '*Ultrasaurus*'), Diplodocidae (*Amphicoelias, Apatosaurus, Barosaurus, Diplodocus, Nemegtosaurus*)

order Ornithischia[†]

suborder Ornithopoda: Fabrosauridae, Hypsilophodontidae (*Dryosaurus, Hypsilophodon, Nannosaurus, Othnielia*), Iguanodontidae (*Camptosaurus, Iguanodon*), Hadrosauridae (*Edmontosaurus*), Pachycephalosauridae

suborder Stegosauria: Stegosauridae (*Stegosaurus*)

suborder Ankylosauria: Nodosauridae (*Panoplosaurus*), Ankylosauridae (*Euoplocephalus*)

suborder Ceratopsia: Protoceratopsidae (*Leptoceratops*), Ceratopsidae (*Styracosaurus, Triceratops*)

CLASS MAMMALIA

subclass Eotheria[†]

order Docodonta: Docodontidae (*Docodon*)

subclass Allotheria[†]

order Multituberculata: Ectypodontidae (*Ectypodus*)

subclass Theria

infraclass Metatheria

order Marsupialia: Didelphidae (*Amphiperatherium*[†], *Glasbius*[†], *Peratherium*[†]), Phalangeridae (*Dactylopsila, Phascolarctos, Pseudocheirus*), Macropodidae (*Hypsiprymnodon, Potorous*)

infraclass Eutheria

order Proteutheria[†]: Pantolestidae (*Cryptopithecus, Dyspterna, Palaeosinopa*)

order Lipotyphla: Nyctitheriidae[†] (*Leptacodon, Pontifactor, Saturninia, Scraeva,* Amphidozotheriinae undet.), Plesiosoricidae[†] (*Butselia*), family uncertain ('*Adapisorex*'[†]), Talpidae (*Eotalpa*[†]), Adapisoricidae[†] (*Paschatherium*), Dormaaliidae[†] (*Macrocranion, Neomatronella*), Amphilemuridae[†] (*Gesneropithex*), Erinaceidae (*Tetracus*[†]) Chrysochloridae (Golden Moles)

order Taeniodonta

order Chiroptera

order Dermoptera

order Primates: Paromomyidae[†] (*Arcius*), Plesiadapidae[†]

(*Plesiadapis*), Omomyidae[†] (*Microchoerus, Nannopithex, Pseudo-loris, Teilhardina*), Adapidae[†] (*Cantius, Europolemur, Lepta-dapis*), Daubentoniidae (*Daubentonia*), Indriidae (*Indri, Propithe-cus*), Cercopithecidae (*Cerocebus, Miopithecus*), Lorisidae (*Galago*)

order Edentata: Dasypodidae (Armadillos)

order Pholidota (Pangolins)

order Rodentia: Paramyidae[†] (Paramyinae: *Paramys*; Micropara-myinae: *Microparamys*; Manitshinae: *Plesiarctomys, Pseudo-paramys*; Ailuravinae: *Ailuravus, Meldimys*), Pseudosciuridae[†] (*Paradelomys, Sciuroides, Suevosciurus, Treposciurus*), Therido-myidae (*Ectropomys, Isoptychus, Pseudoltinomys, Thalerimys*), Gli-ridae (*Bransatoglis[†], Gliravus[†]*), Sciuridae (*Belomys, Iomys, Ratufa, Trogopterus*), Eomyidae[†] (*Eomys*), Hydrochoeridae (capybaras), Cricetidae (*Eucricetodon[†]*), Castoridae (*Steneofiber[†]*), Thryonomyidae (*Thryonomys*)

order Didelphodonta[†]: Didelphodontidae (*Didelphodus*)

order Apatotheria[†]: Apatemyidae (*Apatemys, Heterohyus*)

order Tillodontia[†]: Esthonychidae (*Esthonyx*)

order Pantodonta: Coryphodontidae (*Coryphodon*)

order Creodonta[†]: Oxyaenidae (*Oxyaena, Palaeonictis*), Hyaenodon-tidae (*Hyaenodon, 'Prototomus', Pterodon*)

order Carnivora: Miacidae[†] (*Miacis, Quercygale*), Viverravidae[†], Canidae (*Cynodictis[†]*), Viverridae (*Arctictis*)

order Condylarthra[†]: Arctocyonidae (*Landenodon*), Periptychidae (*Lessnessina*), Hyopsodontidae (*Hyopsodus, Microhyus*), Phenaco-dontidae (*Phenacodus*), Paroxyclaenidae (*Vulpavoides*)

order Hyracoidea: Procaviidae (*Heterohyrax*)

order Proboscidea: Elephantidae (Elephants, *Mammuthus[†]*)

order Perissodactyla

    suborder Hippomorpha: Palaeotheriidae[†] (*Anchilophus, Cymbalo-phus, Hyracotherium, Lophiotherium, Palaeotherium, Plagio-lophus, Propalaeotherium*)

    suborder Ancylopoda: Lophiodontidae (*Lophiodon*)

    suborder Ceratomorpha: Amynodontidae[†], Helaletidae[†] (*Hyrachyus*), Rhinocerotidae (*Ronzotherium[†]*) Tapiridae (*Tapirus*)

order Artiodactyla: Dichobunidae[†] (*Diacodexis, Dichobune, Hyper-dichobune*), Mixototheriidae[†] (*Mixtotherium*), Cebochoeridae[†] (*Acotherulum, Cebochoerus*), Haplobunodontidae[†] (*Amphirhaga-therium, Haplobunodon, Rhagatherium*), Choeropotamidae[†] (*Choe-*

ropotamus), Anthracotheriidae[†] (*Anthracotherium, Bothriodon, Diplopus, Elomeryx, Prominatherium*), Anoplotheriidae[†] (*Anoplotherium, Catodontherium, Dacrytherium, Tapirulus*), Cainotheriidae[†], Xiphodontidae[†] (*Dichodon, Haplomeryx, Xiphodon*) Amphimerycidae[†] (*Pseudamphimeryx*), Entelodontidae[†] (*Entelodon*), Hippopotamidae (hippopotamuses)

suborder Ruminantia: Traguloidea (chevrotains), Cervoidea: Cervidae (deer) (*Cervus*), Giraffidae (giraffe), Bovoidea: Bovidae (antelope, buffalo, cattle, duiker, eland, gerenuk, Grant's gazelle, impala, kudu, oryx, Thomson's gazelle)

Stratigraphic table

The stratigraphic table includes Mesozoic and Cenozoic time divisions and stratigraphic units mentioned in the text. For major early Mesozoic and Palaeozoic stratigraphic divisions see Figure 1.3. (Time scale based on Harland *et al.* (1982) *A Geologic Time Scale*, Cambridge University Press.)

Era	Period	Epoch	Age	Age (MY BP)	North America	Localities and Formations in Europe and Asia
Cenozoic	Tertiary — Neogene	Pliocene	Piacenzian	2.0	Burge Fauna	
			Zanclian			
		Miocene	Messinian			
			Tortonian		Valentine Fm	
			Serravallian			
			Langhian			
			Burdigalian			
			Aquitanian			
	Tertiary — Paleogene	Oligocene	Chattian	24.6		
			Rupelian		Chadronian Flagstaff Rim	Bouldnor Fm
						Bembridge Lst Fm, Headon Hill Fm, Becton Sand Fm
		Eocene	Priabonian			Barton Clay Fm,
						Creechbarrow Lst
			Bartonian			Selsey Div
						Marsh Farm Fm
			Lutetian			Earnley Fm
						Wittering Div
			Ypresian		Willwood Fm	London Clay Fm
						Blackheath Beds
		Paleocene	Thanetian		Fort Union Fm	Woolwich and Reading Beds
						Thanet Fm
			Danian		Puercan Composite	
				65.0	Lance Fm, Hell Creek Fm, Lower Medicine Bow, Fox Hills, Olmos Fm, Judith River,	
			Maastrichtian			Deccan Intertrappean Beds (basal)

Stratigraphic correlation chart

Localities / Formations	Stage	Epoch	Series	Period	Era	Age (Ma)
Scania	Santonian			Cretaceous	Mesozoic	
Mgachi	Coniacian					
	Turonian					
Czechoslovakia (Peruč)	Cenomanian					97.5
Neuse River, Blackhawk Fm, Nanaimo Group						
Tuscaloosa Fm, Gordo Fm, Raritan Fm, Dakota ss, Linnenberger locality, Aspen Shale, Fall River Fm, Cheyenne ss, Cedar Mountain Fm, Blairmore Group						
Potomac Group: Elk Neck Beds	Albian		Early			
Kazakhstan, Kolyma Basin						
Patapsco Fm	Aptian					
Arundel Clay	Barremian	Neocomian				
Patuxent Fm	Hauterivian					
Horsetown Group	Valanginian					
	Berriasian					
Suchan, Mongolia; Wealden Beds	Tithonian		Late	Jurassic		144.0
	Kimmeridgian	Malm				
Morrison Fm	Oxfordian					163.0

Glossary

The glossary includes, in alphabetic order, terms and higher taxa mentioned in the text (animals to the level of families, plants generally to the level of subclasses). The terms are included in one grammatical form only (e.g. noun, anemophily; adjective anemophilous omitted etc.). The definitions are generally restricted to the sense used in the present text.

achene   A dry indehiscent single-seeded fruit.

acrodromous   Applied to leaves having veins that arise near the base and gradually curve towards the leaf apex.

actinodromous   Applied to leaves having three or more primary veins radiating from a single point; palmate.

actinomorphic   Applied to flowers having radial symmetry; cf. zygomorphic.

Aculeata   The stinging Hymenoptera including many of the wasps, ants and bees. Lower Cretaceous–Recent.

acyclic   Applied to flowers having a spiral arrangement of all floral parts.

aeolian   Applied to sediments deposited by the wind.

aestivation   The arrangement of perianth parts within the flower bud.

alkaloids   A group of nitrogen-rich organic compounds, produced by the secondary metabolism of some plants, that includes caffeine, morphine etc.

allometric   Pertaining to differential growth in an organism.

alluvial   Applied to sediments transported by water.

Amentiferae   Group of dicotyledonous families that contains mostly shrubs and trees with simple, anemophilous flowers often borne in catkins, e.g. alder and birch. Corresponds approximately to the subclass Hamamelididae. Middle Cretaceous–Recent.

amphitropous   Applied to ovules having a strongly curved nucellus and embryo sac. Chalaza and hilum positioned near the micropyle.

amynodont rhinoceroses   Rhinoceroses of the family Amynodontidae. Upper Eocene–Oligocene.

317

anatropous   Applied to ovules having a straight nucellus and embryo sac that are inverted 180° and usually fused to the funicle (raphe).

androecium   Collective term for the stamens in a flower.

anemophily   Pollination by the wind.

ankylosaur   Member of the Ankylosauria, the suborder of armored dinsoaurs. Cretaceous.

anther   The part of the stamen that comprises the connective and two, four or several microsporangia (pollen sacs) in which the pollen develops.

anthophilous   Applied to animals attracted to, or feeding on, flowers.

anthophorid bee   Member of the family of predominantly solitary bees the Anthophoridae. Oligocene–Recent.

anthophytes   Seed plants with reproductive organs arranged in flowers or flower-like structures. Commonly used as synonymous with angiosperms; used by Doyle and Donoghue (Chapter 2) in a broader sense to include also the Bennettitales, *Pentoxylon*, and Gnetales.

Apatotheria   Order of placental mammals which comprises animals ranging in size from that of a mouse to that of a cat; characterized by enlarged incisors and curiously shaped cheek teeth. Middle Paleocene–Middle Oligocene.

apetalous   Applied to flowers that have no petals.

Apidae   The family of bees that includes the eusocial subfamilies Bombinae and Apinae (honeybees and stingless honeybees). Eocene–Recent.

apocarpous   Applied to gynoecia having carpels that are free and not fused to each other; cf. syncarpous.

Apocrita   Suborder of Hymenoptera defined by the abdomen being narrowly constricted at the attachment to the thorax. Lower Cretaceous–Recent.

Apoidea   Superfamily of Hymenoptera comprising the bees. Eocene–Recent.

apomorphy   A uniquely derived character-state used in cladistic analyses; cf. plesiomorphy.

arboreal   Applied to animals living in trees, to which they are mainly restricted.

archegonium   The structure enclosing the egg, in the gametophyte of plants.

archetype   Original or ancestral type; prototype.

archosaur   Group of animals, including the dinosaurs, pterosaurs, crocodiles and sometimes birds. Triassic–Recent.

Arcto-Tertiary   Applied to mid- to high-latitude plants of the Northern Hemisphere.

aril   A pulpy or hairy outgrowth that develops following fertilization from funicle or other part of the ovule.

Asteridae   Subclass of dicotyledonous families that contains many taxa with derived features, e.g. sympetaly and zygomorphy in mint and foxglove. Paleocene–Recent.

autapomorphy   A uniquely derived character-state characteristic of a single species or other taxon; cf. synapomorphy.

autochthonous   Applied to accumulations of sedimentary particles or remains of organisms that have not been transported far from their point of origin.

autotroph   An organism capable of synthesizing organic substance from entirely inorganic compounds.

basitarsus   The basal or proximal segment of the tarsus (foot) of an insect. In bees the basitarsus is modified into a pollen comb.

bauxite   Aluminium-rich weathering product formed in tropical or subtropical regions with high year-round rainfall.

Bennettitales   Gymnospermous plants with pinnate or simple cycad-like leaves, reproductive organs borne in bisexual or unisexual flower-like structures and seeds surrounded by sterile scales; cycadeoid. Upper Triassic–Upper Cretaceous.

benthic   Applied to organisms inhabiting the bottom of seas, rivers or lakes.

berry   A fleshy indehiscent fruit, without a stony layer in the fruit wall.

bilophodont   Applied to teeth dominated by two transverse crests.

biofacies   Biological (fossil) characteristics of a rock unit.

biostratigraphy   Characterization and correlation of rock units by their fossil content.

biota   Collective term for the animal and plant life of a given area.

biped   An animal with two feet.

bipinnate   Applied to compound leaves in which the primary leaflets are pinnate.

bisaccate   Applied to pollen grains having two air sacs; characteristic of many conifers.

bisexual flower   A flower possessing both female and male reproductive organs (pistils and stamens); hermaphroditic flower; cf. unisexual flower.

bisporangiate   Applied to anthers having two microsporangia (pollen sacs). Also used for bisexual reproductive structures.

bitegmic   Applied to ovules having two integuments; cf. unitegmic.

Bombyliidae   A family of flower-visiting flies comprising the beeflies. Recent (no fossil record).

bordered pit   Pit in a xylem element in which a thin area of primary (cellulosic) wall is overhung by a flange of lignified secondary wall.

Brachycera   Suborder of flies that contains two important flower-visiting families: the Empididae and Bombyliidae. Jurassic–Recent.

brush inflorescence   Applied to inflorescences having flowers with elongated and conspicuous filaments extending beyond the perianth.

bryophytes   Small, non-vascular plants with an annual or perennial gametophyte and a short-lived sporophyte; Bryophyta; mosses and liverworts. Lower Carboniferous–Recent.

bunodont   Applied to teeth having low, rounded cusps.

calyx   Collective term for the sepals in a flower.

campylotropous   Applied to ovules having a curved nucellus, but straight embryo sac. Chalaza and hilum positioned near the micropyle.

canopy   The uppermost stratum of a forest consisting of the crowns of trees.

capsule   A dry dehiscent fruit formed from a syncarpous ovary.

carnivore   Animal feeding on other animals.

carpel   A unit of an angiosperm gynoecium, bearing ovules.

Caryophyllidae   Subclass of dicotyledonous families that comprises herbs, shrubs or climbers, usually with curved embryos, e.g. carnation, *Cactus* and beetroot. Upper Cretaceous–Recent.

cataphyll   Scale leaf.

catkin   A pendulous inflorescence of small simple flowers.

Cebochoeridae   Family of small pig-like, even-toed ungulates. Eocene.

ceratopsid   Member of the family Ceratopsidae, the horned dinosaurs. Upper Cretaceous.

chalaza   The region in an ovule where the integuments arise.

champsosaur   Member of Champsosauridae, a family of small crocodile-like reptiles. Upper Cretaceous–Eocene.

charophyte   Green, generally freshwater alga characteristically encrusted with calcium carbonate and bearing leaf-like branches in whorls along the main axis. Upper Silurian–Recent.

circinate   Applied to leaf tips that are inwardly coiled.

clade   A group of taxa (represented by lines or branches in a cladogram) derived from a single common ancestor.

cladistic analysis   A method for analyzing relationships using shared derived character-states (synapomorphies).

cladogram   A diagrammatic presentation of a cladistic analysis.

Coleoptera   Order of insects that contains the beetles. Permian–Recent.

Colletidae   A family of primitive solitary bees. Recent (no fossil record).

colpus   An elongated aperture in angiosperm pollen.

columellae   Rod-like elements that separate the tectum from the footlayer in an angiosperm pollen wall.

community   A group of organisms co-existing and interacting in the same area.

companion cell   A small cell with dense protoplasm associated with sieve elements in the phloem.

compitum   An interconnected portion in the gynoecium that serves as a pathway for regular pollen-tube distribution between several carpels.

compression   A fossil or process in which an organism is flattened by compaction of the surrounding matrix.

condylarth   Member of the Condylarthra, an order of archaic ungulates. Upper Cretaceous–Lower Oligocene.

conifers   Gymnospermous plants often with needle or scale-like leaves, reproductive organs in unisexual cones, and dense pycnoxylic secondary wood; Coniferopsida; Coniferales. Carboniferous–Recent.

connation   The fusion of organs.

connective   The tissue connecting the microsporangia (pollen sacs) of the anther.

coprolite   Fossilized excrement.

cordillera   A system of mountain ranges.

corolla   Collective term for the petals in a flower.

corpus   Cells of the apical meristem beneath the tunica.

corystosperms   Group of Mesozoic seed ferns with pinnate leaves and reproductive organs borne on branched axes. Triassic.

cotyledon   A leaf-like appendage of the embryo in seed plants.

craspedodromous   Applied to leaves having secondary veins that run directly to the leaf margin.

cricetid   Member of the Cricetidae, a family of rodents that includes hamsters and voles. Oligocene–Recent.

cupule   A structure in seed ferns that enclosed one or more seeds.

cursorial   Applied to animals adapted for running.

cycads   Gymnospermous plants typically with large, pinnate leaves, reproductive organs in unisexual cones and soft, manoxylic wood; Cycadales. Permian–Recent.

cycadophytes   Heterogeneous group of seed plants comprising the cycads, Bennettitales and seed ferns. Upper Devonian–Recent.

cycadopsids   Heterogeneous group of seed plants comprising the cycads and Bennettitales. Permian–Recent.

cyclic   Applied to flowers having a whorled arrangement of all floral parts.

decussate   Having opposite pairs of leaves or floral parts that are alternately positioned at 90° to each other at successive nodes.

deltaic   Pertaining to sediments deposited at the mouth of a river.

dermopteran   Member of the order Dermoptera, which comprises gliding mammals restricted today to southeast Asian rainforest; also known as colugos, cobegos or 'flying lemurs'. Paleocene–Recent.

diagenetic   Pertaining to post-depositional chemical and physical changes in a sedimentary rock.

diaspore   Collective term for propagative plant organs (fruits, seeds, spores) involved in dispersal; disseminule; propagule.

dicotyledon   An angiosperm that has an embryo with two cotyledons and floral parts often in fives or fours; cf. monocotyledon.

Dilleniidae   Subclass of dicotyledonous families comprising herbs, shrubs, climbers and trees that exhibit a mixture of primitive and advanced characters, e.g. *Dillenia*, *Camellia* and violet. Upper Cretaceous–Recent.

dinosaurs   Members of the extinct reptilean orders Saurischia and Ornithischia. Middle Triassic–Cretaceous.

dioecious   Applied to plants having staminate and pistillate flowers borne on different individuals; cf. monoecious.

diploid   Applied to an organism or cell having two sets of chromosomes.

Diptera   Order of insects that include the true flies. Triassic–Recent.

disseminule   Collective term for propagative plant organs (fruits, seeds, spores); diaspore; propagule.

docodont   Member of the Docodonta, an order of small primitive mammals. Upper Triassic–Upper Jurassic.

double fertilization   Fertilization process characteristic of angiosperms in which one male nucleus contributes to the formation of a diploid zygote, and the other to the formation of a triploid endosperm nucleus.

drupe   A fleshy indehiscent fruit with a stony inner layer (endocarp) enclosing a single or few seeds.

ecosystem   Collective term for a community of organisms and their associated physical environment.

ectexine   The outer region of the exine of the pollen wall.

ectotherm   A cold-blooded animal that has internal body temperature controlled by the temperature of the environment; cf. endotherm.

elaiophore   Organ in plants that bears oily appendages that attract insects and may promote dispersal or pollination.

Elateridae   A family of flower-visiting beetles. Triassic–Recent.

embolism   An obstruction in a vascular element.

embryo   The early stages in the development of an organism.

embryo sac   The megaspore in seed plants giving rise to the female gametophyte.

embryogeny   The formation of the embryo.

Empididae   A family of brachyceran flower-visiting flies. Lower Cretaceous–Recent.

endarch   Applied to xylem having the first-formed elements (protoxylem) nearest to the center of the axis, inside the metaxylem.

endemic   Applied to taxa occurring within a specified area.

endexine   The inner layer of the exine of the pollen wall.

endocarp   The inner layer of fruit wall, forms the stony layer in a drupe.

endosperm   A nutritive tissue formed after fertilization that surrounds the embryo in angiosperm seeds.

endothecium   A cell layer in the anther, usually with characteristic thickenings of the cell walls.

endotherm   A warm-blooded animal that maintains constant internal body temperature by high metabolic rate; cf. ectotherm.

entomology   The study of insects.

entomophily   Pollination by insects.

eomyid   Member of the Eomyidae, a family of rodents related to the Cricetidae. Upper Eocene–Pleistocene.

epeiric   Applied to shallow seas either on, or at the margins of, continents.

epigynous   Applied to flowers having sepals, petals and stamens inserted above the ovary. See also inferior; cf. hypogynous.

epiphyllous   Applied to organisms living on the surface of leaves, e.g. a moss or alga living on a higher plant leaf.

epiphyte   A plant growing on other plants, but not parasitic.

estuarine   Applied to sediments or organisms of an estuary, where a river is entering the sea.

eustele   A cylindrical vascular system with discrete strands of primary xylem arranged either as a discontinuous cylinder or in a scattered arrangement.

evaporite   Saline or carbonate deposit formed by the evaporation of a lake or restricted sea.

exarch   Applied to xylem having the first-formed elements (protoxylem) farthest from the center of the axis, outside the metaxylem.

exine   The sporopollenin layer that forms the outer part of a typical pollen wall. Includes ectexine and endexine.

fabrosaur   Member of the Fabrosauridae, a family of the ornithopod (beaked) dinosaurs. Upper Triassic.

faithful pollinator   Animal visiting flowers of only one or a few taxa at a given time; fidelity.

feeder   A finger-like structure in a gnetalean embryo.

femur   The proximal bone of the hindlimb in vertebrates.

fern   Spore-producing plant with a sporophyte that usually has large compound leaves (fronds) on which the sporangia are borne; Filicopsida; Pteropsida. Carboniferous–Recent.

fidelity   Applied to pollinators visiting flowers of only one or a few flower taxa at a given time; faithful pollinator.

filament   The stalk bearing the anther in a stamen.

fluvial   Applied to sediments or organisms of a river; fluviatile.

folivore   Animal feeding on leaves.

follicle   A dry dehiscent fruit formed from a single carpel, usually containing several seeds, and opening along a single suture.

foraminifer   Unicellular and usually microscopic, marine animal with a shell of one to many variously arranged chambers.

fossorial   Applied to ground-digging or burrowing animals.

frugivore   Animal feeding on fruits.

fruit   The mature ovary, containing the seeds.

funicle   The stalk of the ovule.

fusainized   Turned into charcoal by high temperatures, usually by wildfire.

gamete   The reproductive cell which fuses with another gamete to form the zygote.

gametophyte   The gamete-producing haploid phase in a plant life cycle.

gamoheterotopy Positional change in plant ontogeny in which pollen sacs are replaced by ovules, or vice versa.

Gelinden flora Applied to Early Tertiary subtropical to tropical floras of the Northern Hemisphere.

genotype The set of genes of an organism; cf. phenotype.

ginkgophytes Gymnospermous plants with dense wood, long and short shoots and characteristic fan-shaped leaves. Triassic–Recent.

glade Clearing in woodland vegetation.

glauconite A green mineral of the illite group, formed exclusively in marine environments.

glirid Member of the rodent family Gliridae, known as the dormice. Lower Eocene–Recent.

glossopterids Group of seed plants with reproductive organs attached to tongue-shaped leaves. Commonly assigned to the seed ferns; Glossopteridales. Upper Carboniferous–Triassic.

Gnetales Gymnospermous plants with reproductive organs in compound unisexual cones and vessels in the secondary xylem. Upper Triassic–Recent.

Gondwana Hypothetical ancient supercontinent, comprising South America, Africa, Madagascar, India, Antarctica and Australia, that began to fragment during the Mesozoic.

granivore Animal feeding on seeds.

guard cells The pair of cells forming the stomatal apparatus.

gynoecium Collective term for the carpels and ovules in a flower; pistil or pistils.

gymnosperms Seed plants with naked seeds, includes all seed plants except angiosperms. Upper Devonian–Recent.

hadrosaur Member of the Hadrosauridae, the family of duck-billed dinosaurs. Upper Cretaceous.

Hamamelididae Subclass of dicotyledonous families that contains mostly shrubs and trees with simple, anemophilous flowers, e.g. plane (am.: sycamore), birch and walnut. Corresponds approximately to the Amentiferae. Albian–Recent.

haplocheilic Applied to stomata in which guard cells and associated subsidiary cells are derived from different mother cells; cf. syndetocheilic.

haploid Applied to an organism or cell having one set of chromosomes.

harmomegathic Pertaining to volume changes encountered by pollen grains during dehydration and rehydration.

haustellate Lepidoptera Lepidoptera with mouthparts adapted for sucking fluids. Upper Cretaceous–Recent.

helaletid Member of the Helaletidae, a family of odd-toed ungulates related to modern tapirs. Eocene–Oligocene.

hemicyclic Applied to flowers having floral parts arranged partly spirally and partly in whorls.

herbivore   Animal feeding on plants; phytophage; used by Collinson and
Hooker in a restricted sense to include animals feeding mainly on leaves
and non-woody stems of plants.

hermaphroditic flower   A flower possessing both female and male
reproductive organs (pistils and stamens); bisexual flower; cf. unisexual
flower.

heterochlamydeous   Applied to flowers having a perianth differentiated into
distinct calyx and corolla; cf. monochlamydeous.

heterochrony   Divergence from normal timing in development of organs.

heteromerous   Applied to flowers having a different number of floral parts in
each whorl; cf. isomerous.

heterosporous   Applied to plants producing two kinds of spores, microspores
and megaspores, that germinate to produce male and female gametophytes,
respectively.

heterostylous   Applied to flowers in which the length of styles and stamens
in different flowers of the same species varies and impedes self-pollination
or promotes outcrossing.

hexamerous   Applied to flowers having floral parts in sixes.

hilum   The point of attachment of an ovule to the placenta.

homoiotherm   A warm-blooded animal that maintains constant internal body
temperature by high metabolic rate; endotherm.

homology   Applied to characters having the same phylogenetic origin.

homoplasy   Resemblance of structure due to convergent or parallel
evolution.

horsetails   Spore-producing plants with the sporophyte usually having
whorled leaves and sporangia borne on peltate sporangiophores;
Sphenopsida. Middle Devonian–Recent.

humerus   The proximal bone of the forelimb in vertebrates.

hydromorphic   Applied to organisms adapted to living in water.

hygromorphic   Applied to organisms adapted to wet and humid habitats.

Hymenoptera   Order of insects that includes the bees, ants, wasps, sawflies,
and ichneumonids. Upper Triassic–Recent.

hypogynous   Applied to flowers having sepals, petals and stamens inserted
below the ovary. See also superior; cf. epigynous.

hypsilophodont   Member of the Hypsilophodontidae, a family of small
cursorial, bird-hipped dinosaurs. Upper Triassic–Cretaceous.

iguanodont   Member of the Iguanodontidae, a family of large bipedal
bird-hipped dinosaurs. Cretaceous.

impression   External imprint of a fossil organism in the sediment.

inaperturate   Applied to pollen grains lacking apertures.

incompatibility   Condition in which fertilization between two individual
plants cannot occur.

indriid primate   Member of the Indriidae, a family of primates that inhabits
Madagascar, e.g. indri and sifakas. Pleistocene–Recent.

inferior Applied to ovaries that are borne below the sepals, petals and stamens. See also epigynous; cf. superior.

inflorescence Aggregation of several or numerous flowers.

infructescence Aggregation of several or numerous fruits. Derived from an inflorescence.

insectivore Animal feeding on insects and other invertebrates.

integument The tissue in an ovule that envelops the megasporangium (nucellus).

interseminal scales Sterile appendages between the ovules in bennettitalean seed plants.

isomerous Applied to flowers having an equal number of floral parts in each whorl; cf. heteromerous.

kaolinite White, aluminium-rich, clay mineral formed in acid freshwater environments in warm, wet climates.

K-selected species Species characteristic of stable environments, usually with superior competitive abilities and late reproduction; cf. r-selected species.

lacustrine Applied to organisms or sediments of lakes and ponds.

lagenostome The pollen-capturing portion of the nucellus in some Palaeozoic seed ferns.

laterite Red, aluminium–iron-rich residual soil formed in tropical regions with high rainfall.

Laurasia Hypothetical ancient super-continent, comprising North America, Europe and Asia, that began to fragment during the Mesozoic.

leaf gaps Gaps in the vascular cylinder where the vascular bundles for the leaf depart.

leaf-margin analysis A method for determining past climate by studying the ratio of leaves with incised margins to leaves with entire margins in a fossil flora.

Lepidoptera Order of insects that includes butterflies and moths. Lower Cretaceous–Recent.

leptocaul Applied to trees having thin branched stems; cf. pachycaul.

lithofacies Physical characteristic of the lithology of a rock unit.

lithology The physical and petrographic features of a rock.

lithostratigraphy Characterization and correlation of rock units by their physical and petrographic features.

loculicidal dehiscence Dehiscence in capsules breaking into the loculi.

lophiodontid Member of the Lophiodontidae, a family of odd-toed ungulates with tapir-like teeth. Eocene.

lycopod Spore-producing plant with a sporophyte bearing spirally arranged scale-like or linear leaves, and sporangia borne on, or in, the axils of sporophylls; Lycopsida. Devonian–Recent.

macropodid marsupial   Member of the Macropodidae, a family of marsupials that includes kangaroos and wallabies. Pliocene–Recent.

Magnoliidae   Subclass of dicotyledonous families containing mostly trees and shrubs that retain many primitive features, e.g. *Magnolia*, tulip tree and bay laurel. Barremian–Recent.

mammal   Warm-blooded (homoiothermic) tetrapod with hair, mammary glands and a dentary-squamosal jaw articulation. Upper Triassic–Recent.

mangrove   Plant community of muddy tidal areas along tropical coasts.

manoxylic   Applied to soft, loose wood with wide parenchymatous rays; cf. pycnoxylic.

marsupial   Group of pouched mammals. Upper Cretaceous–Recent.

mastixioidean flora   Applied to Tertiary subtropical to tropical floras that contain many palaeotropical genera; Gelinden.

Maüle reaction   Method for detecting the presence of syringaldehyde and vanillin in lignin.

megaflora   Assemblage of macroscopic plant fossils, typically leaves, fruits and seeds; megafossil flora; cf. microflora.

megagametophyte   The female gametophyte in heterosporous plants.

megaspore   The larger spore of a heterosporous plant giving rise to the female gametophyte; macrospore; cf. microspore.

megaspore membrane   The megaspore wall in an ovule.

megasporophyll   A leaf-like organ producing megaspores or ovules; cf. microsporophyll.

megathermal   Applied to organisms living in a warm climate with mean annual temperature greater than 20 °C.

meristem   A plant tissue of unspecialized cells that are involved in the formation of new cells.

mesarch   Applied to xylem having the first-formed elements (protoxylem) surrounded by the metaxylem.

mesomorphic   Applied to plants having the structure characteristic of plants from warm, equable habitats; cf. xeromorphic.

mesophyll   Leaf-size class comprising leaves with a surface area between 80.0 and 364.5 cm^2.

mesophytic   Applied to plants adapted to living in temperate climate with moderate rainfall.

mesothermal   Applied to organisms living in a climate with mean annual temperature between 13 and 20 °C.

metabolism   The biochemical processes of assimilation and expenditure of energy.

metaxylem   Primary xylem differentiated after the protoxylem and after elongation of the organ.

microflora   Assemblage of microscopic plant fossils, typically pollen and spores; palynoflora; cf. megaflora.

microgametophyte   The male gametophyte in heterosporous plants.

microphyll   Leaf-size class comprising leaves with a surface area between 4.5 and 40.5 cm². Also used of small leaves with no leaf gaps.

micropterigid moth   A primitive moth of the family Micropterigidae characterized by having jaws with interlocking teeth as adults. They feed predominantly on pollen. Lower Cretaceous–Recent.

micropyle   An opening at the apex of an ovule through the integument (or integuments) that permits access to the nucellus.

microsporangium   Organ in heterosporous plants that produces microspores.

microspore   The smaller spore of a heterosporous plant giving rise to the male gametophyte; cf. megaspore.

microsporophyll   A leaf-like organ producing microspores or pollen; cf. megasporophyll.

microsynangium   A group of coherent sporangia that produces microspores.

microthermal   Applied to organisms living in a cold climate with mean annual temperature less than 13 °C.

molar   Back cheek tooth that is not replaced.

mold   Cavity left in the sediment when a fossil organism is dissolved or removed.

monadelphous   Applied to androecia having stamens united into a single tube.

monochlamydeous   Applied to flowers having a simple perianth composed of more or less similar undifferentiated parts (tepals); cf. heterochlamydeous.

monocotyledon   An angiosperm that has an embryo with one cotyledon and floral parts often in threes; cf. dicotyledon.

monoecious   Applied to plants having staminate and pistillate flowers borne on the same individual; cf. dioecious.

monophagous   Applied to animals feeding on one kind of food material.

monophyletic   Applied to a group of organisms derived from a single common ancestor.

monoporate   Applied to pollen grains having a single pore-like aperture.

monospecific   Applied to organisms belonging to the same species, or to a genus, comprising only one species.

monosulcate   Applied to pollen grains having an elongate distal furrow.

multilacunar   Applied to nodes having numerous leaf gaps.

multituberculates   Small, specialized, early mammals adapted for herbivory. Upper Jurassic–Upper Eocene.

mutation   The process (or result of process) that produces a heritable change in the genetic material.

Mycetophilidae   A family of flies that includes the fungus gnats. Jurassic–Recent.

MYBP   Million years before present.

nectary   A glandular organ, usually in a flower, secreting nectar.

neotropical   Pertaining to the tropical regions of the New World (North, South and Central America and adjoining islands); cf. palaeotropical.

Nitidulidae   A family of flower-visiting beetles. Triassic–Recent.

node   The part of the stem at which the leaves arise.

nodosaur   Member of the Nodosauridae, a family of armored dinosaurs with shoulder spines. Cretaceous.

Normapolles   Group of angiosperm pollen taxa that includes tri-aperturate pollen grains with complex, often protruding apertures. Middle Cenomanian–Upper Eocene.

notophyll   Leaf-size class comprising leaves with a surface area between 40.5 and 81.0 cm^2.

nucellus   The tissue of an ovule in which the megaspore is formed and develops into a female gametophyte.

nut   A dry, often woody, indehiscent fruit.

Nyctitheriidae   Family of insectivorous mammals that comprises shrew-like animals, e.g. *Saturninia*. Paleocene–Lower Oligocene.

omnivore   Animal feeding on a variety of both plant and animal material.

ornithischian   Member of the Ornithischia, the order of bird-hipped dinosaurs. Upper Triassic–Cretaceous.

ornithopod   Member of the Ornithopoda, the beaked dinosaurs. Upper Triassic–Cretaceous.

orogeny   The formation of mountains. Major earth movement.

orthotropous   Applied to ovules having a straight nucellus and embryo sac. Chalaza and hilum positioned opposite the micropyle.

ostracod   A small, bivalved crustacean animal that occurs either in marine or freshwater environments.

outcrossing   Fertilization of ovules by pollen from a different individual.

ovary   The lower region of the gynoecium that contains the ovules.

ovipositer   A hardened tube-like structure in insects, derived from various abdominal segments and used for the deposition of eggs.

ovule   A megasporangium (nucellus) containing a single functional megaspore and enclosed by one or two integuments. The female gametophyte develops within the megaspore and after fertilization the ovule matures into a seed.

oxygen isotope analysis   A method for determining palaeotemperatures by the ^{16}O to ^{18}O ratio in marine fossils with calcium carbonate shells, the ^{18}O content being higher in shells from warmer water.

pachycaul   Applied to trees having a thick, sparsely branched stem; cf. leptocaul.

pachycephalosaur   Member of the Pachycephalosauridae, the family of dome-headed dinosaurs. Upper Cretaceous.

paedomorphic   Pertaining to the appearance of juvenile characters in mature organisms, as a result of different rates of development of vegetative and reproductive organs.

palaeobathymetric   Relating to depths of ancient oceans.

palaeolatitude   Latitude estimated for different areas in the past; may be established by analyses of magnetic polarization of rocks.

palaeosol   An ancient soil horizon.

palaeotropical   Pertaining to the tropical regions of the Old World (Africa, Madagascar, India, Southeast Asia, Indonesia, Melanesia, Polynesia); cf. neotropical.

palinactinodromous   Applied to leaves having three or more primary veins that diverge from different points near the leaf base.

palmate   Applied to leaves having three or more leaflets or veins radiating from a single point; actinodromous.

palynoflora   Assemblage of acid-resistant plant microfossils (pollen, spores, algae); microflora.

palynological   Pertaining to the study of pollen and spores and other acid-resistant microscopic plant remains.

palynomorph   Acid-resistant plant microfossils such as pollen, spores and certain algae.

palynoprovince   A palaeofloristic province defined by the content of pollen in contemporaneous strata.

Pangaea   Hypothetical ancient supercontinent comprising all continents existing today; fragmented in the Permian and Triassic into Gondwana and Laurasia.

pantodont   Member of the order Pantodonta comprising archaic ungulate herbivores. Middle Paleocene–Middle Oligocene.

paraphyletic   Applied to a group of organisms derived from a single common ancestor, but which does not comprise all of that ancestors descendants.

paratropical   Applied to climate having mean annual temperature between 20 and 25 °C.

pathogen   A disease-producing organism.

peltate   Applied to leaves having a petiole or stalk that is more or less centrally attached, resulting in an umbrella-like arrangement.

pentamerous   Applied to flowers having floral parts in fives.

perforation plate   The portion of the wall of a vessel element through which it is directly connected to another vessel element.

perianth   Collective term for sepals and petals, or tepals, of a flower.

periporate   Applied to pollen grains having pores evenly distributed over the whole surface.

perissodactyl   Member of the Perissodactyla, an order of placental mammals that comprises odd-toed ungulates, e.g. horses, rhinoceroses and tapirs. Eocene–Recent.

permineralization   A fossil or process in which the original tissues of plants and animals are impregnated by minerals; petrifaction.

petal   Sterile appendage of a flower forming the inner part of the perianth (corolla).

petiole   The stalk of a leaf.

petrifaction   A fossil or process in which the original tissues of plants or animals are impregnated by minerals; permineralization.

phalangerid marsupial   Member of the Phalangeridae, a family of marsupials that includes possums and the koala. Pliocene–Recent.

phenology   The study of temporal biotic phenomena and their relation to climatic factors.

phenotype   The characteristics of an organism resulting from interaction between its genotype and the environment; cf. genotype.

phloem   The food-conducting tissue of vascular plants, composed mostly of sieve elements.

phyllotaxy   The arrangement of leaves or similar organs on an axis.

phylogeny   The evolutionary relationships of organisms.

physiognomy   Characteristic external features of an organism, vegetation or community.

phytophagous   Applied to animals feeding on plants; herbivorous.

pinnate   Applied to compound leaves that have leaflets borne along either side of a common midrib.

pinnatifid   Applied to leaves that are pinnately divided although the lamina of adjoining lobes is continuous and not completely separated.

pistillate flower   A flower possessing carpels (pistils) but no stamens; cf. staminate flower.

placentation   Attachment of the ovules to the ovary.

planktonic   Applied to floating organisms.

platyspermic   Applied to seeds having a bilateral symmetry; cf. radiospermic.

plesiomorphy   An ancestral or primitive character-state recognized in cladistic analyses; cf. apomorphy.

plutons   Igneous rocks formed by solidifaction of magma beneath the ground surface.

pollen-chamber   Chamber formed in the apical region of the nucellus in gymnosperm ovules in which the pollen germinates.

pollen grain   The microspore of seed plants that contains the male gametophyte and transports the gametophyte and gamete to the ovule (gymnosperms) or stigma (angiosperms).

pollenkitt   Lipidic, sticky substance that occurs on the pollen grains in many insect-pollinated flowers.

pollen sac   The pollen-producing cavity in seed plants; microsporangium.

pollen tube   A tube that extends from the pollen grain on germination, and through which the male nuclei may pass to the female gametophyte.

pollination   The transfer of pollen from anthers to the receptive surface of carpels or ovules.

polymerous   Applied to flowers having numerous parts.

polyphagous   Applied to animals feeding on a variety of food material.

polyphyletic   Applied to a group of organisms derived from several ancestral forms.

polyploidy   The condition of having more than two sets of chromosomes.

postcranial   Applied to the skeleton apart from the skull.

premolar   Front cheek teeth in mammals that replace the earlier deciduous (milk) teeth.

primate   Member of the Primates, an order of placental mammals including man, apes, monkeys, lemurs and tarsiers. Paleocene–Recent.

primordial   The first formed; primitive.

progenesis   Paedomorphosis resulting from the attainment of precocious reproductive maturity.

progymnosperms   Group of early land plants with gymnospermous wood, frond-like leaves and free-sporing reproductive organs. Middle Devonian–Lower Carboniferous.

propagule   Collective term for propagative plant organ (fruits, seeds, spores); diaspore, disseminule.

prosauropod   Member of the Prosauropoda, an infraorder of primitive lizard-hipped dinosaurs. Middle–Upper Triassic.

protandrous   Applied to flowers in which the pollen is released before the stigma is receptive.

prothallial cell   A cell of the male gametophyte in gymnosperms and some lycopods and pteridophytes.

protogynous   Applied to flowers in which the stigma becomes receptive and subsequently non-functional before the pollen is released.

protostele   A vascular system comprising a solid xylem core with peripheral phloem.

prototype   Original or ancestral type; archetype.

protoxylem   The first differentiated elements of the primary xylem.

pseudanthium   A condensed inflorescence of pistillate and staminate flowers that functions as a single bisexual flower.

pseudosciurid   Member of the Pseudosciuridae, a family of rodents including e.g. *Suevosciurus*. Middle Eocene–Middle Oligocene.

pteridophytes   Spore-producing plants with a sporophyte that usually has large compound leaves. By some authors used in a broader sense to include lycopods, sphenopsids and ferns. Lower Devonian–Recent.

pteridosperms   Heterogeneous and unnatural group of gymnospermous plants often with fern-like leaves on which the seeds were borne; seed ferns. Upper Devonian–Upper Cretaceous.

pulvinus   Swollen base of petiole.

pycnoxylic   Applied to dense wood with sparse parenchyma; cf. manoxylic.

quadruped   An animal with four feet.

radiospermic   Applied to seeds having a radial symmetry; cf. platyspermic.

ramentum   A thin covering of membranous scale-like structures in ferns.

ramus   Proximal part of a vertebrate mandible.

Ranunculidae   Subclass of dicotyledonous families mainly consisting of herbs that have retained a variety of primitive features, e.g. buttercup and water lily. Upper Cretaceous–Recent.

receptacle   The part of the floral axis on which the reproductive organs are borne.

reticulate   Applied to tissues or structures arranged in a net-like pattern.

rhinocerotid   Member of the family Rhinocerotidae, the rhinoceroses. Middle Eocene–Recent.

riparian   Applied to organisms or sediments of rivers and streams.

rodent   A member of the Rodentia, an order of placental mammals that comprises animals with chisel-shaped front teeth, e.g. dormice, squirrels, hamsters and beavers. Upper Paleocene–Recent.

Rosidae   Subclass of dicotyledonous plants that contains herbs, shrubs and trees with a mixture of primitive and advanced characters, e.g. rose, pea, saxifrage and myrtle. Middle Cretaceous–Recent.

rostrum   A beak or beak-like process.

r-selected species   Species characteristic of unstable environments, generally reproducing early and producing abundant offspring; cf. *K*-selected species.

saccate   Applied to pollen grains having one or more air sacs.

samara   A dry, indehiscent, winged, usually single-seeded fruit.

saprophyte   A plant that absorbs nutrients from dead or decaying organic matter.

saurischian   Member of the Saurischia, the order of lizard-hipped dinosaurs. Middle Triassic–Cretaceous.

sauropod   Member of the Sauropoda, the suborder of large quadrupedal dinosaurs. Jurassic–Cretaceous.

sawfly   Member of the order Symphyta. Upper Triassic–Recent.

scalariform   Ladder-like configuration usually in conducting cells formed by parallel arrangement of elongated perforations or pits.

scansorial   Applied to animals that exhibit some modification for locomotion in trees in which they spend much time foraging.

schizocarp   A dry fruit formed from a syncarpous ovary that splits at maturity into its constituent, usually single-seeded parts.

sclerophyllous   Applied to plants having thick fibrous leaves often adapted to dry conditions.

sclerotesta   A hard stony layer of a seed coat.

seed   A mature ovule containing an embryo.

seed ferns   Heterogeneous group of gymnospermous plants often with fern-like leaves on which the seeds are borne; pteridosperms. Upper Devonian–Upper Cretaceous.

seismic   Pertaining to earth vibrations and earthquakes.

selenodont   Applied to teeth dominated by half-moon-shaped crests.

selenolophodont   Applied to teeth dominated by both straight transverse and half-moon-shaped crests.

self-compatibility   The ability of a hermaphroditic plant to reproduce sexually by self-pollination.

self-incompatibility   The inability of a hermaphroditic plant to reproduce sexually by self-pollination.

semi-hypsodont   Applied to teeth showing trends towards being high crowned.

semi-inferior   Applied to flowers having perianth and androecium fused to the ovary for part of its length.

semi-tectate   Applied to pollen grains having an open reticulate tectum.

sepal   Sterile appendage of a flower forming the outer part of the perianth (calyx).

sieve element   An elongated phloem cell that lacks a mucellus and is involved in the transport of nutrients.

sieve tube   A tube in the phloem formed from a series of sieve elements.

siphonogamy   The transfer of male gametes by a pollen tube to the egg cells.

sister groups   Two monophyletic groups (clades) that are more closely related to each other than to any other group under consideration.

sphecid wasp   Member of the superfamily Sphecoidea thought to be ancestral to the bees (Apoidea). Lower Cretaceous–Recent.

sphenopsids   Spore-producing plants in which the sporophyte has whorled leaves, and sporangia borne on sporangiophores. Middle Devonian–Recent.

sporangium   Spore-producing organ.

spore   A small, usually one-celled, reproductive body produced by a wide range of plants; in bryophytes and vascular plants, germinates to produce a gametophyte.

sporophyll   A leaf-like organ bearing the sporangia.

sporophyte   The spore-producing diploid phase in a plant life cycle.

sporopollenin   Resistant material (carotenoid polymer) forming the exine of spores and pollen.

stamen   Organ in the flower producing the pollen, typically consisting of filament, connective and anthers.

staminate flower   A flower possessing stamens but no carpels; cf. pistillate flower.

staminode   A sterile stamen.

stegosaur   Member of the Stegosauria, the suborder of plated dinosaurs. Jurassic–Cretaceous.

stenophyllous   Applied to elongated leaves.

stigma   The receptive part of the carpel upon which the pollen germinates.

stigmatic mucilage   A slimy liquid secreted at the surface of the stigma.

stomatal apparatus   An opening and associated cells in the epidermis of plants through which gaseous exchange is regulated.

stratigraphy   The study of the origin and nature of stratified rocks.

strobilus   An aggregation of reproductive structures along a common axis, e.g. cone, flower.

style   The extension of the gynoecium between ovary and stigma.

sulcus   Elongate distal aperture of a pollen grain.

superior   Applied to ovaries that are borne above the sepals, petals and stamens. See also hypogynous; cf. inferior.

suspensor   A basal extension of the embryo.

sympatric   Applied to species occupying the same geographical area, or with overlapping geographical ranges.

sympetalous   Applied to flowers having connate (fused) petals.

Symphyta   Suborder of the Hymenoptera defined by the abdomen being broadly attached to the thorax; sawflies. Upper Triassic–Recent.

sympodial   Branching system produced by the successive development of lateral buds.

synangium   A group of adherent sporangia.

synapomorphy   A shared derived character-state used in cladistic analyses; cf. autapomorphy.

syncarpous   Applied to gynoecia having carpels united (fused); cf. apocarpous.

syndetocheilic   Applied to stomata in which guard cells and associated subsidiary cells are derived from the same mother cell; cf. haplocheilic.

synergism   Combined activity of organs or substances resulting in greater effect than the sum of the different organs or substances acting alone.

syrphid fly   Member of the insect family Syrphidae that are important and highly modified nectar feeders. Eocene–Recent.

taeniodont   Member of the Taeniodonta, an order of archaic mammals. Paleocene–Eocene.

tannins   Strongly astringent phenol derivatives.

taphonomic   Pertaining to the processes of deposition and fossilization of dead organisms.

taxon   A group of organisms characterized by a number of common features; species, genus, family, order, etc.

tectate   Applied to pollen grains having a continuous tectum (outermost layer of ectexine).

tectum   The outermost layer of the ectexine in the pollen wall.

terrestrial   Applied to sediments or organisms of the land.

Tethys   Sea or embayment that separated Europe and Africa and extended to southern Asia during the Mesozoic and Early Tertiary.

tetramerous   Applied to flowers having floral parts in fours.

tetrapods   The higher, predominantly terrestrial, vertebrates, comprising amphibians, reptiles, birds and mammals. Upper Devonian–Recent.

tetrasporangiate   Applied to anthers having four microsporangia (pollen sacs).

theridomyid   Member of the Theridomyidae, a family of rodents that includes animals with semi-hypsodont teeth, e.g. *Thalerimys*. Upper Eocene–Oligocene.

thermophilic   Applied to organisms that thrive in warm habitats.

thrips   Members of the order Thysanoptera. Permian–Recent.

Thysanoptera   An order of relatively primitive minute–small exopterygote insects. Permian–Recent.

Tipulidae   A family of flies that contains the craneflies. Jurassic–Recent.

topology   The morphology of geometrical figures, including branching diagrams.

tracheid   A water-conducting lignified cell of the xylem without perforation plates.

tracheophyte   A vascular plant.

triaperturate   Applied to pollen grains having three apertures (furrows or pores) that are usually radially arranged.

tricolpate   Applied to pollen grains having three elongated furrows (colpi) that are usually radially arranged.

tricolporate   Applied to pollen grains having three elongated furrows (colpi) each with a pore-like aperture.

tricolporoidate   Applied to tricolporate pollen grains in which the pores are poorly developed.

trilacunar   Applied to nodes having three leaf gaps.

trimerophytes   Group of early vascular plants that includes small leafless spore-producing forms, sometimes included in the Psilophytopsida. Devonian.

trimerous   Applied to flowers having parts in threes.

triploid   An organism or tissue with three sets of chromosomes.

triporate   Applied to pollen grains having three pore-like apertures that are radially arranged.

trophic   Pertaining to nutrition.

tropical   Applied to climate having mean annual temperature greater than 25 °C.

tunica   Superficial layer or layers of the apical meristem with anticlinal cell divisions.

Turgaian flora   Applied to Early Tertiary temperate floras of the Northern Hemisphere.

ungulate   Hoofed mammal.

unilacunar   Applied to nodes having a single leaf gap.

unisexual flower   A flower possessing either female or male reproductive organs; cf. bisexual (hermaphroditic) flower.

unitegmic   Applied to ovules possessing one integument; cf. bitegmic.

vasicentric   Applied to wood having parenchyma surrounding the vessels in angiosperms.

versatile   Applied to anthers in which the connective is attached near the middle, permitting free movements.

vertebrates   Group of animals having visceral clefts, dorsal nerve cord, post-anal tail and vertebral column of bone or cartilage; includes tetrapods and fishes.

vespoid wasp   Member of the superfamily Vespoidea that includes all social wasp species. Upper Cretaceous–Recent.

vessel   A tube in the xylem formed from a series of vessel elements.

vessel element   A water-conducting lignified cell of the xylem with perforation plates.

viverrid carnivore   Member of the Viverridae, a family of carnivores that includes often nocturnal and scansorial animals, e.g. mongooses, civets and genets. Upper Eocene–Recent.

viviparous   Applied to plants in which germination takes place before dispersal of the seed.

water ferns   Small, heterosporous ferns that grow in, or near water. Includes Marsileales and Salviniales. Lower Cretaceous–Recent.

Western Interior Seaway   Narrow sea that separated eastern and western North America during the Cretaceous extending from Texas to the Arctic Ocean.

xeromorphic   Applied to plants having the structural characteristic of plants from dry habitats, such as thick leaf texture and thick cuticles; cf. mesomorphic.

xerophytic   Applied to plants of dry habitats.

Xyelidae   Family of the Symphyta that are the most primitive and earliest occurring in the fossil record. Upper Triassic–Recent.

xylem   The water-conducting tissue of vascular plants, composed mostly of tracheids and, in some plant groups, vessel elements.

zygomorphic   Applied to flowers having bilateral symmetry; cf. actinomorphic.

zygote   A cell formed by the fusion of two reproductive cells (gametes).

Index

•

Bold type indicates a main text reference; italics refer to figures.